GEOGRAPHY OF ASSAM

Editors

A.K. Bhagabati

B.K. Kar & A.K. Bora

RAJESH PUBLICATIONS

NEW DELHI - 110 002

GEOGRAPHY OF ASSAM

Published by :
Sanjay Gupta
RAJESH PUBLICATIONS
1, Ansari Road, Darya Ganj, New Delhi - 110002
Ph.: 011-23274550, 43084725
Email : rajeshpublications@gmail.com,
rajeshpubdelhi@gmail.com
Website : www.rajeshpublications.com

First Published : 2001
IInd Edition : 2002
IIIrd Edition : 2007
IVth Edition : 2011
Vth Edition : 2012
Reprint : 2015
VIth Edition : 2018
VIIth Revised & Enlarged Edition : 2022

ISBN-10 : 81-85891-41-9
ISBN-13 : 978-81-85891-41-5

Printed in India by :
R.K. Offset Process
Delhi-110095

Dedicated to

Late Professor S.P. Mandal

and

Late Professor S.D. Roy

of

Darrang College, Tezpur

Preface to Revised and Enlarged Edition

Geography of Assam, edited by us and written by a galaxy of scholars, has received immense response from readers ranging from students, research scholars, candidates of competitive examinations to college and university teachers. However, there was a growing demand for revision of certain chapters of the book, especially with respect to updating the database and incorporating new information. We are happy to express here that most of the chapters of the book are more or less revised and enlarged keeping in view the changing scenario of the State's geography. More importantly, four completely new chapters on some pivotal issues including Earthquake, Forest Management and Tea Industry have been included in this edition in order to make the book more acceptable to all. We express our gratefulness to the learned authors, more particularly the authors of the new chapters.

With heavy heart we would like to inform our readers that Professor S.P. Mandal, to whom we dedicated the book, had left this world on 14th of October, 2020 at the age of 83. We take the opportunity here to pay our respectful homage to the departed soul of Professor Mandal.

We are thankful to Mr. Sanjay Gupta, proprietor of Rajesh Publications, New Delhi for coming forward to publish this revised and enlarged edition in a much better form.

With all humility and sincerity, like in the earlier edition, we invite constructive criticism and suggestions from our beloved readers so that the quality of the book may be improved further in the next edition.

August, 2022 — **Editors**

Preface of First Edition

The geography community of the north-east, more particularly of the state of Assam has been in urgent need of an exhaustive book on geography of the state since the late seventies of the last century when the highly acclaimed book by Professor H.P. Das, an eminent scholar in the field, went out of print. During the last quarter of the century, unlike many other parts of the country, Assam witnessed tremendous changes in all aspects of its geography including the most delicate ecological system. The evergrowing population resulting mainly from large-scale immigration, the highly disturbed water regime, 'threatened biodiversity, relative stagnation of the economy, growing unemployment and the volatile ethnic and political issues have made the contemporary geography of the state highly diverse and complicated. Serious kind of systematic studies incorporating all important aspects can only reflect the regional realities of such a spatio-socially diverse area like Assam. It is a task really very difficult for a single hand to tackle. We have, therefore, evolved a joint venture in which both young and matured minds could join hands and blend their ideas and expertise to finally arrive at the goal of publication of a new book on geography of the state.

Two different events of unavoidable concern to us, one being really unbearable, prompted us to bring out the book within a short time. While the basic plan for the book was yet to acquire a concrete shape Professor S.P. Mandal, a reputed geography teacher, retired from his long service as the head of the department of geography, Darrang College, Tezpur. Subsequently, Professor S.D. Roy, a devoted teacher and social worker, who was heading the department, suddenly left this world for heavenly abode after a brief illness. Incidentally, both of them were class mates in the Patna University, became colleagues in Darrang College and immediate neighbour in an urban enclave called Rubber Bagan at Tezpur for more than thirty years. The authors and the editors of this book, most of whom, are either their direct students or colleagues, then considered it to be befitting to dedicate their works in memory of late Professor Roy and in honour of Professor Mandal. We all are really happy to have Professor Mandal among us still as a great source of inspiration to culture geography in this part of the country.

We have tried our best to make the book complete and do all possible justice to the proposed title. Even then, some of the important issues relating to occupational diversity, unemployment, trade and commerce, insurgency, diffusion of information technology, which have enough geographical relevance, could not be addressed directly. These issues need more interdisciplinary effort for their correct and comprehensive presentation. We are, however, very glad to announce here that most of our contributors are young geography teachers in the colleges of Assam and interestingly, for a few of them this would probably be the first attempt to express their views as responsible authors. We hope, the young authors in future will be more enthusiastically rise to the cause of geography education and research in this otherwise less-advanced part of the country.

The editors are greatly indebted to Professor M. Taher of the department of geography, Gauhati University who agreed to take the pain of writing the introductory chapter of the book in spite of his old age. Incidentally, all the authors of the book are his direct and ardent students. Let us all take the opportunity to extend our deep sense of gratitude to Professor Taher, Dr. A.U. Choudhury, a well-known wildlife expert of the region and Dr. S.K. Phukan, a reputed police officer as well as a serious researcher, a rare combination indeed, deserve our hearty thanks for their valuable and highly relevant contribution. Thanks are also due to all other authors for their active participation in this venture. A special word of thanks goes to P. Bhattacharya, Cartographer, department of geography, Gauhati University, for his constant help and assistance.

M.L. Gupta, proprietor of Rajesh Publications, New Delhi willingly came forward to shoulder the responsibility of publishing this book. We are really thankful to him.

Lastly, with all humility and sincerity we invite constructive criticism, advice and suggestions from our learned readers so that necessary improvement may be made in the next edition of the book.

January, 2001 **–Editors**

Contents

Contributors

1. **Dr. A. K. Bhagabati**
Professor of Geography (Retd.)
Gauhati University
GUWAHATI - 781014
2. **Dr. M. Taher***
Professor of Geography (Retd.)
Gauhati University
GUWAHATI- 781014
3. **Dr. A.K. Bora**
Professor of Geography
Gauhati University
GUWAHATI - 781014
4. **Dr. D. Deka**
Asstt. Professor
Department of Geography
Gauhati University
GUWAHATI - 781014
5. **Dr. K. Sarma**
Assoc. Professor
Dept. of Geography (Retd.)
Jagiroad College
JAGIROAD - 782410
6. **Dr. A. U. Choudhury**
Commissioner to the
Govt. of Assam (Retd.)
Dispur
GUWAHATI- 781006
7. **Dr. H.N. Bora**
Associate Professor (Retd.)
Department of Geography
Govt. College
BOMDILLA
8. **Dr. B.K. Kar**
Professor of Geography
Gauhati University
GUWAHATI - 781014
9. **D.C. Bhuyan***
Town Planner
Tezpur Town
TEZPUR - 784001
10. **Dr. M. Dutta**
Assoc. Professor in Geography
Cotton University
GUWAHATI - 781001
11. **Mrs. A. Keot**
Assoc. Professor in Geography
B.P. Chaliha College
NAGARBERA - 781127
12. **Dr. B. Saikia**
Ex-Principal
Dr. B.K.B. College
PURANIGUDAM
13. **Dr. R. Chetry**
Asstt. Commissioner
Govt. of Assam
14. **Dr. B. Barah**
Ex-Principal
J.B. College
JORHAT- 785001
15. **Dr. P. Bhattacharya**
Associate Professor
Department of Geography
Gauhati University
GUWAHATI - 781014
16. **Dr. J. Hazarika**
Ex-Principal
Darrang College
TEZPUR -784001
15. **Dr. S.K. Phukan**
Retd. Sr. Police Officer
ASSAM

**Unfortunately these two learned contributors left the world forever.*

List of Tables

List of Figures

1 Assam: A Geographical Outline

Introduction

Assam, one of the seven states of North-East India, popularly known as the *Land of Seven Sisters*, is a unique geographical entity. It represents one of the richest geographical regions that the country nurtures in terms of nature and human culture. The hills, vales, beels, streams and floodplains that exist in the state are interwoven in such a way that Assam spontaneously symbolizes a great and rare assemblage of diverse natural and cultural landscapes. It is basically a mosaic of landscapes which has provided suitable habitats for a large variety of culture groups with diverse heritages and traditions that help Assam to emerge remarkably as a land of physical heterogeneity and socio-cultural homogeneity. A territory of its parallel is, in fact, difficult to find practically within the country's political limit (Mukerji, 1984).

Locationally Assam holds a unique position. Extending latitudinally from 24°8' N to 27°56' N and longitudinally from 89°42' E to 96°0' E, the state spreads over an area of 78,438 sq. km. It shares the basic characteristics of two important river basins- the Brahmaputra and the Barak- and forms a part of the Indian zoogeographic sub-region within the broad Oriental Zoogeographic Region. Being situated at the center of the country's north-eastern periphery, the state holds enough strategic significance, in addition to its historical importance as a trade route for many of the South and South-EastAsian nations. Surrounded by seven Indian states and two foreign nations, Assam's geographical position in the context of India and South-East Asia naturally ensures great potentialities and promise for future development, and also encounters complicated problems needing immediate attention from the state and the centre as well. The peculiar geographical location in the North-East bordering South-East Asia may be boon or a bane depending

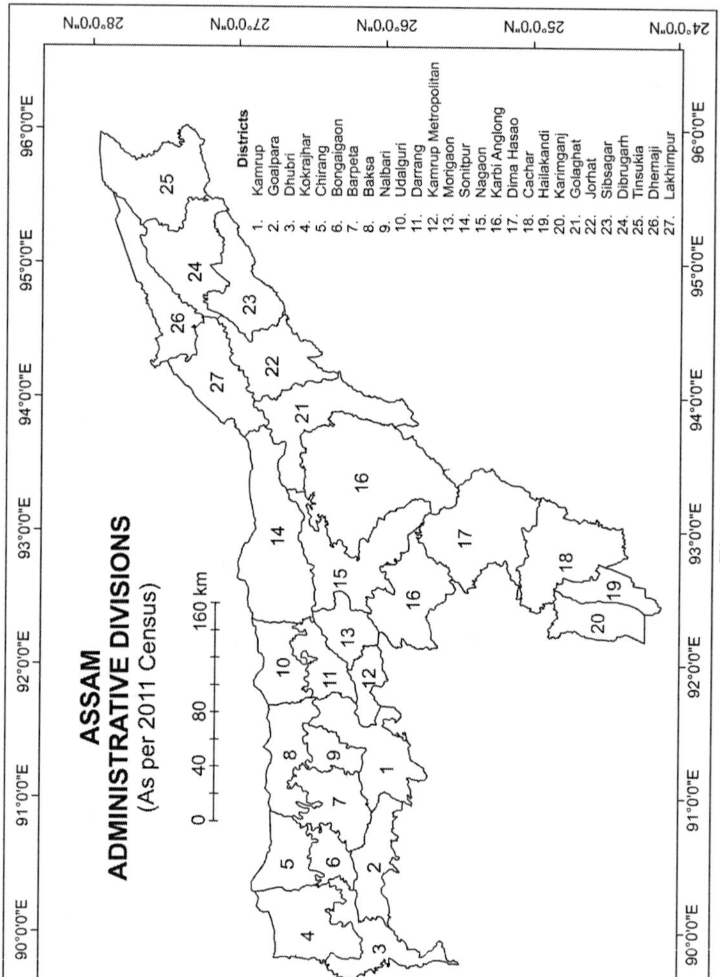
ASSAM
ADMINISTRATIVE DIVISIONS
(As per 2011 Census)
0 40 80 160 km
Districts
1. Kamrup
2. Goalpara
3. Dhubri
4. Kokrajhar
5. Chirang
6. Bongaigaon
7. Barpeta
8. Baksa
9. Nalbari
10. Udalguri
11. Darrang
12. Kamrup Metropolitan
13. Morigaon
14. Sonitpur
15. Nagaon
16. Karbi Anglong
17. Dima Hasao
18. Cachar
19. Hailakandi
20. Karimganj
21. Golaghat
22. Jorhat
23. Sibsagar
24. Dibrugarh
25. Tinsukia
26. Dhemaji
27. Lakhimpur
90°0'0"E
91°0'0"E
92°0'0"E
93°0'0"E
94°0'0"E
95°0'0"E
96°0'0"E
24°0'0"N
25°0'0"N
26°0'0"N
27°0'0"N
28°0'0"N

Fig. 1.1

upon the country's capability to appreciate the potentialities and to manage the problems amicably from sustainable perspectives.

Physical Make-up

The physical make-up of Assam is characterized by diverse features such as floodplain, natural levee, marshes and beels, scattered hillocks, folded hill ranges, old plateau and a dense network of streams. On the basis of the topographic characteristics, the state is broadly divided into the following three major divisions.

1. **The Brahmaputra Valley :** Extending east-west for about 725 km with an average width of 80 km, the Brahmaputra Valley forms an important physiographic unit within the territory of Assam. The valley is wide (80-100) in its upper and lower parts, but narrow (55-60 km) in the middle. It is dotted with occasional hillocks of hard rocks locally called *tilla*. While the valley as a whole has a general slope from east to west, the immediate north bank and the south bank have a tendency to slope down gently to the Brahmaputra river channel. The average east-west slope of the valley stands at 12.5 cm. per km.

The width of the Brahmaputra channel in its long profile varies considerably. At few places where the towns like Tezpur, Guwahati and Goalpara are situated, the river bank is relatively stable as it is over the granitic outcrops. Being overburdened with huge sediment loads, the channel tends to be wider in the portions immediately downstream of these places. Thus, for example, at Tezpur the river is 3.60 km wide, whereas immediately west of it the river widens to nearly 12.5 km. In the wider portion, the velocity of the river water decreases causing increased sinking of sediments to the river bed. This leads to the formation of braided channels chocked with sandbars (Pal and Bagchi, 1975). From Dibrugarh to Dhubri, numerous river islands locally called *chars* are found to occur. Majuli, one of the biggest river islands of the world, now with an area of around 700 sq. km. lies in between the channels of the Brahmaputra as a remarkable geomorphic feature.

Numerous small and big tributaries from the north and south join the Brahmaputra to form a well-developed drainage

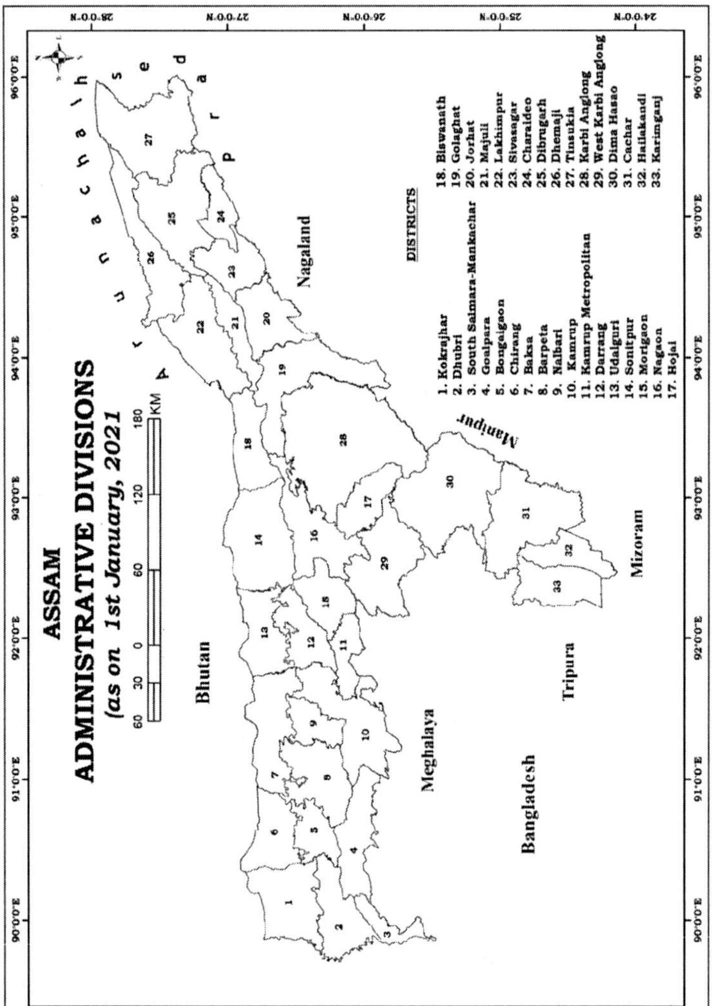
ASSAM
ADMINISTRATIVE DIVISIONS
(as on 1st January, 2021)
60 30 0 60 120 180 KM
Bhutan
Arunachal Pradesh
Nagaland
Manipur
Meghalaya
Bangladesh
Tripura
Mizoram
DISTRICTS
1. Kokrajhar
2. Dhubri
3. South Salmara-Mankachar
4. Goalpara
5. Bongaigaon
6. Chirang
7. Baksa
8. Barpeta
9. Nalbari
10. Kamrup
11. Kamrup Metropolitan
12. Darrang
13. Udalguri
14. Sonitpur
15. Morigaon
16. Nagaon
17. Hojai
18. Biswanath
19. Golaghat
20. Jorhat
21. Majuli
22. Lakhimpur
23. Sivasagar
24. Charaideo
25. Dibrugarh
26. Dhemaji
27. Tinsukia
28. Karbi Anglong
29. West Karbi Anglong
30. Dima Hasao
31. Cachar
32. Hailakandi
33. Karimganj
90°0'0"E
91°0'0"E
92°0'0"E
93°0'0"E
94°0'0"E
95°0'0"E
96°0'0"E
24°0'0"N
25°0'0"N
26°0'0"N
27°0'0"N
28°0'0"N

Fig. 1.2

network. The important north bank tributaries include the Gai, Subansiri, Bharali, Puthimari, Pagladiya, Manas, Ai and Sonkosh. The sources of these tributaries lie in the Eastern Himalayan highlands and are fed with heavy monsoonal downpour as well as melting of snow. The major south bank tributaries on the other hand are Burhi Dihing, Dhansiri, Kopili, Digaru, Kulsi, Krishnai and Dudhnoi. Having their sources in the hills of Nagaland and the northern slopes of the Meghalaya plateau, these tributaries receive water mainly from rain only. However, the age-old practice of shifting cultivation in these highlands causes accelerated erosion generating huge sediments to be carried by these tributaries. Thus, the tributaries from both north and south bring down enormous sediments to get deposited over the Brahmaputra River floor and as a result the river channel has been increasingly rendered shallower and flat. It is interesting to note that the sediment yield of the north bank tributaries is on an average three times higher than that of the south bank ones (Goswami, 1985). Whatever may be the change in the river regime, the people of Assam for generations have watered their fields with its life-giving floods and drunk of its blessed water, their whole history and culture are intimately connected with the Brahmaputra (Barua, 1951).

2. **The Hills** : The hills of Assam are mainly confined to the districts of Karbi Anglong and North Cachar (now Dima Hasao). The Karbi Hills are basically parts of the Meghalaya plateau and the Rengma Hills and North Cachar Hills on the other hand are parts of the Barail Range. While the Kopili river separates the Karbi Hills from the Meghalaya plateau, the Dhansiri on the east separates the Karbi Hills from the hills of Nagaland. The Karbi Hills are drained and immensely denuded by the Kopili system and also the tributaries of the Kolong and the Dhansiri to give the area a plateau character.

Stretching from the state of Nagaland, the Barail Range covers the whole North Cachar district. It divides the district into two distinct parts. The northern part with relatively gentle slope is drained to the Brahmaputra, while the southern one rolls down to the Barak. The watershed in between them rises to above 1,350 meters at places.

3. **The Barak Valley** : Named after the river Barak, the Barak

valley or the Cachar plain lies in between the north Cachar hills on the north and Mizo Hills on the south. In the east of the valley stands the Manipur Hills and on the west it merges with the Sylhet plain of Bangladesh. The Barak along with its tributaries Jiri, Diksha, Chiri, Madhura and Jatinga flows through the Cachar plain in an east-west direction. The central part of the valley is a saucer-shaped depression dotted with low isolated hillocks, where the Barak exhibits a highly meandering course with a number of cut-off wetlands. The immediate floodplains of the river get regularly inundated and it is characterized by the presence of many ox-bow depressions.

The diverse topography, soil type and climatic pattern through their long-continued interplay have composed an environmental condition in the state that favoured the growth of a variety of natural vegetations belonging to as many as six distinct types: tropical wet evergreen, tropical semi-evergreen, deciduous, subtropical broadleaf, subtropical pine and riparian forests. Unfortunately, the state has experienced large-scale deforestation and destruction of wildlife habitats during the recent past and the consequent conflicts between human and wild animals, especially elephants and monkeys. The fast-shrinking base of rich biodiversity in the state due to deforestation calls for prompt attention and action from the authorities concerned for its protection and sustainable development.

Table 1.1
District-wise Area, Total Population, Population Growth Rate and Population Density in Assam, 2011

District	Area (in km^2)	Total Population	Decadal Growth Rate of Population during 2001-11 (%)	Population Density (Persons/km^2)
1. Kokrajhar	3,296	887,142	5.21	269
2. Dhubri	2,176	1,949,258	24.44	896
3. Goalpara	1,824	1,008,183	22.64	553
4. Barpeta	2,282	1,693,622	21.43	742
5. Morigaon	1,551	957,423	23.34	617
6. Nagaon	3,973	2,823,768	22.00	711
7. Sonitpur	5,204	1,924,110	15.55	370
8. Lakhimpur	2,277	1,042,137	17.22	458
9. Dhemaji	3,237	686,133	19.97	212
10. Tinsukia	3,790	1,327,929	15.47	350

11. Dibrugarh	3,381	1,326,335	11.92	392
12. Sivasagar	2,668	1,151,050	9.44	431
13. Jorhat	2,851	1,092,256	9.31	383
14. Golaghat	3,502	1,066,888	12.75	305
15. Karbi Anglong	10,434	956,313	17.58	92
16. Dima Hasao	4,888	214,102	13.84	44
17. Cachar	3,786	1,736,617	20.19	459
18. Karimganj	1,809	1,228,686	21.90	679
19. Hailakandi	1,327	659,296	21.45	497
20. Bongaigaon	1,093	738,804	20.59	676
21. Chirang	1,923	482,162	11.34	251
22. Kamrup	3,105	1,517,542	15.69	489
23. Kamrup Metro	955	1,253,938	18.34	1313
24. Nalbari	1,052	771,639	11.99	733
25. Baksa	2,457	950,075	10.74	387
26. Darrang	1,585	928,500	22.19	586
27. Udalguri	2,012	831,668	9.61	413
Assam	**78,438**	**31,205,576**	**17.07**	**398**

*Gazette Notification on March, 2007.
Source: Census of India, Primary Census Abstract, Assam, 2011.

The People

Assam provides homes for a large number of population groups. The state had to receive people coming from different directions, with varied racial and cultural backgrounds at different points of time and thus the state continues to harbor a highly complicated social structure (Taher, 1993). The state has around eighty socio-cultural groups in the form of tribes, castes and religious sects. Languages spoken in the state belong to Assamese, Bengali, Tibeto-Burman and Siamese-Chinese groups. The tribal people of the state, who constitute 12.4 % of the total population (2011 Census) belong to as many as fifteen major tribes - nine plain tribes and six hill tribes. The plain tribes include the Bodo, Mising, Rabha, Tiwa, Deori, Sonowal Hajong etc, while the hill tribes comprise tribes like Karbi, Dimasa, Kuki, Garo, Mizo and the Hmar.

The geographical distribution of population groups is really interesting to note. The Dimasas are concentrated in the districts of Dima Hasao and Cachar, the Karbis in Karbi Anglong, the Bodos in the north bank of the lower and central Brahmaputra Valley, the Rabhas and the Tiwas in the south bank of the lower and central Brahmaputra valley. Interspersed among various tribal groups are the caste Hindu and indigenous Muslim population groups. The Muslims of immigrant origin are mainly concentrated in the floodplains ofthe Brahmaputra in its lower and central parts.

Nearly 85% of the population of Assam lives in the villages. It is needless to say that even today the most important means of livelihood for majority of the rural people is agriculture, and agriculture in Assam basically means rice farming. Since agriculture remains the main occupation, the common interest grown around it has kept different groups of people in constant interaction andco-operation. The methods and pattern of cropping practised through generations have given rise to an atmosphere of mutual dependence and cooperation leadingto harmonious living among the communities. Though the Assamese society apparently seems to be a highly divided one, the age-old undercurrent of cultural and economic interdependence among the people and areas evolved through sharing common ecological settings and economic systems have given it in long-run a unified character.

The state, however, experienced remarkable change in its socio-economicand political environment due to very high rate of population growth all throughout the last century. It is important to note that during the period 1901-2001, the population of the state increased from 3.28 million to 26.65 million, an eight-fold increase against the country's four and a half fold during the period. It is worth mentioning that the Brahmaputra valley in Assam is one of the most populous regions of India. In 1901 the valley had a population of 2.2 million which increased to 22.6 million by 2001, a more than ten times increase to be precise. Similar type of growth was experienced by Barak Valley also. So far population density is concerned, the state recorded an average density of 42 persons/sq. km in 1901, which increased to 102 persons in 1951 and 340 persons in 2001. Interestingly, some of the marginal areas like the river islands and the immediate river banks, the foot hills in the border areas, and most parts of the two hill districts, which were once repulsive for human habitation because of inhospitable natural environment, are now under growing density of population. The impact of high growth rate and density of population may be far-reaching in all spheres of environment, society and economy. Significantly, as an immediate consequence of it, among others, the exposed lands, wetlands and the grasslands of the state are getting increasingly degraded and the natural forests are fast getting transformed into agricultural lands and other uses (Bhagabati, 2007). If this trend continues unabated, the possibility of encountering more complicated environmental problems in near future cannot be denied.

Economy

The economy of Assam is basically agrarian in character. The practice of agriculture is, to a large extent, traditional and even today primitive type of farming (jhumming) covers nearly 2 per cent of the state's total area. However, during the recent period, the state has witnessed significant positive change in its agricultural sector, particularly in terms of technification and productivity. Interestingly, the food crops continue to dominate the cropping pattern by occupying more than three-fourths of the total cropped area. Among the foodcrops, rice alone covers around 65 per cent the total cropped area. Rice meant, and still means the source of subsistence, nourishment, security and hope for future of the people of Assam (Dutta and Bhagabati, 2007).

The cash crops cultivated in the state include mainly tea, jute, sugarcane and tobacco. With an area of nearly 2 lakhs hectares, tea plantations occupy about 2.5% of the total area and contribute substantially to the state's total income. The recent appearance of certain horticultural crops including vegetables has made the agricultural landscape of the state more diversified and economically efficient.

In spite of being blessed with a high potential for industrial development, the pace of industrialization in Assam has been still slow. Before independence, there was no other industry worth the name except the colonial tea industries and a refinery at Digboi. It is now encouraging to note that over the past few years, in spite of great difficulties, efforts are being made towards improving and increasing various infrastructural facilities necessary for industrialization in the state. At present major industrial establishments in the state include refineries and petrochemical complex, fertilizer plant, paper mill, jute mill, sugar mill, cotton mill, cement factory, etc. The state has more than 100 medium scale industries and around 17,500 small scale industries to produce various consumer and durable goods.

Assam is well-known for its variety of handloom and handicraft product ssuch as Eri (*Philosamiaricini*), Muga (*Antheracaassama*) and bamboo and cane products, bell metal and brass metal products, etc. The production of golden yellow muga silk which is secreted by an indigenous multivoltine caterpillar is the monopoly of Assam. However, most of these local products have faced unprecedented competition before the modern technological products mainly from outside due to inadequate infrastructural facilities, dearth of competent entrepreneurial motivation and low level of private and public sector investment.

Compared to some other states of India, Assam is still rich in forest resources. The reserved forests in the state cover about 20 % of the total area. The state earns every year a considerable amount of revenue from various forests products such as timber, plywood, fuel wood, cane, thatch, etc. However, time has come to take all sorts of care in using the forests products, which are now considered as environmental resources, the cost of which is much more than what is generally estimated in monetary terms.

In the economic background as narrated above, there is a need to reorient the state's economy towards sustainable use of biological resources and to generate employment opportunity, particularly for the weaker section and women in eco-friendly occupations. The present occupational structure clearly indicates that the secondary sector is still very poor and near-stagnant. There is enough scope to increase the output of the agro-based and forest-based products in particular to generate useful employment opportunities, provided the available and potential biological resources are locally and judiciously utilized (Bhagabati, Kalita and Barua, 2006).

Prospects

The development that integrates production with resource conservation and enhancement and that links both to the provision for all of an adequate livelihood and equitable access to resources, it is desirable one. To think of it, to do for it and to achieve it are the real call of the time.

Endowed with relative locational advantage, rich natural resources, both renewable and nonrenewable, diverse human resources, rich tradition and cultural heritage, Assam has immense prospects for development. No doubt, when a traditional land like that of Assam is exposed to the forces of globalization and modernization, there may be many unforeseen problems, and also certain unexplored prospects, if the emerging situations could be managed properly. Till date, Assam has come through a long process of socio-cultural assimilation facilitated basically by its unique geography and eventful history. The land, water, geographical and biological resources scattered in different parts of the state hold keys for a kind of sustainable and diversified development. The traditional knowledge system and indigenous skills that its people have been sharing historically are also well indicative of the region's great potentiality for sustainable prosperity. The reduction of the degree of misuse and underuse and the promotionof appropriate sustainable use of resources, require scientific management of the existing and forthcoming situation. Various dimensions of environment, society and economy need to be objectively integrated and rationally appreciated in order to ensure socially and ecologically acceptable development. The emerging situations in Assam are so diverse and sensitive that in no case their geographical and historical roots can be ignored. The structures (natural and social), processes (cultural and economic) and the patterns (spatial and temporal) are inherently inter-linked needing scientific understanding for their rational management. However, one can hope that the day is not too far when all the spatial, social and structural realities would be understood in their right earnest and all the prospects for development of the area would be translated into reality through best possible ways and means.

References

Barua, B.K. (1951): *Cultural History of Assam*, Vol. 1. Bina Library, Guwahati. p. 2.

Bhagabati. A.K. (2007): Population pressure and degradation of riverine environment in the Brahmaputra valley. Assam. *Journal of Geography*, Vol. 5. pp. 21-33.

Bhagabati A.K., Kalita, M.C. and Baruah, S. (2006): *Biodiversity of Assam: Status, Strategy and Action Plan for Conservation*, RBH Publishers (India). Guwahati.

Dutta, M. and Bhagabati, A.K. (2007): Place of rice farming in Assamese life and culture, in B. Thakur. George Pomeroy et al (ed), *City, Society and Planning*, Vol. II. Concept Publishing Co., University of Akron, USA and American Association of Geographers, pp 359-373.

Goswami, D.C. (1985): Pattern of sediment yield from river basins of the Brahmaputra system. N.E. India, *North Eastern Geographer*, Vol. 17. p. 6.

Mukerji, A.B. (1984): The Region of the Seven Sisters: A Distant View, *North-Eastern Geographer*, vol. 15, No. 1&2.

Pal. S.K. and Bagchi, K. (1975): Recurrence of floods in the Brahmaputra and the Kosi basin: a study in climatic geomorphology, *Geographical Review of India*, Vol. 37, No.3. pp. 246-57.

Taher, M. (1983): The peopling of Assam and contemporary social structure, in A Ahmad (ed), *Social Structure and Regional Development,* Rawat Publication, Jaipur.

2 Historical Evolution and Peopling

Assam, located in the tropical latitudes (24.3^{0}N and 28^{0}N) and eastern longitudes (89.5^{0}E and 96.1^{0}E), is the most populous state in North-East India. It is surrounded on three sides by hills and mountains. The rivers Brahmaputra and Barak, in the north and south respectively, carve out deep valleys, that represent the major part of the state. Between the two valleys there lies a strip of highland made up by hills and plateaus. The two plains form the easternmost parts of the north Indian plain, which are apparently shut to the south-East Asia by the Patkai Range and its southward extensions. In the north, the Brahmaputra plain rises steeply to the Himalayas, while to the east, the parallel hill ranges of Arunachal, Nagaland and Manipur attain height as one moves eastward, culminating at the Patkai- Manipur-Arakan Yoma hills. Although the hills and mountains mentioned above, appear to provide physical blockade, they are not inaccessible. There are a large number of mountain passes across both the Himalayas and the eastern hills. Situated as such, it is a transitional territory between South Asia and South- East Asia. The transition is reflected not only in the region's flora and fauna, but also in human types, languages and cultures.

Assam is surrounded by seven Indian states and two foreign countries. There are only a few Indian states which have such a strategic location. The state has an area of 78,438 km^2 representing 2.39 per cent of the Indian landmass and a population of 31,205,576 (2011) accounting for 2.58 per cent of the total population of the country.

Historical Evolution

Assam is an ancient land with a long history. The region is referred to as Pragjyotishpura in the earlier epic period. Its capital was also named Pragjyotishpura. The old Sanskrit literature seems to use both Pragjyotishpura and Kamrupa to refer to the same territory. The boundary of the ancient Kamrupa used to fluctuate in different periods. At one stage it extended from Dikkar River (that flowed in the Sadiya region) in the east to the Karatowa River (a river, consisting in the past, of now abandoned channels of the Koshi, Mahananda and Tista).

Between the fourth and the twelfth centuries A.D., Kamrupa was ruled by the Barman, Salastambha and Pala dynasties. Kumar Bhaskar Barmana (594 A.D.- 650 A.D.), a contemporary of the north Indian emperor, Harsha Vardhana, was a famous king of Kamrupa. It was during his reign that the famous Chinese traveller, Huen Tsang, visited Kamrupa (643 A.D.) and wrote in praise of beauty and prosperity of the country. At the beginning of the thirteenth century, Muslim invasions started (1205 A.D. onward) from the west, and the Ahoms entered from the east under the leadership of Seu-ka- pha (1228 A.D). With the entry of these two forces into the political arena of the territory, there occurred a series of political changes. The western part of Kamrupa became an independent kingdom called Kamata, with its capital at Kamatapur (located 32 km south of the present Cooch Bihar Town), the middle part of the Brahmaputra Valley came under the sway of the Bhuyan feudal lords, and the Kacharis and Chutiyas carved out their Kingdoms in the southern and north-eastern parts of the Valley respectively. The Ahoms, meanwhile, established their kingdom over the area between the Burhi-Dihing and Disang and Dikhou rivers. While they continued their effort to consolidate and expand their rule, the Koches, in the fifteenth century, created a kingdom of their own (in 1485 A.D.) in the western part, based on the erstwhile Kamata territory. In course of time, the Koch Kingdom was divided (1584 A.D.) into two portions, the western portion becoming a protectorate of the Mughals and the eastern part becoming vassal of the Ahom king.

The origin of the name 'Axom' (Asom or Assam) is not fully clear. Baharistan - I - Ghaybi written by Mirza Nathan, depicting the geo-political conditions of eastern India during the period 1608-1624, referred to the Ahom territory as the 'Country of Asom' . It is mentioned here that Saiyid Abu Bakr was appointed to command the expeditionary force to conquest Asom, i.e. the territory ruled by the Ahoms (Nathan, 1992). Banikanta Kakati points out that Suryakhari Daivajna, who compiled the 'Darang Rajvanshawali, in the sixteenth century, referred to the Ahom community (and not the land they held) as Axom (Kakati, 1996). Sankardeva, in one of his verses, composed in the early sixteenth century, while referring to the castes and communities of the then Assam, used the term Axom to mean the Ahom (Dutta, 1969). Daityari Thakur, in his biography of Sankardeva, written in the seventeenth century also referred to the Ahom community as Axom (Kakati, 1996). Sihabuddin Talish, who accompanied Mir Zumla as a scribe in the latter's invasion of the Ahom Kingdom in 1662-63, referred to the land east of Guwahati up to Sadiya as

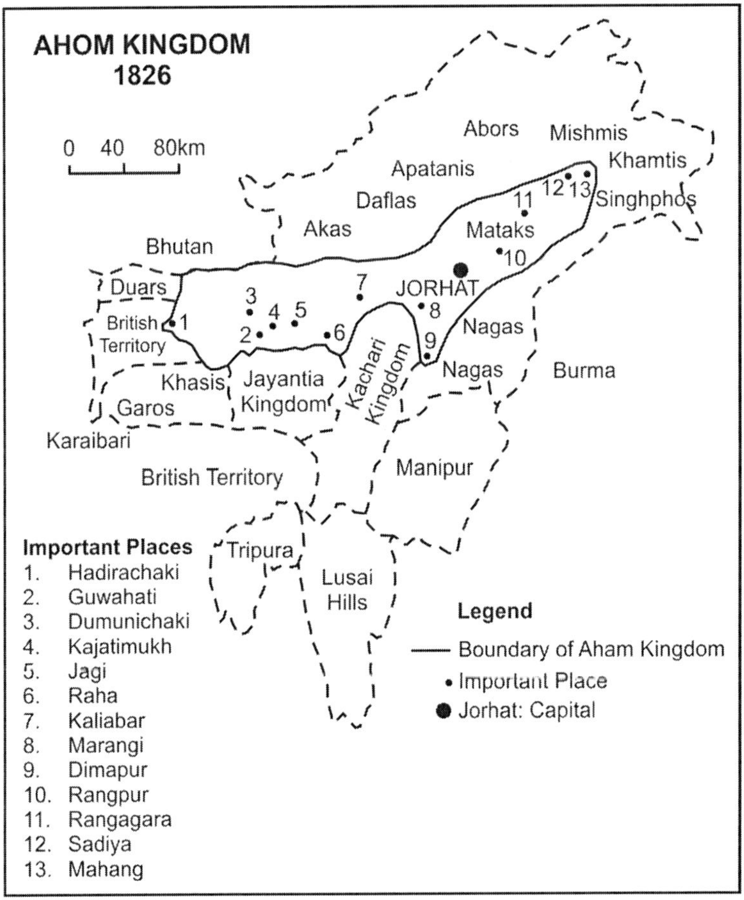

Fig. 2.1

Asom (Talish, 1872). Incidentally, that exactly was the territory then held by the Ahom king.

These references show that the term 'Axom or Asom' was earlier used to mean the Shan community (the term by which the Ahoms initially introduced themselves), who established their Kingdom in eastern Assam. Subsequently, the term came to be used to mean also the territory they ruled. With the fall of the Koch power, when the western part of the Brahmaputra Valley also came occasionally under Ahom administration, the term 'Axom' or 'Asom' received a wider connotation to mean whole of the valley. Later, when the British annexed the territory in 1826

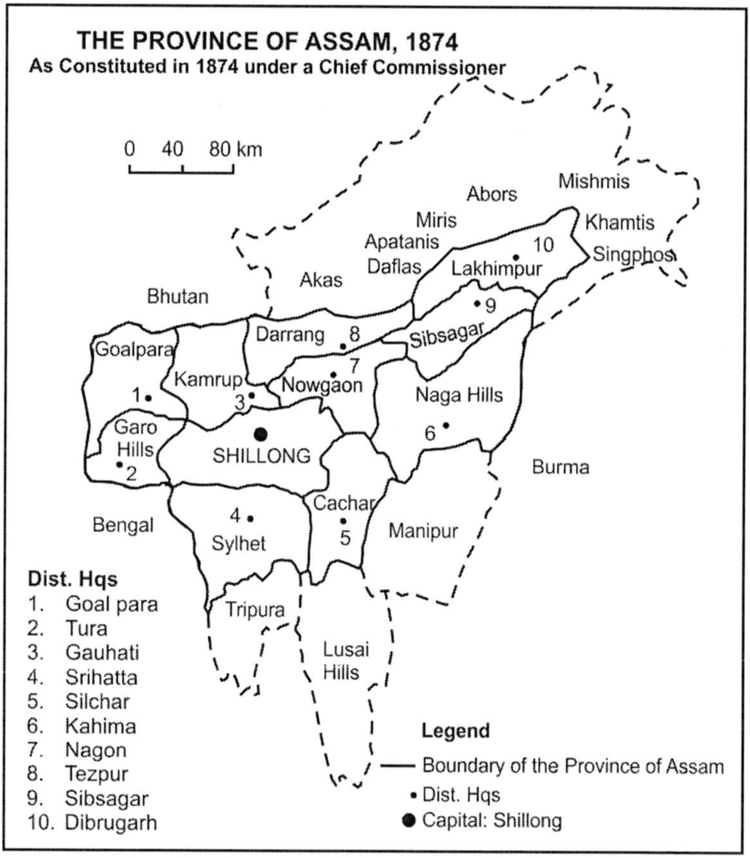

Fig. 2.2

(Fig. 2.1) and constituted it into an administrative division under a Commissioner, they retained the name Assam. Further they added to it the southern hills, plateaus and plains, which they subsequently annexed. The whole territory was constituted a Province on February 6, 1874, as the Province of Assam (Fig. 2.2) under a Chief Commissioner.

The formation of Assam as an administrative province under British India is a story of annexation of small native kingdoms and principalities in North-East India and their merger into one unit. The undivided Goalpara district (comprising the present districts of Dhubri, Bongaigaon, Goalpara and the southern part of Kokrajhar) and (Karaibari, western part of Garo Hills) came to be controlled by the East India Company following the transfer of 'Dewani' from the Mughal emperor to the former on August 12, 1765. In fact, a district named Goalpara was constituted by the British, comprising the above mentioned areas, under the Commissioner of Cooch Bihar in 1821. The Brahmaputra Valley between the Manas River in the west and Kumatia River (on the north bank) and the Burhi Dihing River (on the south bank) in the east was annexed by the company following the Treaty of Yandaboo with the invading Burmese army on February 24, 1826. Between 1828 and 1838, the area covering the present districts of Lakhimpur, Sibsagar,Jorhat and Golaghat was allowed to be administered by the Ahom King Purandar Singh. It was finally merged with the British territory in November 1838, on the plea of incompetent but ruthless administration of the king. The area to the east of the Ahom territory, i.e. the portion comprising present Dhemaji, Dibrugarh and Tinsukia districts (except Sadiya Sub-Division) was taken over from the Matak prince by the British in November, 1839, on the plea of the former's inability to pay the annual tribute. The area to the farther east (east of the Matak territory), now covered by Sadiya Sub-Division and foothills of the Lohit district was annexed by the British in 1836 from the Khamti prince due to alleged rebellion of the latter. Subsequently, the area in the southern part of the present Lohit district was also taken over by the British in 1843 from the Singphos on the plea of alleged insubordination of the latter, thus completing the annexation of whole of the Brahmaputra Plain and its immediate foothills in the east.

The Plain was then divided into two administrative parts and named them as Upper Assam (east of the Jia Bharali on the north bank and the Dhansiri on the south-bank) and Lower Assam (west of the rivers named, up to the Sonkosh and Karaibari). The two parts were placed under two Assistant Commissioners. The headquarters of the upper Assam part was Sibsagar (till 1912, when it was shifted to Jorhat), and that of the lower Assam part was Guwahati. By about 1834, the area of the Brahmaputra Valley annexed by the British, was divided into four districts, viz. Goalpara, Kamrup, Darrang and Nowgong on the one hand, and the protectorates of Upper Assam, on the other. Subsequently, the protectorates were also taken over, as mentioned above, by 1843 and the whole Valley was constituted by a Division, called Assam Division under a Commissioner with its headquarters at Guwahati.

Simultaneous with the annexations in the Brahmaputra Valley, there were annexations also in the south. The territory covering the present Cachar- Hailakandi districts and North Cachar Hills, was under the Kachari King, Govind Chandra. He died heirless and his territory was annexed by the British on August 14, 1832. The territory in the Barak Valley thus annexed, was first made a district in the name of Cachar District under the Commissioner of the Brahmaputra Valley ('Assam'), but later on it was transferred to the Dhaka Division of Bengal. The hilly part of the Kachari Kingdom, i.e. the present North Cachar Hills District and a large part of the present Diphu Sub-division, was then under Tularam Senapati, a rebel Kachari leader, again as a Company's protectorate. This portion (surrounded by the Jamuna river in the north, the Dhansiri in the east, Barail Range in the south and the Diyung and Kapili rivers in the west) was annexed by the British in 1834 on the plea of Tularam's depredation on the former's territory. Tularam was, however, allowed to remain as its feudal lord. After his death in 1844 the land was divided between his two sons. They could not properly administer their respective parts. The East India Company finally took over the region. The northern part inhabited mainly by the Karbi people was joined with Nowgong district, while the present North Cachar Hills was constituted an administrative sub-division and attached to Cachar District.

It is noticed that by about 1828, the East India Company virtually became the master of both the Brahmaputra and the Barak Valleys. They wanted to construct a road linking the two Valleys across the Khasi Hills. The road began to be constructed from Sylhet northward and ultimately reached Nongkhlaw (in West Khasi Hills District), where they wanted to construct a sanatorium also. This step was not kindly taken by the Siem of the area, Tirot Singh, who started a resistance against the British on may 5, 1829. The resistance continued for long six years. Tirot Singh was finally defeated and Khasi Hills was annexed by the British on February 11, 1835.

The Jaintia territory was still a separate kingdom under Ram Singh III, alias Rajendra Singh, albeit as a Company's protectorate. The king abdicated his throne on March 15, 1835 and the territory was annexed by the Company. The Khasi and Jaintia territories were merged into one unit and was constituted a district under the Assam administrative Division in the name 'United Khasi and Jaintia Hills District'.

There was little geo-political change in this part during the period 1843-1865. Then in 1866, the northern part of the present Kokrajhar district (comprising Bijni Duar, Sidli Duar, Chirang Duar, Rifu Duar and Guma Duar areas), which was earlier under the dual control of the Bhutanese King and Koch feudal lords, was annexed by the British and constituted a district, called Eastern Duars under Kuch Bihar Division. The district was, however, dissolved in 1867 and made a part of the then Goalpara district.

In 1866, the present Wokha District of Nagaland was attached to the British territory on the plea of depredation of the Angami Nagas on the British areas. Like other Naga foothill areas, this tract was firstly made a part of Sibsagar district. But later on , all the areas of the Naga hills, annexed by the British, were constituted into a district called Naga hills District.

While the foothill part of Garo hills bordering the Bengal plain and the Brahmaputra Valley was already under the British, comprising an estate called Karaibari, the interior higher hills, bordering the Khasi Hills in the east, still remained under the tribal chiefs. This interior part was taken over by the British and a new district named Garo Hills District was constituted in 1869. This district also included the western Karaibari area and was placed under the Commission of Kuch Bihar.

While political consolidation continued in the Brahmaputra Valley, the local people kept insisting on introduction of Assamese, their mother-tongue, as the medium of instruction and administration since the late eighteen-thirties in the districts of the Valley. This was agreed to by the British administration in 1871 and Assamese became the medium of instruction in the primary schools, courts and offices of the region. The identity of Axom or Asom or Assam may have received a boost from this politico- social decision also.

On February 6, 1874 the British government constituted a Province under a Chief Commissioner in North-East India and named it Assam, which included their Assam administrative Division (the Brahmaputra Valley) and the hills and plains to the south of it, annexed by the British between 1826 and 1866. The Province initially included ten districts, viz. Goalpara, Kamrup, Darrang, Nowgong, Sibsagar, Lakhimpur, Naga Hills, United Khasi and Jaintia Hills, Garo Hills and Cachar (which included a part of the Barak plain and the present North Cachar Hills District). However, seven months later, on September 12, 1874, the Sylhet District of the then Bengal was added to it, raising the number of districts in Assam to eleven.

There was a period of geopolitical lull after the formation of the province of Assam. However, The British administration used to make excursions to the northern mountainous territory allegedly to contain the depredations of the hill and plain tribes like the Mishmi, Khamti, Singpho, Adi, Apatani, Nishi, Hrusso, Aka, etc. In fact, the British government, based in Assam, started marking out Inner Line and Frontier Tracts, from September 30, 1875, partly to regulate the activities of the said tribes and partly to consolidate their administration in Assam.

In 1895 almost whole of Lusai Hills (now Mizoram) was annexed by the British government and in 1898 the region was tagged to Assam as its twelfth district.

With the turn of the century, direct and indirect efforts, aimed ultimately at attaining independence of India, began to emerge among the educated Indians and the British government started taking counter-measures covertly and overtly to check the tide. As a part of such a design, they divided the then Bengal in 1905 and its eastern part, comprising 16 districts, was joined with the 12 districts of Assam to form a new Province named 'Eastern Bengal and Assam' under a Lieutenant Governor. This was done partly perhaps to break down the strength of the united Bengali nationality and partly to divide it on the basis of religion (for eastern part of Bengal had Muslim majority, while the western part had Hindu majority population). However, this arrangement of administration had to be abandoned in 1912, under tremendous public opposition when the province of Eastern Bengal and Assam was dissolved. Assam once again became a separate province

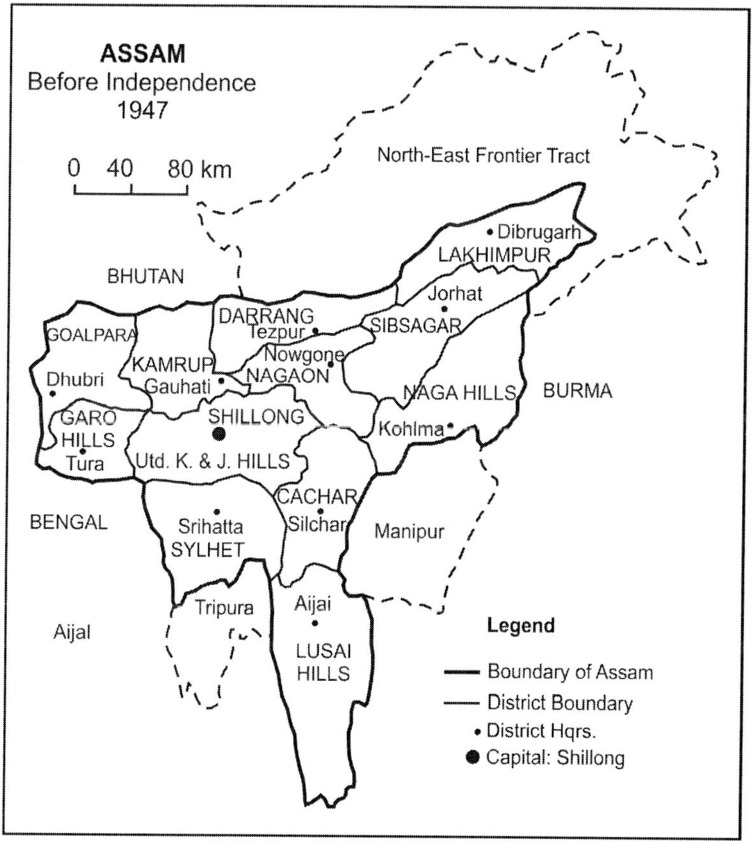

Fig. 2.3

under a Chief Commissioner. Meanwhile minor administrative adjustment were made within Assam. For example, in 1913-14, the Dimapur area, which was in Sibsagar district, was transferred to Naga Hills district. In 1921, under the Montague-Chelmsford Reform suggested in 1919, Assam was made a full-fledged province under a Governor. There was no geopolitical change thereafter till 1937, when the four plain areas of the then North-East Frontier Tracts, adjoining the Brahmaputra Valley, namely the Balipara area, Jonai area, some portions of Sadiya area and Margherita area were merged with Assam and the area of the province remained so till 1947 (Fig. 2.3).

In 1947 when India was partitioned, Sylhet district was separated and tagged to the then East Pakistan, except the Thana areas of Karimganj, Badarpur, Ratabari and Patharkandi, which were merged with the undivided Cachar district.

After independence, there was a long period of administrative reorganisation of the areas of North- East India. On November 17, 1951 the district of United North Cachar and Mikir Hills was created by merging 4421.12 km^2 of hill area from Nowgong district, 1539.65 km^2 of hill area from Golaghat Sub-Division of the then Sibsagar district and 4371.23 km^2 of hill area from Jaintia Sub-Division of United Khasi and Jaintia Hills district. In the following year, an Autonomous District 'Hills district Council' was set up for the district as per Sixth Schedule of the Constitution. In 1963, Naga Hills district was separated from Assam for creating a state out of it. On January 2, 1970, United North Cachar and Mikir Hills was divided into two Autonomous district, viz. North Cachar Hills and Mikir Hills (which was renamed Karbi Anglong in 1976). In 1970 Garo Hills and United Khasi and Jaintia Hills were seceded from Assam to form an Autonomous state, which was accorded full-fledged statehood as per North Eastern Areas (Reorganisation) Act 1971 of the Union Government. This Act also took away Lusai Hills from Assam, which was constituted as Union Territory under the name Mizoram. By then, the number of administrative districts in Assam came down to only nine.

However, the exigencies of socio-economic development, under the regime of planning, necessitated the creation of smaller districts. On October 2, 1971, old Lakhimpur district was divided into Dibrugarh and Lakhimpur districts. While the part lying to the south-east of the Brahmaputra was made Dibrugarh district, the socio- economically backward north-western part comprising Dhemaji sub-division and Lakhimpur sub-division, was made the district of Lakhimpur. Assam, thus, came to have ten districts from late 1971. Following the above principle, the number of districts was raised to 17 between 1981 and 1987. The districts of Dhubri and Kokrajhar were carved out of the old Goalpara district. Similarly, Barpeta and Nalbari were carved out of the old Kamrup district. Besides, the old Darrang district was divided into Darrang and Sonitpur districts, old Sibsagar into Sibsagar and Jorhat districts and old Cachar into Cachar and Karimganj districts. Thus, during the early nineteen eighties, the number of

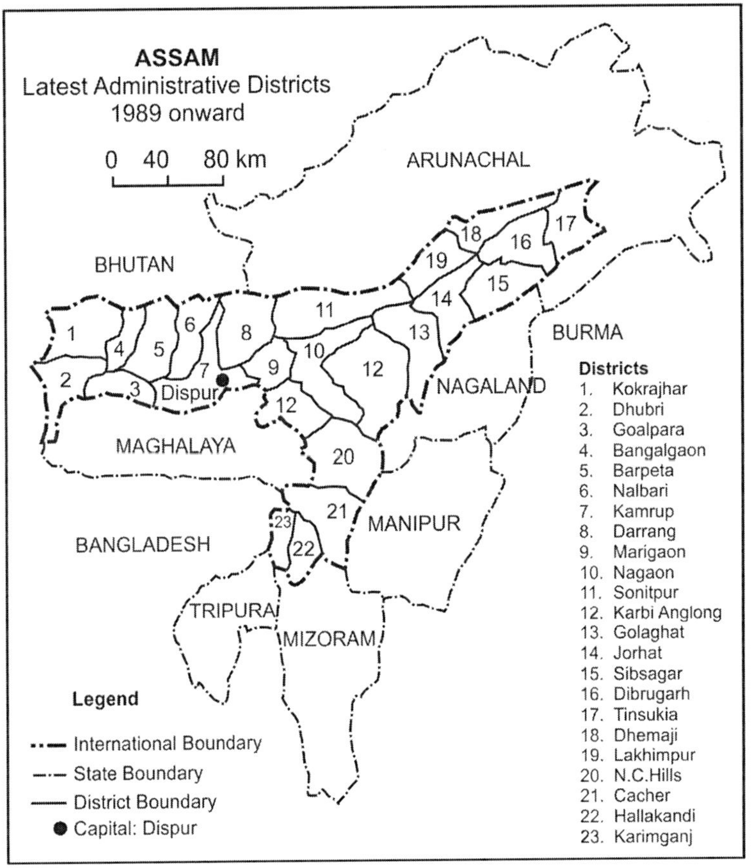

Fig. 2.4

districts in Assam rose to 17. Again between 1988-89, Golaghat was carved out of Jorhat, Dhemaji out of Lakhimpur, Marigaon out of Nagaon, Tinsukia out of Dibrugarh, Bongaigaon out of Kokrajhar and Goalpara and Hailakandi out of Cachar. Thus by 1989, the number of administrative districts rose to 23 (Fig. 2.4). By 2017 the number of districts in Assam reached 34. The aim of creation of smaller districts, as mentioned above, was to bring administration nearer to the people, so that the benefit of socio-economic development can be quickly and directly delivered to them.

A new administrative change was introduced in Assam during the 1990s. The plain tribes having sufficiently large population in each and living more or less in contiguous areas, were granted autonomy, in respect of certain civil administrative aspects, following agitation by such groups as the Bodo, Rabha, Tiwa and Mising, who are not only socio-economically backward, but also occupy relatively backward areas. The Bodo Autonomous Council was constituted on March 7, 1993, comprising 2570 villages of Kokrajhar, Barpeta, Nalbari and Darrang districts along the Himalayan foothills with headquarters at Kokrajhar. Subsequently, between 1994 and 1997, Rabha-Hasong Autonomous Council was formed over the south-eastern foothills of Goalpara and south-western foothills of Kamrup districts with headquarters at Dudhnoi. Similarly, the Tiwa Autonomous Council was constituted over the south-western part of Marigaon with headquarters at Jagiroad and Mising Autonomous Council came up over the Mising dominated villages of Dhemaji and Lakhimpur districts and Majuli sub-division of Jorhat district, with headquarters at Gogamukh (in Dhemaji District).

The above discussion gives an outline as to the process of evolution of Assam into a British Province and finally to an Indian State from the Ahom Kingdom and its neighbouring principalities-large and small. A close look at the geopolitical events reveals that the period of evolution spread over the nineteenth and twentieth centuries, can be divided into two periods : Firstly, the period of geopolitical consolidation, from 1826 to 1947 and secondly, the period of geopolitical reorganisation from 1947 onward.

Process of Peopling

Because of its transitional location between South and South-East Asia, Assam has been receiving migrants from the north, east, south and west, across the high Himalayas, over the tangled mass of ridges and valleys of the Patkai Range, Manipur Hills, Lusai Hills and Arakan Yoma and over the plains of the Ganga-Brahmaputra Doab since time immemorial. The state had a population of more than 31 million as per the census of 2011. The natural environment of Assam with fertile river valleys, abundant rainfall, luxuriant vegetation and a rich collection of fauna, had been such that it could attract groups of people to settle. There had been, therefore, streams of migrations into it even during the palaeolithic period. An analysis of the sequence of peopling of Assam is difficult to make, firstly because, no systematic and tangible evidence has so far been found as to the human occupation of the land during the prehistoric and protohistoric periods; and secondly, there has been such an intermixture of both blood and cultures of different groups of people in this transitional part, that little can be traced to establish a succession of peopling.

However, a close look at the ethnolinguistic composition of the people of the state reveals that there are people speaking Austro- Asiatic, Tibeto-Chinese and Indo-European languages. An analysis of peopling of Assam on the above ethnolinguistic basis, coupled with whatever scanty palaeolithic, neolithic and historical evidences are available, shows that there are as many as eleven 'waves' and 'streams' of migration into Assam (The term 'wave' is used to mean a migration that occurred once at a particular point of time and the term 'stream' is used to depict a migration of a group of people that continues for long time). The 'waves' and 'streams' of migration are identified as follows:

1. The first group of migrants to settle in this part of the country is perhaps the Austro-Asiatic language speaking people who came here from South-East Asia a few millennia before Christ. They had perhaps earlier lived in the foothills bordering the Brahmaputra Valley and then moved to the hills of Meghalaya, perhaps yielding place to the new migrants of Tibeto-Burman stock. These Austro-Asiatics are now represented by the Mon-Khmer speaking Khasis (both the Khasis proper and Pnars taken

together), some of whom also live in Karbi Anglong and North Cachar Hills of Assam.

2. The second group of migrants came to Assam from the north, north-east and east. They are mostly the Tibeto-Burman language speaking people and include such major groups as the Eastern Himalayan (Bhutias, Monpas, etc.), North Assam (Aka, Hrusso, Nishi, Adi, etc.), Bodo, Naga, Kuki-Chin and Kachin. They are racially Mongoloid, and unlike the Austro-Asiatics have set up a stream of migration which has been continuing still today, although in a trickle. Cross border migration of some Tangsas, Nagas and Kuki-Chins from Myanmar is not altogether unknown even today. Assam has a very large number of Tibeto-Burman speaking people comprising both indigenous and recent migrants.

3. From about the fifth century before Christ, there started a trickle of migration of the people speaking Indo-Aryan language from the Gangetic plain. Their migration has continued ever since. In the past. Indo-Aryan migration was even occasionally sponsored by the local kings and Chieftains of Assam, although in most cases, there were voluntary streams with migrants moving in search of economic opportunity in the Brahmaputra and Barak plains. They are originally of Caucasoid racial descent , who mixed with the subsequent migrants of the same stock and the locals of other stocks and have now formed the majority of the Hindu population of the state. They are also the ones, who spread and built-up the Vedic socio-cultural basis of the state. The spread of the Indo-Aryans into the Barak valley was contemporaneous to that in the Brahmaputra Valley, their settlements being an extension of that in the Gangetic Delta of Bengal.

4. While the migration of the old Indo-Aryan socio-cultural group was still continuing, there started another stream of disguised migration of another group of Indo-Aryans belonging to Islamic faith. Mohammad Bin Bakhtiar Khilji, a Muslim General of Kutubuddin, in about 1205 AD, reached Assam with his force of army in his attempt to go to Tibet. His movements within Assam enraged the local king. He was defeated and some of his captive soldiers were settled in Assam, especially in the area between Hajo and Sipajhar in the Brahmaputra Valley. Subsequent to that, there were several invasions by the Nawabs of Gaur and Dhaka and Sultans and emperors of Delhi. In most of their

invasions, whether they won or lost, they had to leave behind some of their men and materials. It is important to note in this connection, that the persons connected with the Muslim invasions of Assam were not only soldiers and generals, there were also artisans, craftsmen, technicians, apart from scribes, linguists and travelogue writers. All these personnel were necessary for the invading party, because invasion of Assam, then a terra incognita, was a long-term project. Muslim captives and left-outs, therefore, consisted of the types of people mentioned above and they were found to be useful by the local kings. They were, therefore, allowed to settle. Thus, there grew up in Assam, a Muslim community, which grew over time with local conversions mostly perhaps through marriages. This disguised migration of the Muslims was feeble and halted and the settlers mostly married locally. Thus, they became a part of the greater Assamese community. The introduction of Muslim element into the Barak Valley was also contemporaneous to that described above, for after the conquest of Bengal by the Muslim in 1203-1204 AD, they reached this valley through the Gaur Kingdom.

5. About twenty-three years after the first invasion by the Muslims, there came in another group of migrants from Shan State of northern Myanmar. Under the leadership of an able and mighty general, Seu-ka-pha, in 1228 AD, there came in a group of Tai or Shan people belonging to the Siamese-Chinese linguistic sub-family, who were locally known as Ahom in Asom. They established a kingdom in eastern Assam between the rivers Burhi-Dihing and Disang, which grew into a large one, occasionally covering whole of the Brahmaputra Valley. The Ahoms married locally, learnt the local language (Assamese) in course of time and adopted Hindu religion and culture. The kingdom they built lasted for long six hundred years from 1228 to 1826. They made positive contributions to consolidate Assam's regional socio-cultural identity.

6. After the consolidation of administration by the Ahoms in upper Assam and by the Koches in lower Assam, there prevailed a period of peace and prosperity in the Brahmaputra Valley and the stream of Indo-Aryan migration from Bengal and other parts of the Gangetic Valley received a fillip under the sponsorship of these kings, for religious and economic reasons. Similarly, the

invasion of the Nawabs and Sultans also continued to leave behind some Muslim settlers. Apart from these, there was a migration of some Tai people from upper Myanmar, which added a new element to the population mosaic of Assam. These 'Later-day' Tai people, entering Assam during the period from seventeenth century to early nineteenth century, are the Khamtis, Khamyangs, Aitons, Phakes and Turungs. Each of these groups came at different time from different areas across the Patkai. They are Buddhist by religion and speak Tai language and unlike the Ahoms, have still retained their religion and culture. They are to be found in small groups in upper Assam today.

7. It was the above six groups of people who inhabited Assam until it was taken over by the British. The British annexation of Assam brought about not only political change, but also profoundly changed the population structure of the region. The British introduced new administrative and economic order. They needed English educated personnel to help them in administration. They also needed trademen with the introduction of full-scale monetised economy in Assam. Further, they needed policemen, security personnel, technicians, professionals and artisans for safety and security and for public works and various other services. Thus started a new stream of migration with the coming in of Bengali government service-holders and professionals, Rajasthani tradesmen and north-Indian labourers and artisans, Nepali security men, etc. The trend of immigration, thus gaining momentum, continued unabated till recently, giving a new dimension to the population base of Assam.

8. Some British explorers and prospectors, who had known of the presence of natural tea bushes in Assam, started tea plantation in the region once it was annexed by the East India Company. As the local people were found to be unwilling to work in the tea gardens as labourers, the planters recruited indentured labourers from the tribes of Chotanagpur of Central India. These tribals are enthnolinguistically Austro- Asiatic speaking Mundari group of languages. Most of these belong to such tribes as Munda, Ho, Santal, Savara, Oraon, Gond, etc. Thus a new wave of Austro - Asiatic migration started in about 1848 and continued till 1940. Assam today has about two million people of the origin of this group of immigrants.

9. After the annexation of the Brahmaputra Valley by the British in 1826, they consolidated their rule over almost whole of North-East India, including the hills. But, inter-tribal feud, political conflict and social unrest still prevailed in upper Myanmar, across the border of British India. As such, a large number of Kachin, Naga and Kuki-Chin people migrated to the contiguous areas of North-East India across the border. Some of them, especially the migrants of the Kuki-Chin stock, migrated to Assam beyond the hills of Mizoram and Manipur. Some of the Singphos, living in the Changlang border of Tinsukia district and the various Kuki tribes, who live in North Cachar Hills and Karbi Anglong, are of the origin of these migrants of the nineteenth and early twentieth centuries.

10. Still another stream of immigration started in the last two decades of nineteenth century with the coming in of the landless Muslim peasants from the then eastern Bengal (now Bangladesh), firstly, at the instance of the landlords of the undivided Goalpara district in connaivance of the British administrators. The stream became large during the period between 1901 and 1947, the immigrants occupying the active flood-plains of the Brahmaputra, where they took to jute, vegetable and Aus and lowland rice growing. This stream has brought in one of the largest volumes of immigrants which have affected even the demographic scenario of Assam. It is further noticed that the stream of this immigration continued even after the partition of India. The people of the origin of this group of immigrants now number about 4 million in Assam. They are racially perhaps Aryo-Dravidian and speak a dialect of Indo-Aryan Bengali. Their descendents are fast adopting Assamese language and culture.

11. The partition of India, at the time of independence, provoked unofficial transference of population and millions of Hindu refugees from the then East Pakistan entered India. Assam being contiguous to it, received a large volume of Hindu refugees, from 1947 over a few decades, numbering about a million. Unlike the Muslim immigrants from East Bengal, who generally went to the flood-plains and took to agriculture, the Hindu refugees of East Pakistan generally settled around the towns, service centres and railway junctions and stations and took mostly to tertiary activities.

The eleven waves and streams of immigration, recorded above, are the ones that stand out in temporal perspective. Besides these, there were other minor streams and waves also, as evidenced by the people coming in from all over India and beyond in search of job, trade opportunity and for other economic activities. They normally settle in the town and industrial locations, and being small groups get themselves merged with the local population in course of time. Be that as it may, the present population of Assam, who are permanent residents, whether of indigenous or immigrant origin, are largely the descendents of the eleven main streams and waves of migrants. Their intermixture, over the years, has contributed to the rich socio-cultural heritage of the region.

References

Barpujari H.K. (ed. (1977): *Political History of Assam*, Vols. 1 & II, Government of Assam, Dispur, Guwahati.

Bhattacharjee, K.K. (1983): *North-East India : A Study,* Cosmo Publication, New Delhi.

Dutta, H.N. (ed.) (1969): *Shrimadbhagavat, Skandha*-2, Dutta Baruah and Co., Nalbari, p.38.

Kakati, B.K. (1996): 'Asamiya Bhasar Gathan Aru Bikash', Translation from *Assamese Language, Its Formation and Development* by B. Kakati Birth Centenary Celebration Committee, Barpeta, pp. 1-6.

Nathan, Mirza (1935): *Baharistan-I- Ghaybi,* tr. by M.I. Borah, DHAS, Govt. of Assam, Vol. I, Guwahati, p. 319.

Taher, M. (1993) : 'The Peopling of Assam and Contemporary Social Structure', in Ahmad, A. ed. *Social Structure and Regional Development*, Rawat Publications, Jaipur, pp 201-219.

Talish, Sihabuddin (1872): 'Fathiyahi- i - Ibriya as' cited by H. Blochman in 'Koch Behar, Koch Hajo and Assam in the 16th and 17th Centuries...', *Journal of Asiatic Society of Bengal,* Calcutta, pp 64-101.

3

Physical Background

Geology

The state of Assam occupies a unique position amidst complex geologic and physiographic make-up of the north-eastern region of India. It is bordered on the north by the Arunachal Himalayas, on the south by the Meghalaya plateau and the hills of Manipur, Mizoram and Tripura, on the east by the Arunachal Himalayas and the Naga-Patkai hill ranges and on the west by the plains of Bangladesh and West Bengal. The state is bounded by highlands of hills and plateaus on the three sides except the western one, where the Brahmaputra valley merges with the Gangetic plain. The geologic formations occurring in Assam belong to the Archaean, Pre-Cambrian, Tertiary and Quaternary periods. As revealed by its geologic history, the formations in Assam may be broadly classified into five. These include (i) the Archaean group of rocks (ii) the Pre-Cambrian rocks (iii) the lower Tertiary sediments (iv) the upper Tertiary sediments, and (v) Quaternary alluvial sediments.

The Archaean rocks comprise the metamorphic rock types of gneisses and schists which are intruded by younger acidic and basic intrusives. These rocks are found to occur in the northern and central parts of the Karbi Plateau and the northern foothills of the Meghalaya Plateau along the Assam-Meghalaya border. The isolated monadnocks like remains consisting of gneisses and granites scattered along the north and south bank of the Brahmaputra in Goalpara, Kamrup, Darrang, Sonitpur and Nagaon districts also belong to the Archaean group of rocks. The Pre-Cambrian rocks consisting of quartzites and phyllities are confined to small areas over the western flank of the Karbi Plateau and the northern part of the North Cachar hills. The lower Tertiary sediments known as the Jaintia Group composed mainly of sandstone and limestone got deposited during the Eocene period and are found to occur in the Kopili valley, the southern and eastern slopes of the Karbi Plateau and some parts of the North Cachar hills. The upper Tertiary sediments of sandstone, sandy shales and siltstones deposited during the Oligocene-Miocene-Pliocene periods are confined to the southern flank of the Karbi Plateau, the North Cachar hills, the hills of the Barak valley of Cachar district, the foothills of the Naga-Patkai ranges bordering the southern margin of undivided Sibsagar and undivided Dibrugarh districts and the narrow belt of the Himalayan foothills along the northern border of Assam. The Quaternary sediments of older and newer alluviums, on the other hand, comprise the recent alluvial deposits of the Brahmaputra and Barak valleys, the red river-bank soils and high level terrace deposits.

Physiography and Relief

The present physiographic configuration of Assam has taken its shape only during the geologically recent times. The geologic and tectonic base of the state has given rise to a variety of landforms under varying climatic conditions and geomorphic processes. The low hill ranges with hot and humid climate and heavy rainfall, concentrated to a few months of the year, experience solifluction, sheet erosion and landslides. The incidence of landslides is high in the Himalayan foothills where heavy rainfall, high seismicity and toe-cutting of hill slopes by the streams are most frequent. Heavy rains often loosen soils and soft Tertiary rocks of the Himalayan foothills as rain water percolates through joints, fractures, foliations and pores of rocks and soils. This eventually leads to heavy slope failure and landslides. Fluvial processes, on the other hand, are significantly dominant on the valley bottoms and plains where alluvial deposition takes place due to erosion of the higher surfaces by rivers and floodings in the valleys. The erosional and depositional processes conspicuously intensified by copious rainfall and frequent seismic movements play dominant role in shaping various physiographic units of the state. However, Assam's physiography may be described in terms of the physiographic elements like plains, floodplains, hills, foothills, plateaus and river valleys. Thus, the broad physiographic divisions of Assam can be delineated as (a) the Brahmaputra Valley (b) the Barak Valley (c) the Karbi Plateau, and (d) the Barail and Southern Hills. (Fig. 3.1).

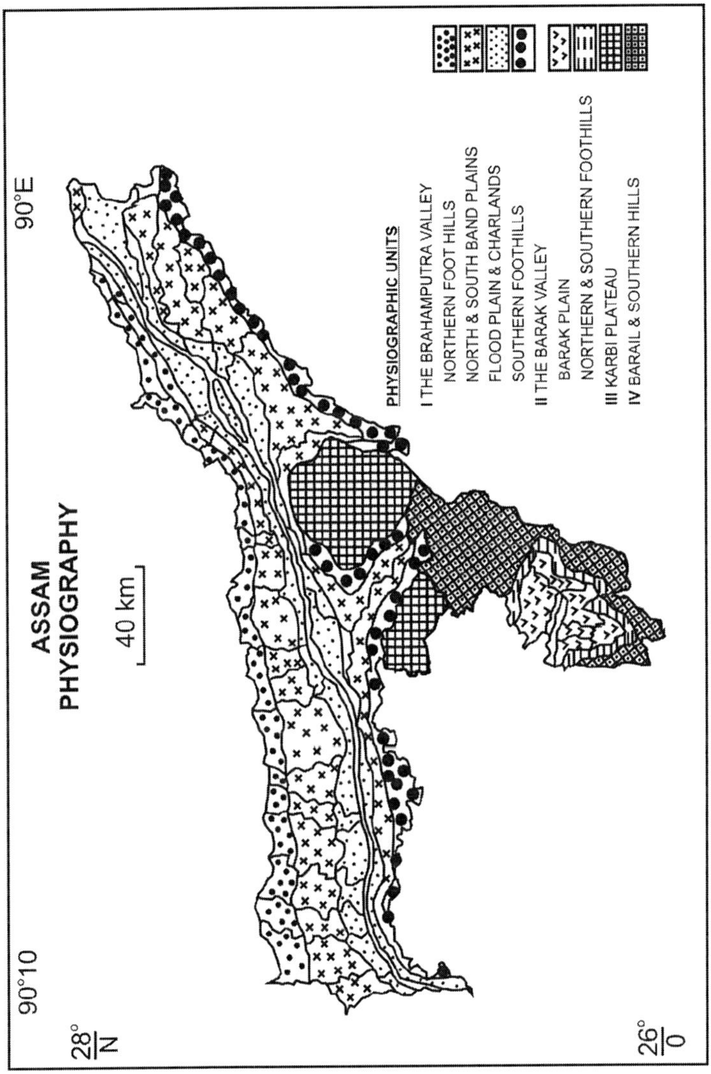
ASSAM
PHYSIOGRAPHY
40 km
90°10
90°E
28°
N
26°
0
PHYSIOGRAPHIC UNITS
I THE BRAHAMPUTRA VALLEY
NORTHERN FOOT HILLS
NORTH & SOUTH BAND PLAINS
FLOOD PLAIN & CHARLANDS
SOUTHERN FOOTHILLS
II THE BARAK VALLEY
BARAK PLAIN
NORTHERN & SOUTHERN FOOTHILLS
III KARBI PLATEAU
IV BARAIL & SOUTHERN HILLS

Fig. 3.1

The Brahmaputra valley is the major physiographic unit of Assam. It is a narrow valley with an approximate east-west extension of about 720 km and average width of 80 km. The valley is girdled by the Eastern Himalayas on the north and east, and the Naga hills, Karbi and Meghalaya Plateaus on the south. It is open in the west merging with the plains of West Bengal and Bangladesh. The 640 km long reach of the Brahmaputra and its 32 major north and south bank tributaries drain the valley which covers about 72 percent of the total area of Assam. The valley as a whole gently slopes from north-east to south-west with an average gradient of 13 cm per km. In its upper reach, near Dibrugarh, the gradient is about 17 cm per km while the same is about 10 cm per km near Guwahati.

Geologically the Brahmaputra is a very young river and its present configuration took shape only during the Pleistocene and Recent times. The valley developed over the foredeep in between the Peninsular mass and the Tethyan Geosyncline. The foredeep, believed to be under the sea till the sub-Recent period, received deposits during all the periods of Tertiary and Quaternary Ages. The Tertiary deposits consisting mainly of sandstones, shale, grit, conglomerate and limestone and the Quaternary ones consisting of alluvial sediments comprising pebbles, sand, silt and clay got deposited over the Archaean basement of the trough. The valley deposits of reddish brown sandy clay with some pockets of unassorted pebbles, cobbles, sand and silt have been identified as older alluvium. The sediments, comprising clay, sand and silt grouped as newer alluvium overlie the older ones with a thickness of 200-300m.

Physiographically the Brahmaputra valley may be divided into two parts near Guwahati (26^{0} 10'N and 91^{0}-45'E): the western lower part and the eastern upper part. The eastern part trending ENE-WSW may be of 'rift valley' origin, bounded by the northern Himalayan thrusts and the southern Naga Hills thrusts. The western lower part trending almost east-west is a 'ramp valley' or a sag in the crust formed tectonically during Miocene period along with the Gangetic depression in front of the rising Himalayas. These two valleys together formed the foreland mentioned earlier.

The Brahmaputra valley in its east-west direction has four

distinct physiographic units, viz. the northern foothills, the north and south bank plains, the floodplain and charlands, and the southern foothills. The northern foothills comprise the sub-Himalayan ranges mainly composed of Tertiary sandstones. The narrow zone with elevation ranging between 150-600 m is relatively wider on the western part and tappers eastward in Sonitpur and Lakhimpur districts. The highlands are characterised by southernly dipping piedmont plains formed as a result of coalescence of alluvial cones. It is a zone of unassorted detritus, where a major portion of stream water percolates down and reappears a few kilometers downstream. The southern part of this zone is a plain of tall grasses and damp soil which provides ideal sites for reserved forests and wildlife sanctuaries.

The built-up zone of the valley lying between the northern and southern foothills and the active floodplain is made up of the recent alluvial sediments carried by the Brahmaputra and its tributaries from the surrounding highlands. This zone consisting of the north-bank and south-bank plains is a region of immense human significance with high population density, rich agricultural fields and a good network of roads and railways. The north bank plain is relatively wider, being about 30 km on the average in the lower Brahmaputra Valley while it narrows down to an average of about 10 km in the upper Brahmaputra Valley. The south bank plain, on the other hand, is narrower on the average of about 5 km in the districts of Kamrup, Goalpara and Dhubri due to jutting out of the Meghalaya Plateau towards the south bank of the Brahmaputra. The plain then starts widening to the east of Guwahati and merges with the Kopili plain. Thereafter, the plain again gets reduced to a narrow strip in Nagaon and in the western part of Golaghat district due to northward extension of the Karbi Plateau. In the districts of Golaghat (eastern part), Jorhat, Sibsagar, Dibrugarh and Tinsukia the plain is fairly wide as the rivers like Dhansiri, Dikhow, Burhi Dihing and Dibru have expanded the plain by their headward erosion. Like the north-bank built-up plain, the south-bank built-up plain also supports high population concentration, fertile agricultural fields and the networks of roads and railways.

The floodplain of the Brahmaputra including the charlands inside the river lies between the north and south bank plains. The floodplain is irregular in its transverse extension due to the occurrence of occasional hillocks and incipient levees on both the banks of the Brahmaputra. On the north bank, the floodplain contains numerous swamps and '*beels*' and is fairly wide in Dhemaji and Barpeta districts. The floodplain on the south bank, which is generally narrower is somewhat extensive near the confluences of Burhi Dihing, Disang, Dhansiri, Kalong- Kopili and Kulsi. The floodplain zone of the Brahmaputra Valley provides *Ahu* and *Bao* rice fields and the islands inside the river support favourable sites for rabi crops and cattle grazing. The continuity of the floodplain is broken by some isolated hillocks of Archaean origin dotted along both the banks of the Brahmaputra. Starting from the easternmost Sirang Pahar (189 m) in Nagaon district, the other important south-bank hillocks towards west include the Kukrakata Pahar (270m) and Kamakhya Hill (239m) of Nagaon district, the Bura Mayang Pahar (348m) of Morigaon district, the Tatimara (160m), Hatisila Pahar (100m), Kharghuli Pahar (216m), Kamakhya-Nilachal (293m) and Dakhla Pahar (138m) of Kamrup district, the Surya Pahar (221m), Hulokanda (131m), Pancharatna Hill (313m), Paglatek (200m), Moronga (221m) of Goalpara district. On the north bank the important hillocks starting from the eastern -most Bhomoraguri hill (140 m) near Tezpur are the Singri hill (172m) of Sonitpur district, the Kurua (203m) of Darrang district, the Dirgheshwari (329m), Agyathuri (205m), Rakhasini (93m) and Hatimura (228m) of Kamrup district, the Mahadeo (110m), Chakrasila (216m), Bhairab Pahar (498m) and Biswakarma Pahar (320m) of Bongaigaon district and the Chandardinga (245m), Dudhnath (210m)and the Tokrabandha (258m) of Dhubri district. All these Archaean hillocks are the remnants of the northerly outcrops of the Karbi and Meghalaya plateaus.

The large channel of the Brahmaputra having an average width of 6-8 km represents a most dynamic fluvial regime and is marked by intense braiding, rapid aggradation and drastic bankline changes. Such a dynamic fluvial regime is highly responsible for the formation of a large number of sandbars on its own bed. The wide alluvial bed is thus dotted with more than 600 small and big sandbars locally called *chars* or *chaparis* which are the characteristic features of a braided river. These chars, with varying sizes and degree of permanency are separated by several distributary channels. During summer most of the chars are submerged by flood water and they change their locations, sizes and other physical parameters. A small number of these chars are more or less permanent in their existence and among them only a few remain visible even during high floods. Some chars, being endowed with fertile alluvial top soil, support a good number of permanent, semi-permanent and also temporary human settlements. The core areas of the permanent chars, which are relatively high and least affected by floods, provide sites for permanent settlements. Away from the cores with the decrease in height, the effect of flood increases rendering the settlements a semi-permanent character. On the other hand, the fringes of the chars are highly susceptible to erosion and frequent inundation and consequently less attractive for human settlements. It is also to be noted here that Majuli, the largest riverine island of the world, bounded between 26°45' and 27°15'N latitudes and 93°39' and 94°35'E longitudes is situated within the river Brahmaputra in its Upper Assam reach. But, unlike the numerous bars and islands formed due to intense braiding processes of the river, Majuli has its own history of genesis. Majuli was formed as a result of the southward migration of the Brahmaputra which is an important fluvio-geomorphic process of the river. In fact, Majuli was an integral part of the present Jorhat district of Assam on its south bank. As revealed from the historical records, southward shifting of the Brahmaputra by way of capturing the lower course of its tributary, the Burhi Dihing flowing parallel to the master stream had caused Majuli to emerge as a riverine island during the first half of the 18th century. Covering presently an area of 645 km^2 with a population of 1,35,378 (1991 census), inhabiting in 155 villages, the island is now an administrative sub-division of the Jorhat district of Assam. But, in recent years, especially after the great earthquake of 1950, recurring floods and severe bank erosion are causing havoc to the island inundating and eroding large areas. A spatio-temporal study of the erosional scenario of Majuli reveals that in 1915 the island occupied an area of 1245sq. km. and its area was reduced to 925 sq. km. and 752 sq. km in 1971 and 1984 respectively. The area of the island was further reduced to 645 sq. km. in 1995, thereby maintaining an average rate of erosion in the order of 7.4 km^2./year (Bora and Das, 1998).

The foothills bordering the southern fringe of the

Brahmaputra valley comprise the foothills of the Patkai-Naga ranges, northern foothills of the Meghalaya and Karbi Plateau and the Barail range. The foothills of the Patkai- Naga range run along the Assam-Arunachal and Nagaland border in the districts of Tinsukia, Dibrugarh, Sibsagar, Jorhat and Golaghat. These foothills with an average elevation of 300-600m are composed of Tertiary sediments and characterised by the presence of some active faults. Rivers like Burhi Dihing, Disang, Dikhow, Janji, Bhogdoi and Dhansiri, flowing into the Brahmaputra have their built-up plain embayments in the foothill region partly through their headward erosion and partly through subsequent deposition. These gently sloping built-up plains adjoining the foothills provide favourable sites for tea gardens in the bordering districts. This foothill zone consists of piedmont plains, anticlinal ridges and synclinal valleys with terraced alluvial fills. The foothills of the Karbi Plateau which belong to Archaean formations, extend along its northern, eastern and western margins with an average elevation of 500m. The rivers like Dhansiri, Kalong and Kopili with their numerous tributaries extend finger- like plain embayments into the plateau along the foothills. These plains, in their turn, descend to the Brahmaputra plain through terraces. Such terraces filled with older alluvium and residual soils support rich tea gardens in Golaghat and Nagaon districts. The foothills lying along the Assam-Meghalaya border form the parts of the Garo and Khasi-Jaintia hills. These hills after losing their heights merge with the Brahmaputra plain through gentle gradients punctuated occasionally by isolated hills and hillocks in the districts of Kamrup, Goalpara and Dhubri. The rivers like Digaru, Kulsi, Boko, Deosila, Dudhnoi, Krishnai, Jinari and Jinjiram flow northward down the plateau and create small embayment valleys by dissecting and denuding the foothills. In general, the southern foothill belt of Nagaon, Kamrup, Goalpara and Dhubri districts is characterised by erosional platforms born out of age-old erosion by the streams alternated with a series of *beels* and swamps. The north-western part of the Tertiary Barail range also forms a section of the southern foothill zone of the Brahmaputra valley. The foothills on the north-western face of the Barail range with elevation between 600 m and 1400 m cover the northern part of the North Cachar Hills district and comprise the catchment areas of the rivers like Dhansiri and Kopili. These two river systems, through their headward erosion, have dissected and denuded the north facing hills of the Barail range to the foothills of lower elevations and gentle slopes.

Like the Brahmaputra valley, the Barak valley also forms one of the important physiographic units of Assam. The valley comprising the Barak plain and the northern and southern foothills is the product of the fluvio-geomorphic processes of the river Barak and its tributaries. The Barak river, while moving westward for a distance of 225 km in Assam through the Cachar district and along the northern border of the Hailakandi and Karimganj districts, builds up its alluvial plain, sometimes called Cachar Plain. This plain is enclosed by hill ranges on three sides -The Meghalaya Plateau and the Barail range of the N.C. Hills district on the north, Manipur hills on the east and the Mizo hills on the south. The western side is open to the Kushiyara-Surma plain of the Sylhet district of Bangladesh. The Barak plain is the headward piedmont part of the Barak-Surma-Kushiyara plain, which lies largely in Bangladesh (Taher, 1986). The Barak plain slopes down gently to the west and the river Barak flows sluggishly in meandering course through it leaving a series of ox--bow lakes and swamps. The immediate neighbourhood of the river represents an active floodplain characterised by extensive marshy tracts and annual inundation. Occurrence of isolated low hillocks in the plain indicates the erosional activities of the Barak and its tributaries. The western and the central parts of the plain are built up of the alluvial deposits. Thus the plain is created by both the aggradational and degradational activities of the Barak river system. Tributaries like Chiri, Diksa, Digli, Jiri, Madhura, Jatinga, Larang, etc. from the Barail range of N.C. hills and the Sonai, Dhaleswari-Katakhal, Singla and Langai from southern Mizo hills join the Barak adding considerable volume of water and load of sediments. The Barak plain is bounded by the foothills of the Barail range on the north and the foothills of the Mizo hills on the south. As in the Brahmaputra valley, this plain also gradually rises to the north and south and merges with the foothills. The terraces and isolated hillocks (tilas) on the interfluves, linked often to the foothills, provide ideal sites for tea gardens.

The Barail hills of Assam covering the N.C. Hills district are the westward continuation of the Barail range stretching from

Tuensang across Nagaland. The Barail range is a Tertiary mountain range topographically bridging the Archaean Meghalaya Plateau with the Tertiary Naga hills which is the south-western projection of the Himalayas. The Barail range divides the N.C. Hills into two parts -the northern part falling under the Brahmaputra basin and the southern part falling under the Barak basin. The range rising from 300 m in the southern part of the Karbi Plateau attains a maximum height of 1866 m in the Theipibung peak of N.C. Hills district. The other notable peaks of the Barail range in the district are the Hemeolowa (1679 m), the Mahadeo (1739m), the Kaukaha (1736 m) and the Tukbai (965 m) peaks. The southern range of the Barail range is steeper than the northern face due to faulting. The north flowing rivers like Kopili and Dhansiri and their headstreams have dissected the range by their headward erosion and thus have subdued the northern face to lower elevations with gentle slopes.

The Karbi Plateau genetically belongs to the Meghalaya Plateau and consists of two hilly lobes of unequal size, separated by the Kopili valley. The eastern lobe covered by the Rengma hills spans over the entire Diphu and Bokajan sub-divisions of the Karbi Anglong district. It is approximately double the size of its western counterpart and is dome shaped with altitudes ranging from 192 m to the highest Dambak peak and Chenghehishon (Singhasan) peak (1357 m). The Rengma hills project northward up to the proximity of the south bank of the Brahmaputra. The western lobe of the plateau continuous to the Meghalaya Plateau, on the other hand, covers the Hamren sub-division of the district. It presents a rugged topography and slopes from south--west to north-east. This part with an average elevation of 900 m is dissected by headstreams of the Kopili and its main tributary the Barapani. The highest peak of this region is the 1219 m high Umlaper peak located at its extreme western border. In general, the Karbi Plateau has assumed a roundish shape, giving out some radially patterned rivers like Dhansiri,Jamuna, Na-noi, Misa, Diju, Diphalu, Kaliani, Doigrung, Nambar, etc. and their headstreams. These rivers often extend finger-like plain embayments into the eastern, western and northern margins of the plateau, which may be referred to as undulating plains with altitudes ranging from 75 m to 200 m above the mean sea level.

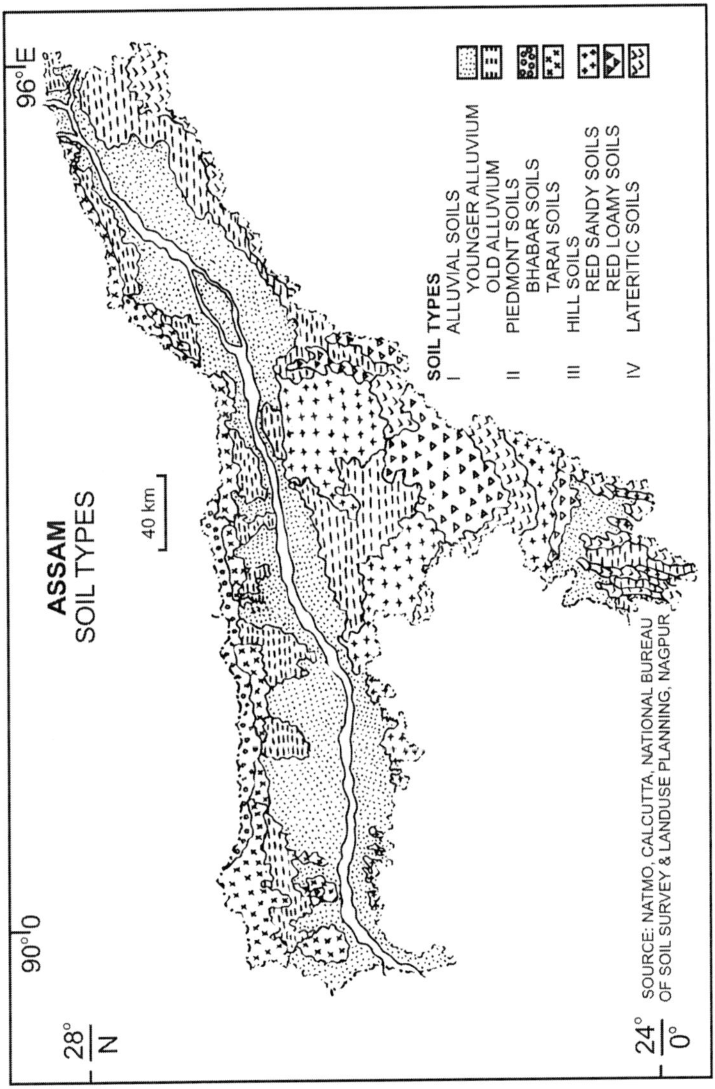
ASSAM
SOIL TYPES
40 km
90°0
96°E
28°
N
24°
0°
SOIL TYPES
I ALLUVIAL SOILS
YOUNGER ALLUVIUM
OLD ALLUVIUM
II PIEDMONT SOILS
BHABAR SOILS
TARAI SOILS
III HILL SOILS
RED SANDY SOILS
RED LOAMY SOILS
IV LATERITIC SOILS
SOURCE: NATMO, CALCUTTA, NATIONAL BUREAU
OF SOIL SURVEY & LANDUSE PLANNING, NAGPUR

Fig. 3.2

Soils

Soil is a vital element of the physical environment. It comprises a complex mixture of mineral (inorganic) and biological (organic) materials and serves as one of the prime requisites for life. Lying over the rocky crust of the earth at different depths, soil is the most valuable natural resource. Soils and in its turn the land through their relative fertility support all agricultural activities. As generally recognised, the parent material, climate, organisms, topography and time govern the soil forming processes. In other words, any given soil can be regarded as the product of the interactions of all these factors, although locally one or a set of factors may exert a strong influence. As regards the soils of Assam, geology (parent material), topography and climate seem to play vital roles in their formations. Therefore, under varying geological conditions, topographical characteristics and agro-climatic situations different types of soils are found to occur in the hills, piedmonts, plateaus and plains. The soils of Assam may thus generally be divided into four groups, viz. alluvial soils, piedmont soils ,hill soils and lateritic soils (Fig. 3.2).

The alluvial soils are extensively distributed over the Brahmaputra and Barak plain. Because of alluvium left by the rivers after leaving the hills and flowing through the plains these soils receive high fertility. There are two types of alluvial soils - the younger alluvium and old alluvium. The younger alluvial soil occurs in an extensive belt of the north-bank and south-bank plains including the active floodplains of the Brahmaputra and Barak rivers. This soil characterised by recent alluvium deposits is moderately deep to very deep and grey to molted grey. It is mostly composed of sandy to silty loams and slightly acidic in reaction. On the river banks it is less acidic, sometimes neutral or slightly alkaline. This soil lacks in profile development and is deficient in phosphoric acid, nitrogen and humus. The old alluvium soil occurs in some patches of Kokrajhar, Barpeta, Nalbari, Kamrup, Darrang, Sonitpur, Lakhimpur and Dhemaji districts between the northern piedmont soil belt and the southern new alluvial soils of the Brahmaputra valley. In the south-bank districts of the valley it occurs in a narrow belt bounded between the southern hill soils and northern new alluvial soils. In the Kopili plain covering Nagaon district the old alluvium finds wider extension. The Barak plain , on the other hand, has some elongated patches of old alluvium confined between the new alluvial soils of the active floodplain and the hill soils bordering Mizoram. Generally the old alluvial soil is very deep, brownish to yellowish brown with texture of fine loams to coarse loams and is slightly to moderately acidic. Some south-bank patches of old alluvium, however, represent pale brown to reddish brown colour with strongly to moderately acidic character.

The piedmont soils are confined to the northern narrow zone along the piedmont zone of the Himalayan foothills. These soils comprise the Bhabar soil and the Tarai soil covering respectively the Bhabar and Tarai belt of the Brahmaputra valley. The Bhabar soil occurs in the narrow Bhabar zone along the Assam-Arunachal boarder extending east up to the river Subansiri. This zone consisting of fairly high ground formed as a result of the coalescence of alluvial cones is characterised by unassorted detritus of boulders, pebbles, cobbles, sands and silts. As developed from the Upper Tertiary sediments, this soil is deep to very deep and fine to coarse loamy in texture. The Tarai soil of the Tarai region occuring just south of the Bhabar soils extends up to Dihang river in some discontinuous narrow patches. This soil varies from sandy to silty loams that remain saturated and support tall grasses in a series of swamps.

The hill soils are generally found in the southern hilly terrains of the state. The fertility of these soils differs greatly in different regions. These soils are rich in nitrogen and organic matters. On the basis of the physical texture and chemical composition, the hill soils may be divided into red sandy soils and red loamy soils. The red sandy soils are distributed covering a narrow belt along the Assam - Meghalaya border, the Karbi Plateau, southern parts of Barail ranges of the N. C. Hills district and some parts of the foothills along the eastern border of the Cachar district. These soils mostly derived from the parent material of Archaean origin are very deep, well drained and brownish to yellowish red in colour. These are strongly to moderately acidic with high organic content. The red loamy soils, on the other hand, occur in the narrow southern foothill belt running along the Assam's border with Arunachal and Nagaland and also in the southern fringes of the Karbi Plateau and the Barail hills of N.C. Hills district. A few patches of these soils trending north-south are also found along the Assam-Mizoram border. These soils are deep, dark greyish brown to yellowish red and fine to coarse loamy. Red loamy soils are lightly to moderately acidic and these lack in nitrogen, phosphoric acid, humus and lime.

The lateritic soils in the state extensively occur almost entirely over the N. C. Hills district covering some parts of southern Karbi Plateau while few patches are confined to eastern margin of the Hamren sub-division of Karbi Anglong district, southern border of Golaghat district and the northern part of the Barak plain along the foothills of the Barail range. These soils are dark and finely textured with heavy loams and deficient in nitrogen, potash, phosphoric acid and lime.

In general, the soils of Assam are rich in contents of nitrogen and organic matter .The alluvial soils of the Brahmaputra and Barak valley are highly fertile and suitable for raising cereals, oilseeds, pulses and plantation crops. These soils are very responsive to manuring and irrigation. Acidic alluvial soils of the upper Brahmaputra valley with a good proportion of phosphoric content are mostly suitable for tea plantation while the alluviums of the lower Brahmaputra valley being deficient in phosphoric content disfavour tea cultivation. New alluvial soils occuring in the charlands of the Brahmaputra are most suitable for growing oilseeds, pulses and rabi crops. The alluviums of the inter-hill plains offer excellent opportunity for cultivating rice and vegetables. Soils occuring in the upper reaches of the hill slopes are favourable for horticultural and plantation crops while the soils of the lower reaches and narrow valleys may be well utilised for paddy and vegetables.

As in the other parts of India the soils of Assam also suffer from a number of soil degradation problems like heavy erosion in the hill slopes, flooding and water logging in the low-lying areas and high acidic nature of the soils. Under heavy precipitation and humid climate loss of top soil through surface runoff is the most common type of soil degradation in the entire state. The problem of top soil erosion is severe in the plains during the flood season. It is estimated that nearly 3.2 million hectares of land of the plains districts of the state are vulnerable to top soil loss with varying severity. Terrain deformation through mass movement is another type of soil degradation which is primarily confined to the hill districts of North Cachar and Karbi Anglong covering an area of about 1.53 million hectares. Waterlogging due to high ground water level in and around the depressed areas and marshy tracts is also a causative factor for soil degradation under high moisture regime. In the state about 0.92 million hectares of land suffer from waterlogging where normal crops and plants other than water-loving ones may not be raised. Besides, most of the soils of Assam irrespective of their types are acidic in nature which leads to chemical degradation of soils. Soil acidity characterised by low P^H value and base saturation affects mostly the hill soils of high rainfall zones of Assam covering an area of 1.72 million hectares.

Climate

The state of Assam lies in the regime of monsoon climate of the sub-tropical belt. It enjoys heavy summer rainfall, winter drought, high humidity and relatively low temperature during a year. The state shows marked spatial variation in the climatic pattern primarily because of its location and physiography, the seasonal change in the pressure condition over the Bay of Bengal and the north-western Indian landmass, the tropical oceanic air masses of south-west monsoon, flow of local mountain and valley winds and presence of water bodies and rivers. The climate of the Brahmaputra and the Barak valleys is normally characterised by orographic lows. During summer the plains become hot and the air over the hills and mountains remains relatively cool. Thus, local low pressure systems develop over the valleys and create a climate of humid sub-tropical type which is slightly different from the Koppen's Cwg type, i.e. the humid meso-thermal Gangetic type.

The Himalayan ranges standing on the north and east of the Brahmaputra valley protect the area from the chilly cold winds of the Tibetan region in winter and obstruct the warm moist winds blowing from the south-west in summer. The cloud brought by the south-west monsoon that passes over the southern hills including the Meghalaya Plateau precipitates in the Brahmaputra valley. Intensity of this precipitation depends on the strength of the winds and the angle at which they meet the mountain barrier. The precipitation decreases on the lee-ward side of the hills, especially

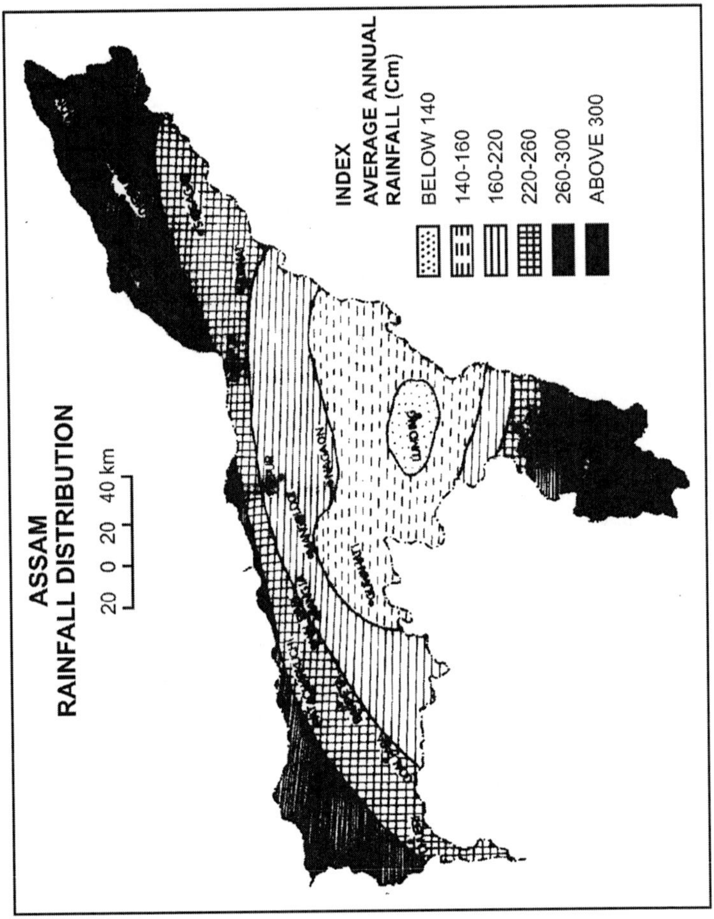
ASSAM
RAINFALL DISTRIBUTION
20 0 20 40 km
INDEX
AVERAGE ANNUAL
RAINFALL (Cm)
BELOW 140
140-160
160-220
220-260
260-300
ABOVE 300

Fig. 3.3

the eastern portions of the hilly areas of Karbi Anglong and North Cachar and increases towards the foothills of the Himalayas. The depression at the eastern end of the Himalayas permits the north-east monsoon during post winter months and has an influence on the precipitation of the valley. The lower Brahmaputra valley which has an opening at the western end receives rainfall derived from the clouds that come across the Meghalaya Plateau and the ones those enter the valley from the western side. The Barak valley, on the other hand, permits the south-west monsoon to penetrate eastward up to the Manipur hills and receives heavy rainfall. The valley being converging and bounded by highlands on its north, east and south is open on the west to orographic and other cyclonic disturbances of the south-west monsoon. The following table 3.1 shows the variation in rainfall in some selected centres of Assam. The table indicates that North Lakhimpur and Silchar receive high rainfall in the order of 335 cm and 323 cm respectively on average annual basis and Lumding records the lowest amount of average annual rainfall of 129 cm.

Table 3.1
Average annual rainfall in some selected places of Assam

Place	Rainfall (cm)	Place	Rainfall (cm)
1. Dhubri	256	11. North Lakhimpur	335
2. Barpeta	229	12. Dibrugarh	273
3. Nalbari	242	13. Sibsagar	250
4. Patacharkuchi	239	14. Gohpur	237
5. Rangia	190	15. Haflong	227
6. Goalpara	242	16. Jorhat	227
7. Guwahati	166	17. Lumding	129
8. Tezpur	182	18. Sadiya	273
9. Mangaldoi	181	19. Silchar	323
10. Nagaon	203		

Source : Department of Environmental Science, Gauhati University, Guwahati.

So far the spatial distribution of rainfall is concerned there is a marked variation within the state (Fig. 3.3). The unique physiographic features of the state and its surrounding mountain and hill ranges affect the areal distribution of rainfall. In general, the hills and foothills receive more rainfall. Besides, the districts of Tinsukia, Dibrugarh, Sibsagar, Dhemaji, Lakhimpur, Kokrajhar and Dhubri in the Brahmaputra valley and the southern most districts of Cachar, Hailakandi and Karimganj enjoy annual rainfall more than 250 cm on average. On the other hand, Nagaon and Karbi Anglong districts receive relatively low rainfall having the lowest in the Lanka-Lumding area (average annual rainfall of 129 cm) due to its location on the leeward side of the Meghalaya Plateau and the Karbi hills.

Under varying intensities of the weather elements and resultant weather conditions Assam normally experiences four climatic seasons, viz. (i) Pre-Monsoon, (ii) Monsoon (iii) Retreating Monsoon and (iv) Dry Winter.

(i) Pre-Monsoon : The pre-monsoon begins in the early part of March and continues up to the end of May. Temperature starts rising gradually from the beginning of the season onward. Pleasant mornings, hot and trying afternoons and occasional thundershowers are some of the important characteristics of the season. In this season marked atmospheric instability develops and severe thunderstorms occur, sometimes preceded by dust-raising squalls. Rainfall increases both in amount and frequency as the season advances which greatly favours the cultivation of tea, jute and *ahu* rice.

(ii) Monsoon : The monsoon sets in by the last week of May or in early June and it lasts up to September or the first part of October. It is the rainy season when the state receives spells of continuous and moderate to heavy rains. June, July and August are the rainiest months when more than 70 per cent of the total annual rainfall occurs. This is the most important season during which *sali* rice, the principal crop of the state is cultivated on the one hand, and the rivers of the Brahmaputra and Barak valleys start rising causing extensive floods on the other.

(iii) Retreating Monsoon : The south-west monsoon withdraws sometimes in between the last part of September and first part of October. Consequently, the intensity of rainfall and the number of rainy days go on decreasing. This season continues up to the middle of November, when fogs commonly occur.

(iv) Dry Winter : The winter season begins in the middle of

November and continues up to the end of February .This season is characterised by low temperature, regular morning fogs and very little amount of rainfall. December and January are the driest months and generally, January is the coldest month.

References

Barthakur, M. (1968) : Some Aspects of Weather in the Brahmaputra Valley, *Jr. of the Assam Sc. Soc., Vol XI*

Barthakur, M (1986) : Weather and Climate of North East India, *North Eastern Geographer, Vol. 18,* No. 1 and 2.

Bhattacharya, S.K. (1969-71) : The Brahmaputra Valley and Tectonics, *Jr. Geol. Soc. of Assam,* Vol. I, Geol Soc. Assam.

Bora, A.K. and Das, D. (1998): Flood and Erosion Hazards in the River Island of Majuli, Assam; paper presented at the 9th Conference of the Indian Institute of Geomorphologists (IGI), Delhi University, New Delhi.

Das, H.P. (1970) : *Geography of Assam,* National Book Trust, New Delhi.

Das, M.M. (1981) : Pattern of Soil Distribution in Assam, *North Eastern Geographer,* Vol. XIII, No. 1 and 2.

Das, M.M. (1985) : Physiographic Frame of Assam, *North Eastern Geographer,* Vol 17 No. 1 and 2.

Dutta , S.K. (1969-71) : Assam Through Ages, *Jr. Geol. Soc. of Assam,* Vol. 1, Geol. Soc. Assam.

Goswami, D.C. (1985) : Brahmaputra River, Assam, India : Physiography, Basin Denudation and Channel Aggradation, *Water Resources Research,* 21,pp. 959-978.

Sarmah, S.K. (1969) : Some Observations on the Rainfall of Assam, *Jr. Assam Sc. Soc.,* 12.

Taher, M (1974) : Fluvial Procces in the Brahmaputra Plain, *Geog. Rev. of India,* Vol. 36, No. 1.

Taher, M. (1986): Physiographic Framework of North-East India, *North Eastern Geographer,* Vol. 18, No. 1 and 2.

4 Drainage and Flood

Drainage

Assam is a land of rivers. By virtue of its unique geographical location amidst the northern, eastern and southern hills and highlands and humid tropical monsoon climate, Assam happens to be a land of innumerable rivers and their tributaries. The entire state is drained by the dense networks of two river systems, viz the Brahmaputra and the Barak systems. Both the river systems are international in their extensions and go out to Bangladesh. In general, the Brahmaputra system is very much extensive, while the Barak system is relatively small, and it is also so in their extensions over the state of Assam (Fig. 4.1).

The Brahmaputra river is one of the biggest rivers of the world. Its extensive river system drains a unique physical setting of diverse environments, such as the cold, dry plateau in Tibet, the steep rain- drenched slopes of the Himalayas, the landlocked alluvial plain in Assam and the vast deltaic plain in Bangladesh. The river is known as the Tsangpo in Tibet (China), the Siang or Dehang in Arunachal Pradesh, the Brahmaputra in Assam and the Jamuna in Bangladesh. Originating from the Chema-Yung-dung glacier mass on the southern slope of the Kailash range in Tibet at an altitude of 5150m, the river traverses a total distance of 2,906 km through the Tibetan Plateau, Himalayan mountains and hills and plains untill reaching the Bay of Bengal in Bangladesh. The Brahmaputra basin covers an area of 5,80,000 km^2 of which 1,94,413 km^2 falls within India. In India, the basin lies in the states of Arunachal Pradesh, Assam, Nagaland, Meghalaya, Sikkim and West Bengal, and the river flows through Arunachal Pradesh and Assam for a distance of 918 km.

The Brahmaputra river of Assam forms a complex river system characterised by the most dynamic and unique water

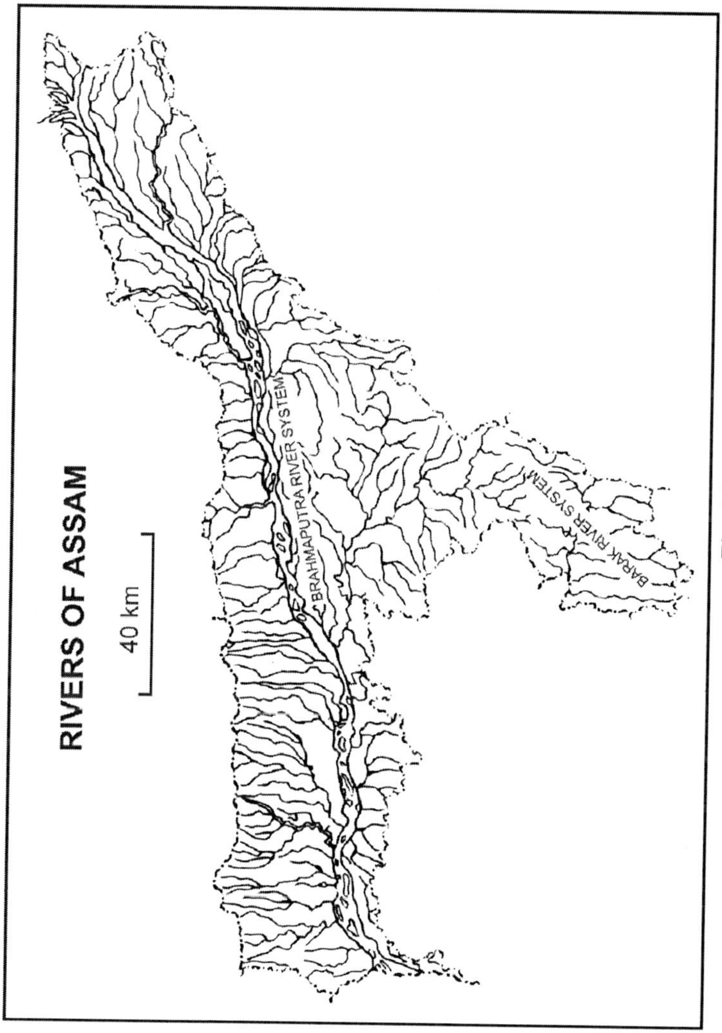
RIVERS OF ASSAM
40 km
BRAHMAPUTRA RIVER SYSTEM
BARAK RIVER SYSTEM

Fig. 4.1

and sediment transport patterns. It is the fourth largest river in the world in terms of average water discharge at the mouth with a flow of 19,830 m^3s^{-1} (Goswami, 1985). The river carries 82 percent of its annual flow at Pandu (Guwahati) during the rainy season (May through October). The water discharge of the river on August 23, 1962 measuring 72,794 m^3s^{-1} has been the highest flood on record. As regards sediment transport, the river is also one of the leading sediment carrying rivers of the world. It is the second river only to the Yellow river of China in the amount of sediment transported per unit of drainage area (Goswami, 1985, 1988). As measured at Pandu, the river carries an average annual suspended load of 402 million metric tons. The high intensity of monsoonal rains, easily erodible rocks, steep slopes and high seismicity contribute a substantial amount of sediment to the Brahmaputra. The river with such a high volume of water discharge and sediment load, represents the most dynamic fluvial regime. Its wide alluvial channel in Assam having an average width of 6-8 km. is dotted with more than 600 small and large sandbars, locally called *chars*. The river is characterised by intense braiding, rapid aggradation and drastic bankline migration. In general, braiding in the Brahmaputra follows the mechanism of central bar formation. However, bar formation and channel division (braiding) in the river are due to a combination of factors like variability of flow, excessive sediment load, easily erodible bank materials and aggradation of the channel. As regards bank erosion and channel migration, the river is found extremely unstable at some vulnerable locations, such as Nagajuli-Maijan, Majuli, Bhairabpur, Balikuchi, Kaziranga, Howlighat and Palasbari. The Brahmaputra, in general, is migrating laterally southward. The southward migration of the river has engulfed the lower most parts of the south bank tributaries, and their confluences are thus shifting eastward. Moreover, most of the tributaries before meeting the Brahmaputra flow sub-parallely from east to west in their lower courses because of incipient levee deposition on both of its banks.

The Brahmaputra rolls down the plains of Assam from east to west for a distance of 640 km. upto the Bangladesh border. Throughout its course, the river receives innumerable tributaries coming out of the northern, north-eastern and southern hill ranges. The mighty river with a well-knit network of tributaries drains an area of 56,480 km^2 of the state accounting for 72 per cent of its total geographical area. The principal north bank tributaries are the Subansiri, Jia Bharali, Dhansiri, Puthimari, Pagladiya, Manas and Champamati. Of these, the Subansiri, Jia Bharali and Manas are the trans-Himalayan rivers. The principal south bank tributaries include the Burhi Dihing, Disang, Dikhow, Dhansiri, Kopili and Krishnai. Certain geomorphological and hydrological characteristics of some important tributaries of the Brahmaputra are presented in table 4.1.

The north and south bank tributaries show marked differences among them in terms of their fluvio-geomorphic characteristics. The wide variations in geology, geomorphology, physiography, relief, precipitation and soils of the northern and southern regions of the Brahmaputra Valley bring about notable differences between these two groups of rivers. The following points of differences may be observed among the north and south bank tributaries of the Brahmaputra.

(a) The northern tributaries have steep channel gradient, but the southern ones have comparatively flatter gradient.
(b) The northern tributaries have shallow braided channels, while the southern tributaries have their meandering channels over the plains.
(c) The northern tributaries have their long courses in the hilly terrain and short courses over the plains, whereas the southern tributaries have their long courses over the plains and short courses in the hills.
(d) The northern tributaries generally have coarse sandy beds with occasional gravel beds up to some distance from the foothills. But, the southern tributaries have beds and banks composed of fine alluvial soils.
(e) The northern tributaries carry enormous sediment load as compared to the southern tributaries. On the average, sediment yield of the north bank tributaries is three times higher than that of the south bank tributaries.
(f) The northern tributaries generally have flash floods, but the southern ones have comparatively less flash floods.
(g) The northern tributaries are characterised by frequent shift of their channels during floods but, the southern tributaries change their courses less frequently. The northern tributaries

Table 4.1
Certain geomorphological and hydrological characteristics of some important north bank and south bank tributaries of the Brahmaputra

Rivers	Catchment area (sq. km.)	Length (km)	Average annual discharge (m^3s^{-1})	Average annual suspended load (ha.m.)	Sediment yield (tons/km^2/year)
Northern tributaries					
1. Subansiri	28,200	442	755771	992	959
2. Ranganadi	2,941	150	74309	186	1598
3. Buroi	791	64	20800	16	529
4. Bargang	550	42	16000	27	1749
5. Jia Bharali	11,716	247	349487	2013	4721
6. Gabharu	577	61	8450	11	520
7. Belsiri	751	110	9300	9	477
8. Dhansiri	1,657	123	26577	29	463
9. Noa Nadi	907	75	4450	6	166
10. Nanoi	860	105	10281	5	228
11. Barnadi	739	112	5756	9	323
12. Puthimari	1787	190	26324	195	2887
13. Pagladiya	1674	197	15201	27	1883
14. Manas-Aie-Beki	41350	215	307947	2166	1581
15. Champamati	1038	135	32548	13	386
16. Gaurang	1379	98	22263	26	506

17. Tipkai	1364	108	61786	31	598
18. Godadhar	610	50	7000	0.21	272
Southern tributaries					
1. Burhi Dihing	8730	360	141539	210	1129
2. Disang	3950	230	55101	93	622
3. Dikhow	3610	200	41892	34	252
4. Jhanzi	1130	108	8797	16	366
5. Bhogdoi	920	160	6072	15	639
6. Dhansiri	10242	352	68746	147	379
7. Kopili	13556	297	90046	118	230
8. Kulsi	400	93	11643	0.6	135
9. Krishnai	1615	81	22452	10	131
10. Jinari	594	60	7783	3	96

have peculiar channel shifting pattern. The Subansiri and all other eastern rivers shift their channels westward, while the rivers between the Subansiri and Pagladiya shift eastward. Again from Manas upto the river Sonkosh in the west, all the rivers migrate westward.

Like the Brahmaputra, the Barak is also an important river of Assam. The Barak system is the second largest river system in North East India as well as Assam. The river with a total length of 900 km from source to mouth drains an area of 52,000 km^2. In India, the river traverses a distance of 532 km upto the Indo-Bangla border. Originating at the border of Manipur and Nagaland, the river flows westward for some distance forming the boundary between the two states. It then suddenly turns southward and flows through Manipur upto Tipaimukh, south of the tri-junction of Assam, Manipur and Mizoram. The river then takes a northward turn and flows for about 60 km. Thereafter it turns westward at Jirimukh and runs through the Cachar plain sluggishly. Near Karimganj, it bifurcates into the northern branch of Surama and the southern branch of Kushiyara.

In Assam, the Barak has a total length of 225 km and it drains the southern part of the state which includes the districts of Cachar, Karimganj and Hailakandi, and the southern portion of North Cachar Hills. The valley has a width of 25-30 km and it covers an area of 6,962 km^2 accounting for about 9 percent of the total geographical area of the state. The plain, usually known as Cachar plain is a narrow one and slopes gently to the west. Through the plain, the river flows sluggishly for a distance of 125 km and exhibits a highly meandering pattern. The meandering channel is characterised by a series of ox-bow lakes, cut-offs and swamps. The important north bank tributaries of the Barak are Jiri,Siri, Modhura, Jatinga and Larang, while the important south bank tributaries include Sonai, Ghagra, Katakhal-Dhaleswari, Singla and Longai. The Barak and its tributaries create recurring floods in the Cachar plain during summer. The water discharge of 6,284 m^3s^{-1} occurring on August 12, 1989 has been recorded as the highest flood, so far observed on the Barak. The floodplain on both the banks is uniquely characterised by depressed tracts developed between the levees along the river and the gradually rising plains merging with the northern and southern foothills.

Floodwater entering into the depressed tracts thus remains stagnant for several days and causes heavy damage.

Floods

Floods in Assam have been a recurring feature since early times. The plains of Assam have been variably inundated by the floods of the Brahmaputra and the Barak river systems in different periods. Although written records of flood in the state do not date back beyond the 13th century, the pedologic and landform characteristics of the plains are clearly indicative of large scale actions of recurring and occasional floodings in the historical past. Historically, the people of Assam are well acquainted with floods and spontaneously they have acquired the art of living with these to a considerable extent. But, in the recent past due to a variety of newly emerged causes, in addition to the generally recognized ones, floods in both the Brahmaputra and Barak valleys sometimes attained such a magnitude that they appeared to be greatly hazardous to a major section of the floodplain dwellers. The nature of floods in terms of frequency of occurrence during a year and the quantum of damages caused has been changing frequently. Now, in certain parts of the Brahmaputra and Barak valleys, floods of different order occur at least 3-4 times in a year. The two valleys surrounded by high hill ranges and plateaus on the north, east and south have high flood potential and thus frequently receive the fury of floods (Fig. 4.2).

Though floods have some beneficial effects, such as fertile silt deposition on agricultural fields, recharge of soil moisture, increase in fish population and washing effect on dirty environment, they cause substantial damages to standing crops, dwelling houses and developmental infrastructures. Generally, the damages during floods are caused by inundation and associated bank erosion. Presently, Assam accounts for nearly 8 per cent of the total flood-damaged area in India.The flood damage data available for the Brahmaputra valley indicate that the cropped area damaged and number of affected people are gradually increasing in the years during 1954 -98 (Table 4.2). It is due to growing human encroachment into the floodplain and increasing severity of floods. As regards annual floods, the flood of 1988, occurring in four severe waves has been recognised as the most devastating one. This flood characterised by high water levels throughout the Brahmaputra from Dibrugarh to Dhubri

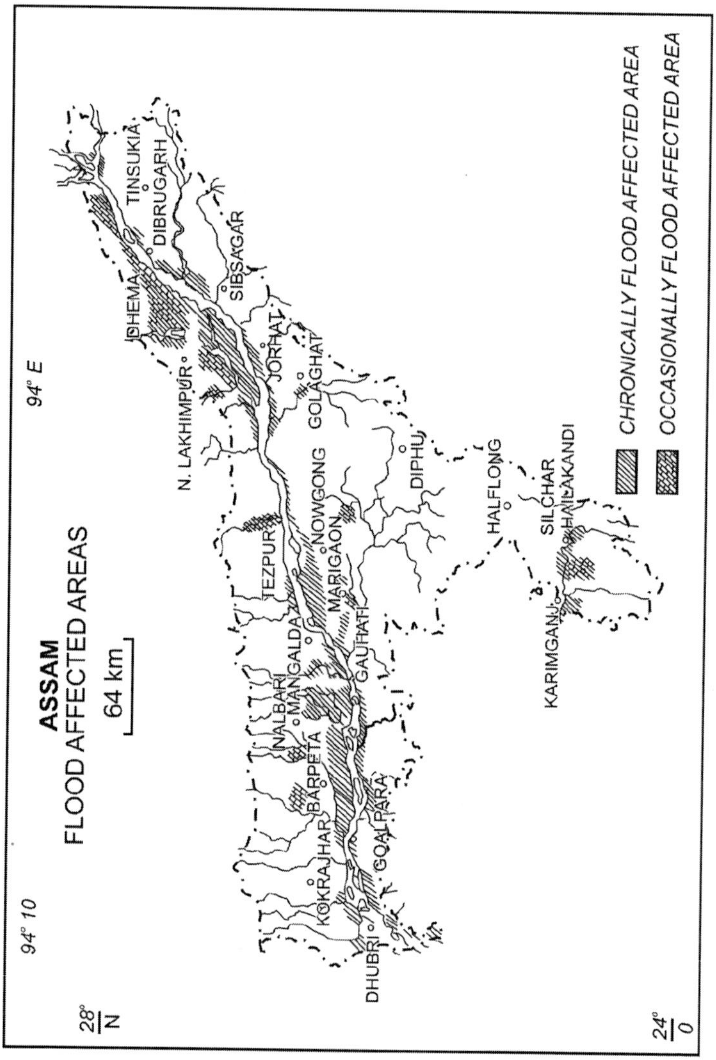
ASSAM
FLOOD AFFECTED AREAS
64 km
94° 10
94° E
28° N
24° 0
DHUBRI
KOKRAJHAR
GOALPARA
BARPETA
NALBARI
MANGALDAI
GAUHATI
MARIGAON
TEZPUR
NOWGONG
N. LAKHIMPUR
GOLAGHAT
JORHAT
DIPHU
SIBSAGAR
DHEMAJI
DIBRUGARH
TINSUKIA
HALFLONG
SILCHAR
HAILAKANDI
KARIMGANJ
CHRONICALLY FLOOD AFFECTED AREA
OCCASIONALLY FLOOD AFFECTED AREA

Fig. 4.2

caused tremendous devastations and extensive inundation accounting for 62 per cent area of the Brahmaputra Valley. So far the spatial extent of flood effect is concerned, normally around 40 per cent of the state's area are affected. The recurring floods of characteristically high magnitudes have assumed devastating dimensions heavily affecting the agricultural lands, crops, cattles and people, and above all, the agrarian economy of the state.

Floods in Assam can be attributed to a variety of factors of natural, hydrometeorological and anthropogenic origin. Of the two river systems - the Brahmaputra and Barak, the Brahmaputra system is highly responsible for causing more devastations. It is so vast and dynamic that the physical and hydrological processes themselves are capable of creating floods in the narrow valley. However, heavy monsoonal rains and devastating landslides coupled with easy erodibility of rocks, steep slopes and high seismicity constitute the major natural causes of Assam floods. The runoff patterns of the Brahmaputra catchment in the Himalayas and the Barak catchment in the hills of Nagaland and Manipur are primarily governed by the quantity and nature of distribution of precipitation. The Barak valley receives on the average an annual rainfall of 280 cm and its hilly catchment receives more than 500 cm. Similarly, the Brahmaputra valley receives an average annual rainfall of 230 cm, while the Himalayan sector of its catchment enjoys annual rainfall in the order of 500 cm. But, interestingly 60-70 per cent of the annual rainfall generally occurs during the summer months of June to September. Since the rainfall normally starts from the month of April, the water holding capacity of the basins gets already lost before the beginning of monsoon rains.

Table 4.2

Flood damages during major floods in the Brahmaputra valley

Years of major floods	No. of flood waves	Area affected (mha)	Cropped area damaged (mha)	Population affected (m)
1954	4	2.90	0.30	1.30
1955	-	1.35	0.07	0.18
1958	-	1.23	0.06	0.40
1962	2	1.59	0.36	3.91
1966	4	1.51	0.06	3.62

1969	-	1.06	0.07	0.89
1972	2	1.01	0.36	2.95
1973	-	2.31	0.17	1.84
1977	3	1.02	N.A.*	4.54
1980	-	1.14	0.26	3.35
1984	4	1.49	0.14	5.61
1987	-	1.62	0.43	N.A.*
1988	4	3.82	1.12	8.31
1993	-	1.25	0.22	5.30
1996	-	1.33	0.28	5.12
1998	3	0.97	0.29	4.69

*N.A. Stands for not available

As soon as the monsoon starts, most of the rain water runs into the streams causing high stage in the Barak and the Brahmaputra and their tributaries. Particularly the Brahmaputra river, carrying high volume of water and sediment load, when additionally charged with heavy concentration of rain during monsoon, obviously becomes capable of creating unbelievable flash floods.

Seismic and geologic factors combined with the hydrometeorological factors also aggravate the flood situation of Assam. The earthquakes of 1897 and 1950 of 8.7 and 8.6 magnitudes in the Richter scale respectively, caused tremendous changes in the fluvial regime of the state, particularly the Brahmaputra valley by suddenly raising the river beds. Changes in river courses and accompanying erosion have together accelerated the silt deposition on the river beds. The aggradation of river beds due to silting has forced the rivers to spill their water overbank, inundating large tracts of land in the valley. The aggressive nature of flooding in the valley made a quantum leap, especially after the great 1950 earthquake. The severe floods of 1954, 1955, 1956 and 1958 are the daring examples.

Besides the natural factors, the human - induced factors are also highly responsible for devastating floods in the state. Deforestation and unwise tree felling in the hilly catchments of the Brahmaputra and Barak are the important causes of flooding in the plains. The absence of protective cover of vegetation accelerates soil erosion, and finally results in silting up of the river beds. The shifting cultivation practised in the hill districts of Karbi Anglong and N.C. Hills has substantially reduced forest cover. As a result, enormous sediments are generated and continually getting deposited on the river beds. The flood problem has become more serious due to some other human-induced factors like large scale human occupance in the floodplains, destruction of wetlands *(beels)* and the poorly managed embankment network. In recent years, increasing encroachment on the floodplain and low-lying areas for habitation and cultivation has significantly aggravated the flood problem. Similarly, the destruction and degradation of *beels* and swamps due to various forms of human interference has also added an adverse dimension to the flood problem. Moreover, the structural measures of flood control, especially the poorly managed embankments also often cause sudden floods. The floodwater coming across the embankments through overtopping or breaching, eventually remains stagnant in a locality for days together and causes gross damage.

However, flood control measures in Assam were started by the government during the early fifties. Till today, embankments are widely adopted as a measure of flood control. In the Brahmaputra valley, the embankments measuring 984 km and 2,567 km have been built along the Brahmaputra and its tributaries respectively. In the Barak valley also, the embankments along the Barak and its tributaries cover a total length of 738 km. Moreover, the other structural flood control measures so far adopted in the state include the construction of 600 km long drainage channel, 421 anti-erosion schemes and 56 major sluices. But, all these short-term and ad hoc measures are now proved to be inadequate and incapable to combat the chronical flood problem of the state. So, there arises a pressing need for evolving some effective strategies to solve this everexpanding problem. In fact, the proper management of flood cannot be achieved unless some long-term measures are undertaken. Measures like development of the vast water resources of the Brahmaputra and the Barak rivers, protection of existing vegetal cover and generation of new growth in the catchments, adoption of an appropriate crop calendar adjusted to the rhythm of floods, and pursuance of the approach of 'living with floods' may be the key factors in the better management of Assam floods.

Table 4.3
Districtwise and typewise distribution of wetlands in Assam

Districts	No. of wetlands of various types: Lake/ Pond	Ox-bow lake/ meander cut-off	Waterlogged (seasonal)	Swamp/ marsh	Reservoir	Tank	Total No. of wetands	Total area under wetlands (ha)
Dhubri	73	68	25	65	-	2	233	6459.5
Goalpara	31	32	31	68	-	3	165	3832.5
Bongaigaon	25	36	17	21	-	1	100	3158.5
Kokrajhar	12	38	20	14	-	1	85	1578.4
Barpeta	17	30	7	42	1	-	97	3301
Nalbari	10	29	-	24	2	3	68	1987
Kamrup	49	65	64	155	1	18	352	11407
Darrang	11	45	22	18	-	7	103	3516
Sonitpur	23	35	110	22	-	16	206	3651
Morigaon	37	41	4	59	-	3	144	11658
Nagaon	68	71	138	92	-	10	379	11295
Golaghat	113	104	80	29	2	2	330	5466.5
Jorhat	39	24	34	9	-	3	109	2108.5
Lakhimpur	38	43	56	7	-	7	151	3033.5
Dhemaji	36	22	61	17	-	3	139	3960

Sibsagar	16	49	16	8	-	20	109	2135
Dibrugarh	13	36	23	12	-	2	86	2752.5
Tinsukia	3	22	31	18	-	-	74	2732.5
Karbi Anglong	6	37	21	3	-	10	77	897
N. C. Hills	-	-	6	-	4	-	10	2552.5
Cachar	57	29	231	19	-	4	340	7188
Hailakandi	7	4	36	-	-	-	47	840
Karimganj	6	1	53	10	-	-	70	5719.5
Assam	**690**	**861**	**1086**	**712**	**10**	**115**	**3474**	**1,01,229.4**

Source : Assam Remote Sensing Application Centre, Guwahati.

Wetlands

Wetlands locally known as *beels* are the most common and integral features of the fluvial landscape of Assam. The riverine tracts of the Brahmaputra and Barak valleys provide favourable geo-ecological conditions for the growth and development of wetlands. The genesis and growth of the wetlands are directly associated with the geo-physical conditions and tectonic evolution of the region, hydrologic and fluvio-geomorphic behaviours of the rivers and the pattern of landuse and human occupance in the river valleys. As revealed from satellite data analysed by the Assam Remote Sensing Application Centre (ARSAC), Guwahati in 1997, the total number of wetlands in the state is 3474 having areal size equal to or more than 2.5 hectares each. These wetlands cover an area of 1,01,229.4 hectares accounting for 1.29 per cent of the total geographical area of the state. The table 4.3 shows the district-wise and type-wise distribution of wetlands in the state.

The wetlands of riverine origin may be categorised as natural ones which include fresh water lakes, ox-bow lakes or abandoned channels, marshy tracts and seasonally water-logged areas. The natural lakes and ponds are generally the body of water held in tectonic depressions or the detached water bodies due to channel shifting of rivers. Most of these wetlands contain water with low turbidity and are non-vegetated or partially vegetated. The important wetlands under this category are Deepar ***beel*** in Kamrup district, Dhir ***beel*** in Dhubri district, Tamaranga and Dalani *beels* in Bangaigaon district. The ox-bow lakes or meander cut-offs are the crescent shaped body of water located along the streams. These are the results of neck cut-offs from the meandering channels. Wetlands of this type sometimes have feeder channels controlling the inflow and outflow of water and are usually attached with typically meandering rivers. Majority of this type of wetlands have low turbidity of water and are mostly non-vegetated. Some important wetlands of this category are the Morikalong and Patoli *beel* in Nagaon district, Mer ***beel*** in Golaghat district, and Guruajan in Morigaon district. The swampy and marshy areas, on the other hand, constitute another major group of wetlands of the state. These are intermittently or permanently covered with water, mostly having shrubs. They usually have no feeder channels to control the inflow and outflow of water. Due to presence of various minerals and organic matters these marshy wetlands show moderate to high turbidity. The notable ones under this category are the Nandani-Sonai *beel* in Morigaon district, Batha *beel* in Darrang district, and Urpad *beel* in Goalpara district. The seasonally waterlogged areas, unlike the other wetlands, constitute a major portion (Table 4.3). The waterspread areas of these wetlands vary in different seasons of the year. Due to heavy and prolonged monsoon rain in the state, most of the waterlogged areas get coalesced to form a single big wetland. Such wetlands with low turbidity generally support only few seasonal aquatic vegetations.

The wetlands play a vital role in the creation of a good ecosystem in the low-lying Brahmaputra and Barak plains. All the wetlands, besides being the economic resource base and rich in bio-diversity help in reducing flood height, and thus flood damage as they act as buffer zones for rivers and other water ways during the period of heavy monsoonal rains. In spite of their well recognised importance, the wetlands of Assam presently suffer from the problems of degeneration and ecodegradation. Rapid eutrophication caused by debris from the floating vegetation and the silt carried by surface runoff as well as streams, large scale human encroachment, growth of human activities on the banks, overuse of the economic resources including extensive fishing, especially during the breeding season, and above all, the mismanagement of the wetlands have all contributed to the destruction of the wetland system of the state. It is, therefore, an imperative need to conserve the wetlands, especially for enriching the geo-environmental quality of the region. The concerned government agencies, NGOs and the local people have their greater role to play in the matter of conservation and management of the dwindling wetlands.

References

Barua, P. and Goswami, D.C. (1997) : Status of Wetlands of Assam - A Study using Remote Sensing Technique; paper presented at the workshop on Conservation of Wetlands of Assam organised by W.W.F. for nature, N.E. India and Assam Sc. Society, Guwahati During October, 1997.

Bhattacharyya, N.N. and Bora, A.K. (1997) : Floods of the Brahmaputra River in India, *Water International,* IWRA, University of New Mexico, U.S.A., Vol. 22, No. 4.

Bora, A.K. and Barman, B. (1998) : A Geo-Ecological Study of the Wetlands of Barpeta District, Assam, *North Eastern Geographer*, Vol. 29, No. 1 and 2.

Coleman, J.M. (1969) : Brahmaputra River : Channel Processes and Sedimentation, *Sedimentary Geology*, Vol. 3, pp. 129-239.

Flood Control Department, Assam (1989) : Report of the committee for examining the causes of flood occurred in Assam in 1988, Govt. of Assam. p. 25.

Goswami, D.C. (1985) : Brahmaputra River, Assam, India: Physiography, Basin Denundation and Channel Aggradation, *Water Resources Research*, 21, p. 959-978.

Goswami, D.C. (1988) : Estimation of Bedload Transport in the Brahmaputra River, Assam, *Ind. Jr. of Earth Sc.*, Vol. 15, No. 1.

Goswami, D.C. (1988) : Magnitude and Frequency of Fluvial Processes in the Brahmaputra Basin, Assam : Some Observations, *Geomorphology and Environment*, Allahabad Geog. Soc., pp. 203-211.

Goswami, D.C. (1989) : Floods and their Impact on the Agriculture of Assam, *Agriculture in Assam*, P.C. Goswami ed., Inst. of Dev. Studies, Guwahati, Assam.

North Eastern Council (NEC), (1993) : *Morphological Studies of River Brahmaputra*, Govt of India, New Delhi.

Sarma, J.N. (1993) : *Rivers of Assam* (In Assamese), Published by Asam Sahitya Sabha, Jorhat, p. 336.

5 Seismic Environment

Introduction

Earthquakes are the most devastating natural seismic hazards causing huge loss of life and property. They are caused by the release of stored energy in the form of seismic waves through the earth's structures resulting in shaking of the ground. They occur mostly along geologic faults, narrow zones where rock masses move in relation to one another. The area where earthquake originates is called the focus and the point on the ground directly above the focus is the epicenter. North-East India including the state of Assam, due to its unique geographical position and geo-tectonic setting, is very much prone to various earthquake events from time to time.

Tectonically, the zone between the Shillong Plateau and the Assam valley is the source area of major earthquakes including that of 1897 occurred in the region (Fig. 5.1) (Kayal et.al, 1992). North-East India has been in the zone V, the zone of highest seismic hazard potential, according to the seismic zonation of India (BIS, 2002). The state of Assam may experience earthquakes with MM intensity of IX. The Global Seismic Hazard Assessment Programme suggests that the state is at a high risk zone and may have a peak gravitational acceleration along the state border with Meghalaya.

Assam along with other parts of N. E. India represents four major tectonic zones, viz. the Eastern Himalayan Zone, the Mishmi Zone, the Shillong Plateau Zone and the Indo-Burmese Zone (Fig. 5.2).

1. Eastern Himalayan Zone includes
   a) MBT (Main Boundary Thrust)
   b) MCT (Main Central Thrust)

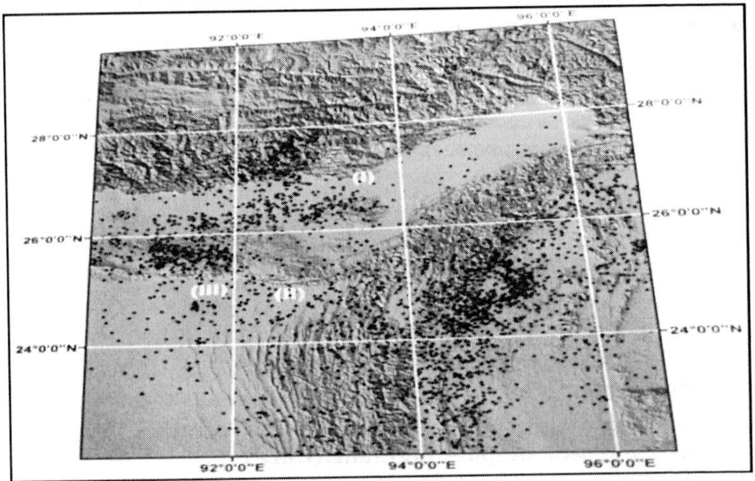

Fig. 5.1: Epicentric pattern in Assam and its adjoining areas

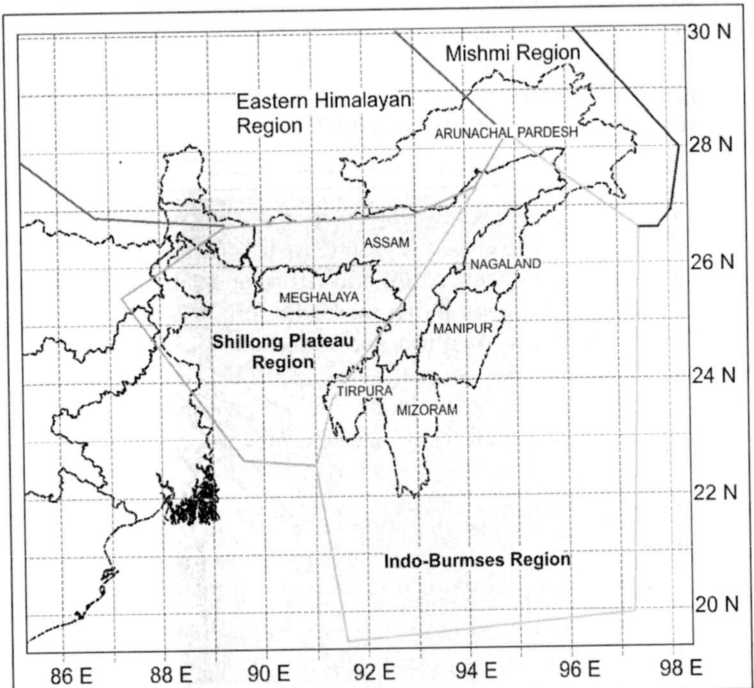

Fig. 5.2: Four Major Tectonic Zones covering Assam and its adjoining areas

Source: Bahuguna and Sil, 2018

2. Mishimi Zone includes
   a) Mishimi Thrust
   b) Lohit Thrust

3. Shillong Plateau zone includes
   a) Oldham Fault
   b) Dauki Fault
   c) Dapsi Thrust
   d) Kopili Fault
   e) Dhubri Fault
   f) Samin Fault

4. Indo Burmese zone includes
   a) Sagging Fault
   b) Chaurachandpur- Mao Fault (CMF)
   c) Chittagong Coastal fault
   d) Kabaw Fault
   e) Kaladam Fault
   f) Disang Thrust
   g) Naga Thrust
   h) Myanmar Central Basin (MCB)

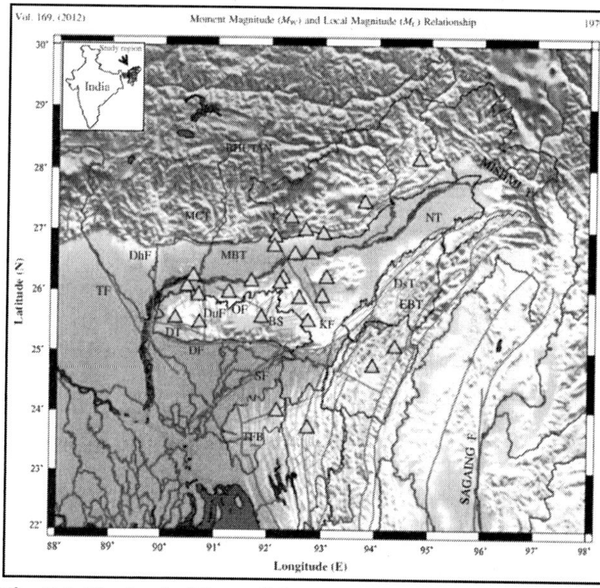

Fig. 5.3: Tectonic Map of Assam and its adjoining areas
Source: after Baruah et. al, 2012.

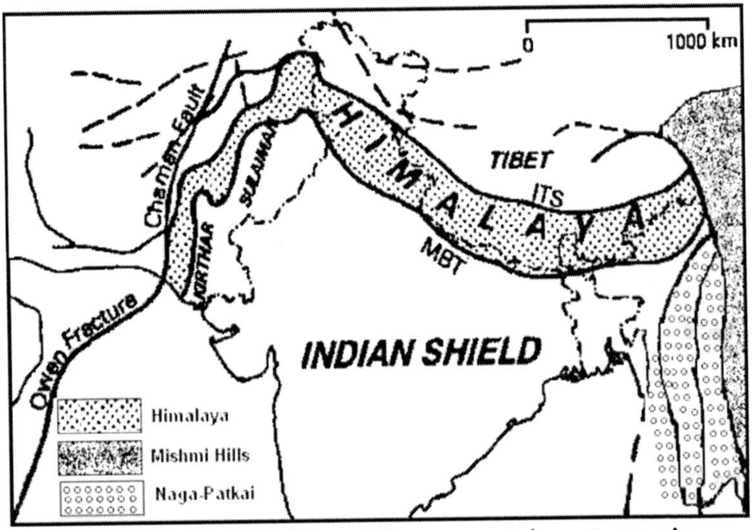

Fig. 5.4(a): Juxtaposition of three mountain systems, viz. Himalayan ranges, Mishmi Hill ranges and Naga-Patkai hill ranges. (after Valdiya, 1998)

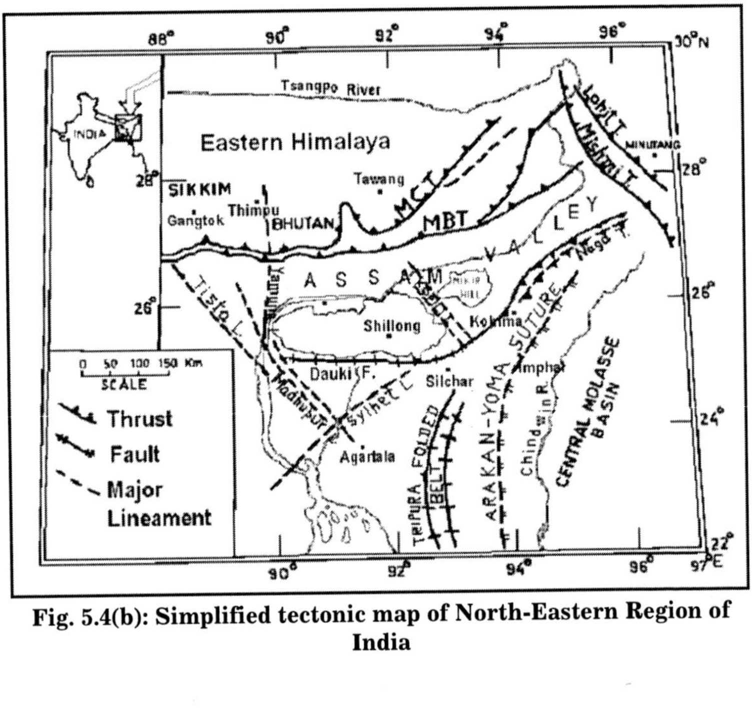

Fig. 5.4(b): Simplified tectonic map of North-Eastern Region of India

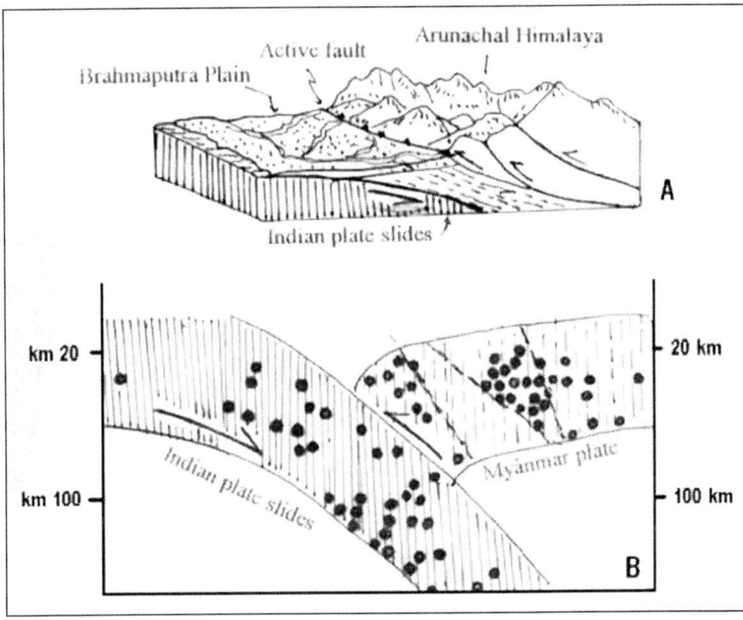

Fig. 5.4(c): Indian Plate sliding under the Arunachal Himalaya (A) and Myanmar Plate (B). (After Valdiya, 2002).

The geological surroundings of Assam valley show a very strategic and vulnerable situation. The notable faults and thrusts are the Himalayan Frontal Fault (HFF), Main Boundary Thrust (MBT), Main Central Thrust (MCT), Bame Fault, Lohit Thrust, Mishmi Thrust, Brahmaputra Fracture, Kopili Lineament, Dauki Fault, Yamuna Lineament, and a number of thrust slices of Arakan-Yoma Folded Belt (Fig. 5.3). These thrusts and lineaments represent zones of crustal weakness and are the source of earthquake activity, which is evident from the distribution of earthquake epicenters. In fact, the Arunachal Himalaya is being pressed by the northward moving Indian shield. The Naga-Patkai-Arakan ranges, on the other hand, are linked with the Andaman-Nicobar Islands Arc, constituting an active zone where Indian Ocean floor is descending under the Malaysian Plate (Fig. 5.4(a), 5.4(b) and 5.4(c) –after Valdiya, 2002).

Distribution of Earthquakes in Assam

Historical and instrumentally recorded earthquake data show that this region has seen several moderate to large earthquakes.

Table 5.1
Number of Earthquakes (Main Shock) reported in the North East India during 1761-2011

From	To	$4 < M_w < 5$	$5 < M_w < 6$	$6 < M_w < 7$	$7 < M_w < 8$	$M_w > 8$	Total
1761	1770	1		2			3
1771	1780	-	-	-	-	-	0
1781	1790	-	-	-	-	-	0
1791	1800	-	-	-	-	-	0
1801	1810	2	-		-	-	2
1811	1820	-	2	-	-	-	2
1821	1830	6	4	1	-	-	11
1831	1840	-	1	1	1	1	4
1841	1850	3	8	1	1	-	13
1851	1860	1	2	2	-	-	5
1861	1870	2	5	2	1	-	10
1871	1880	-	1	-	-	-	1
1881	1890	1	1	1	-	-	3
1891	1900	1	-	-	-	1	2
1901	1910	-	2	-	1	1	4
1911	1920	-	1	1	1	-	3
1921	1930	-	9	14	2	-	25
1931	1940	-	7	17	2	1	27
1941	1950	-	12	7	1	1	21
1951	1960	-	17	8	1	-	26

1961	1970	23	93	20	-	-	136
1971	1980	129	146	8	-	-	283
1981	1990	268	293	16	2	-	579
1991	2000	581	287	22	1	-	891
2001	2010	260	104	5	-	-	369
2011	2015	309	147	6	-	-	462
Total		**1587**	**1142**	**134**	**14**	**5**	**2882**

Source: Bahuguna and Sil (2018).

Table 5.2

List of Latest Earthquakes in Assam (January to July, 2021)

S. No.	Origin Time	Latitude	Longitude	Depth	Magnitude	Location (epicentre)
1	2021-01-03 22:15:28 IST	26.52	93.03	10	2.8	27Km ESE of Tezpur, Assam, India
2	2021-01-04 00:13:07 IST	26.21	90.94	5	2.5	79Km W of Guwahati, Assam, India
3	2021-01-06 04:32:07 IST	26.95	92.93	8	2.5	35 Km NNE of Tezpur, Assam, India
4	2021-01-24 14:01:02 IST	27.79	95.04	10	3	37 Km NNE of Dibrugarh, Assam, India
5	2021-02-08 13:49:15 IST	26.57	93.14	30	3.7	35 Km ESE of Tezpur, Assam, India
6	2021-02-09 04:09:11 IST	26.99	92.54	5	2.3	45 Km NW of Tezpur, Assam, India
7	2021-02-17 17:54:46 IST	26.71	92.63	10	4.7	17 Km WNW of Tezpur, Assam, India
8	2021-02-22 14:01:53 IST	26.298	92.896	10	2.8	40 Km SSE of Tezpur, Assam, India
9	2021-02-25 22:33:21 IST	26.19	92.05	30	3.6	31 Km E of Guwahati, Assam, India
10	2021-03-02 01:32:22 IST	26.41	92.29	15	2.9	56 Km WSW of Tezpur, Assam, India
11	2021-03-02 01:47:58 IST	26.78	93.05	15	2.7	29 Km ENE of Tezpur, Assam, India
12	2021-03-12 16:14:16 IST	26.13	92.71	5	3.1	58 Km S of Tezpur, Assam, India
13	2021-03-21 17:52:41 IST	26.53	92.58	25	4.1	25Km WSW of Tezpur, Assam, India
14	2021-03-21 18:12:55 IST	26.64	92.45	10	3.1	34Km W of Tezpur, Assam, India
15	2021-04-04 22:38:31 IST	26.57	93.89	24	2.9	36Km WSW of Jorhat, Assam, India
16	2021-04-06 03:42:16 IST	27.55	95.54	39	2.7	62 Km E of Dibrugarh, Assam, India
17	2021-04-14 04:34:57 IST	26.44	92.82	10	2.8	46 Km S of Tezpur, Assam, India
18	2021-04-27 12:42:04 IST	26.81	92.73	28	2.6	18Km NNW of Tezpur, Assam, India
19	2021-04-28 07:51:25 IST	26.69	92.36	17	6.4	43 Km W of Tezpur, Assam, India
20	2021-04-28 08:03:01 IST	26.74	92.42	15	4.7	38Km WNW of Tezpur, Assam, India
21	2021-04-28 08:13:21 IST	26.69	92.41	20	4	38Km W of Tezpur, Assam, India
22	2021-04-28 08:25:40 IST	26.54	92.45	26	3.6	36Km WSW of Tezpur, Assam, India
23	2021-04-28 08:44:34 IST	26.59	92.46	25	3.6	33Km WSW of Tezpur, Assam, India

24	2021-04-28 10:05:54 IST	26.39	92.52	27	3.2	39Km SW of Tezpur, Assam, India
25	2021-04-28 10:39:12 IST	26.62	92.46	10	3.4	33Km W of Tezpur , Assam, India
26	2021-04-28 12:32:06 IST	26.36	92.38	23	2.9	52Km SW of Tezpur , Assam , India
27	2021-04-28 14:34:52 IST	26.83	92.50	10	3.4	35Km NW of Tezpur, Assam, India
28	2021-04_28 15:55:48 IST	26.77	92.36	10	2.9	44Km WNW of Tezpur , Assam ,India
29	2021-04-28 17:39:57 IST	26.62	92.40	19	3	39Km W of Tezpur, Assam, India
30	2021-04-28 21:38:11 IST	26.73	92.38	15	2.8	41Km W of Tezpur, Assam, India
31	2021-04-29 00:24:02 IST	26.76	92.93	18	2.6	18Km NE of Tezpur, Assam, India
32	2021-04-29 01;10:24 IST	26.71	92.40	19	2.9	39Km W of Tezpur, Assam, India
33	2021-04-29 01:20:53 IST	26.72	92.43	19	4.6	36Km W of Tezpur, Assam, India
34	2021-04-29 01:41:49 IST	26.78	92.40	11	2.3	41Km WNW of Tezpur, Assam, India
35	2021-04-29 01:52:26 IST	26.73	92.43	16	2.7	37Km WNW of Tezpur, Assam, India
36	2021-04-29 02:38:33 IST	26.73	92.42	11	2.7	38Km WNW of Tezpur, Assam, India
37	2021-04-29 07:13:36 IST	26.86	93.01	30	2.8	31Km NE of Tezpur, Assam , India
38	2021-04-29 13:04:20 IST	26.61	92.33	27	3.6	46Km W of Tezpur, Assam, India
39	2021-04-29 16:17:33 IST	26.59	92.47	26	3.3	32Km WSW of Tezpur, Assam, India
40	2021-04-29 19:14:48 IST	26.74	92.36	10	2.8	44Km WNW of Tezpur, Assam, India
41	2021-04-30 01:08:23 IST	26.90	92.43	10	2.8	45Km NW of Tezpur, Assam, India
42	2021-04-30 16:49:28 IST	26.58	92.51	40	3.5	29Km WSW of Tezpur, Assam, India
43	2021-04-30 18:27:31 IST	26.59	92.39	27	3.2	40Km W of Tezpur, Assam, India
44	2021-05-01 20:09:46 IST	26.72	92.39	15	3.3	40Km W of Tezpur, Assam, India
45	2021-05-01 20:26:58 IST	26.73	92.37	14	2.7	42Km W of Tezpur, Assam, India
46	2021-05-02 03:48:51 IST	26.71	92.40	16	2.7	39Km W of Tezpur, Assam, India
47	2021-05-02 23:02:37 IST	26.82	92.43	13	2.6	40Km WNW of Tezpur, Assam, India
48	2021-05-03 01:23:39 IST	26.43	92.24	40	2.7	59Km ENE of Guwahati, Assam, India
49	2021-05-03 18:13:08 IST	26.74	92.45	10	3.7	35Km WNW of Tezpur, Assam, India
50	2021-05-03 21:55:01 IST	26.71	92.39	17	2.8	40Km W of Tezpur, Assam, India

51	2021-05-04 00:09:03 IST	26.65	92.41	25	2.8	38Km W of Tezpur, Assam, India
52	2021-05-05 19:22:12 IST	26.68	92.39	27	3.5	40Km W of Tezpur, Assam , India
53	2021-05-06 10:30:16 IST	26.55	92.39	10	3.6	41Km WSW of Tezpur, Assam, India
54	2021-05-07 00:19:36 IST	27.01	92.46	17	2.9	51Km NW of Tezpur, Assam, India
55	2021-05-07 00:19:36 IST	26.33	92.34	68	2.8	57 Km SW of Tezpur, Assam, India
56	2021-05-07 19:33:18 IST	26.75	92.42	10	2.8	38 km WNW of Tezpur, Assam, India
57	2021-05-08 01:09:55 IST	26.32	91.44	28	2.9	35 Km WNW of Guwahati, Assam, India
58	2021-06-01 23:48:30 IST	26.77	92.42	10	3.5	39 KM WNW of Tezpur, Assam, India
59	2021-06-30 21:02:17 IST	26.52	92.52	65	2.7	30 Km WSW of Tezpur, Assam, India
60	2021-07-16 08:39:21 IST	26.48	92.44	10	3.4	40 Km WSW of Tezpur, Assam, India

Source: National Centre of Seismology, Ministry of Earth Science, Govt. of India.

The worst of these was the great Assam-Tibet Earthquake that occurred on the Independence Day in 1950. Assam is bounded by the Himalayan belt in its northern and eastern part which is prone to great earthquakes of magnitude exceeding 8.0. Scientific publications have warned that the region would experience several earthquakes, with the possibility of affecting the lives of several million people. Since 1950 the region has been experiencing several moderate earthquakes. On an average, the region experiences an earthquake with a magnitude greater than 6.0 every year. Some of the earthquakes occurred in the past are presented in Table 5.1 and Table 5.2.

Earthquakes in Assam: Types, Magnitude and Frequency

Assam is located in seismic zone V and this area is seismically identified as the most active and hazardous region. This region has experienced some of the country's most powerful shocks with magnitudes exceeding 7.0 and intensity higher than IX. Some of the earthquakes occurred in different seismic source zones of North-East during 1762 - 2015 with different magnitudes are shown in Table 5.3.

Table 5.3
Pattern of occurrence of earthquakes in North-East India during 1762-2015

Seismic Source Zone	No. of Earthquakes occurred with different magnitudes				
	4-4.9	5-5.9	6-6.9	7-7.9	8 and above
Indo-Burmese Zone	874	619	66	7	1
Shillong Plateau	340	223	13	4	1
Eastern Himalayan Zone	175	151	23	0	0
Mishimi Region	86	66	1	1	1

Source: Bahuguna and Sil (2018).

Earthquake Occurrences

Located between two plate boundaries, the Himalayan in the north and the Indo- Burmese in the east, the north-eastern region is one of the seismically most active regions of the world (Fig. 5.1). The complex tectonic and geological set up of the region and intense continental convergence of the northward moving Indian plate at the rate of 20+03mm/year can produce earthquakes of magnitudes 8 and above every few hundred years. Consequently, it has been under continual threat from major earthquakes, a subject being seriously debated in national and international platforms. The two great earthquakes of magnitudes 8.7 in 1897 and 1950 are memorable events in the annals of the history of earthquake in the world. These earthquakes were so intense that the rivers of the region changed their courses and ground elevations got immensely affected. Besides, as many as twenty destructive earthquakes of the magnitudes 6~7 rocked this region during the past century. Some better known damaging earthquakes of N.E. region are described below:

Significant Earthquake Events in Assam

Seismicity is very common in the Assam region. It has frequently been experiencing earthquakes of varying magnitude (M). The seismic record shows that this region had experienced 2 great earthquakes of M >8.7, 20 earthquakes of M >7, 29 earthquakes of M between 6 and 7, and innumerable smaller earthquakes. The data for a period of five months per year (1980-86) further showed a total of 7,000 earthquakes occurred in different parts of the region (Kayal, 1996). In Subansiri district of Arunachal Pradesh alone, nearly 1,000 earthquakes were recorded in 10 months' time in 1990 (Kayal et al., 1992).

Earthquakes in Assam can be considered as occurred during non-instrumental and instrumental periods. Both instrumented and non-instrumented events are listed below. It was only after mid-nineteenth century when the instruments to measure the magnitude were widely used. Before that no scientific measurement of the occurrence were reported. That is why reported magnitudes are listed for instrumented events, while the historical ones are listed as non-instrumented shocks.

Earthquakes during Non-Instrumental Period in Assam:

1548 earthquake during the reign of king Suklenfa of Ahom kingdom
1596 earthquake during the reign of king Sukhamfa of Ahom kingdom
1601 earthquake during the reign of king Susengfa of Ahom kingdom

1642 earthquake during the reign of king Sarufa.of Ahom kingdom
1663 earthquake when Mirjumla fled away from Assam
1696 earthquake during the reign of king Rudra Singha. of Ahom kingdom
1732 earthquake.
1759 earthquake during the reign of king Rudra Singha of Ahom kingdom
1772 earthquake of relatively severe intensity.
1832 earthquake.
1848 earthquake.

Source: CNDM Report

Major Earthquakes during Instrumental Period in Assam

A large number of earthquakes occurred in the last one and half decades in Assam and its adjoining areas. The following are some glimpses of some of the major earthquakes. A major earthquake that struck Cachar, Assam on 10th January 1869 caused heavy damage in the entire region The epicenter of this earthquake was located in the Cachar area of Assam. The impact of the shock was felt over 6,50,000 square kilometers. There was heavy damage in the towns of Silchar, Shillong, Cherrapunji and Sylhet and also in Manipur. Fissures opened on the banks of the Surma river and sand vents threw up great amounts of sand and water. The epicentral tract was 30-45 kilometers long and 5-6 kilometers wide lying on the northern border of the Jaintia Hills. The hypocenter had a depth of 50 kilometers. In 1897, there occurred one of the most powerful earthquakes in the Indian sub-continent and probably one of the largest known anywhere. The quake wreaked havoc across south- west of the present states of Assam, Meghalaya and also Bangladesh. About1542 people were killed and hundreds more injured. The earthquake caused great destruction to many towns in Assam and Meghalaya. The towns of Dhubri, Goalpara, Guwahati and Coach Bihar in Assam and West Bengal were heavily damaged. In 1930, a strong earthquake that occurred in the morning hours greatly affected parts of western Assam, West Bengal and Bangladesh. Most of the buildings in Dhubri and the surrounding areas were destroyed in this shock. It was felt as far away as Dibrugarh, Kolkata, Chittagong and Patna. It was felt nearly all over the northern-eastern region. Again an earthquake occurred on 21st January,

1941 and it had a magnitude of 7.7. This earthquake was felt over larger part of Assam, Bengal (up to Kolkata) and Bihar (up to Purnea). At Jorhat in Assam, water overflowed river banks. At Dibrugarh, Jorhat and Tezpur cracks developed in the building walls causing failure of electricity. This earthquake was also felt at Silchar, Cooch-Bihar, Mymensingh, Dhubri, Rangpur, Tezpur and Guwahati.

The biggest earthquake of the 20th century in this region with a magnitude of 8.7 occurred on 15th August, 1950, which is often referred to as the "Assam Earthquake of 1950". About 1500 people were killed and the drainage of the region was greatly affected by this earthquake. The resultant floods were the cause of most of the fatalities aftermath of this earthquake. The initial shock was followed by thousands of aftershocks, some of which were big enough to be reckoned, and struck a relatively sparsely populated region along the Indo-China border. There was widespread devastation in Upper Assam, the Abor Hills and the Mishmi Hills. The region that suffered the most damage to life and property was 15,000 square miles in area. This included the districts of Jorhat, Lahkimpur, Sibsagar and Sadiya in Assam. Dibrugarh and Saikoaghat were among the worst affected areas. The most dramatic feature of the 1950 event was landslides. The landslides covered 15,000 sq. km of area (Mathur, 1953). Another major earthquake had occurred on 1st July, 1957 with a magnitude of 7.0 and it had the epicenter near Indo-Burma border. This earthquake was widely felt over Assam, Manipur, Tripura, East Pakistan and parts of West Bengal and Bihar. The shock was felt at Tezpur, Halflong, Guwahati, Kailasahar, Silchar, North Lakhimpur, Rowriha, Kumbhirgram, Kohima, Dhubri, Lumding, Pasighat, Imphal, Bindukuri, Shillong, Mazbat, Goalpara, Agartala. After 1950 also many earthquakes have occurred from time to time causing some minor impacts in different parts of Assam, but fortunately not causing widespread destruction (CNDM Report).

In last few years also, many small shocks have been felt in Assam and adjoining areas, but no notable earthquake shocks have been reported. Very recently, on 28th April, 2021, an earthquake of magnitude 6.4 occurred in Sonitpur district of Assam at a focal depth of 17 km. The epicenter was located near the Kopili fault and closer to Himalayan Frontal Thrust (HFT). This earthquake was felt in the entire North-East India and some parts of West Bengal, Bihar and Bangladesh. Damages of buildings from Sonitpur, Nagaon and Guwahati have been reported. Cracks in roads and agricultural fields have also been seen in many rural areas near the epicenter.

Earthquake Vulnerability

Assam and its adjoining areas of north-eastern region is seismically a highly vulnerable region being surrounded by the Himalayas, Arakan Yoma, Hills and plains of Bangladesh, etc. It falls within the zone V of seismic intensity and one of the sixth active zones of the world. Earthquakes in this region mainly occur due to collision between India and Chinese plates in the north and Indian and Burmese plates in the east and south and due to local rift and fault displacements. Apart from the geological and topographic aspects, ever increasing anthropogenic activities may lead this region towards high vulnerability.

Unplanned and unsustainable developmental activities, particularly concrete construction of high rise buildings for enjoying the modern lifestyle, are increasing day by day without considering their vulnerability to the earthquakes. The capital towns, business and other towns of Northeastern region have reached the utmost critical level of their bearing capacity. Tremendous growth of population is continuously putting enormous pressure to construct houses on the hill slopes, thereby increasing the instability in the physical landscape. The point of concern is not just increasing population, but also urban characterization in the delicate eco-system which is severely prone to natural hazards. The people are crazy for constructing concrete buildings, most of which are without engineering design, and even thc houses have paved the way for multistoried masonry buildings. The situation has become vulnerable to disaster, and therefore, there is a strong need for proper planning to ensure sustainable development. It is possible through a long-term program for natural hazards assessment and mitigation with an integrated approach ensuring participation of the society and scientific community (Singh and Tewari, 2005).

Growth of population in the last four decades is more than

three times from the year 1950. Urbanization on an unprecedented scale is being observed almost in all the states of the North-eastern region. Rapid urbanization results in not only the setting up of new settlements, but also leads to overcrowding with buildings and people. Most of the people want to settle down in the urban towns, which have developed and mushroomed over the years. However, the land resources for urban settlement are limited and, thus, almost all the capital towns and other towns and business centers in this region have reached the critical level of their load bearing capacity.

The construction of high rise buildings in Guwahati and other cities of this region has been continually flouting the rules of BIS codes. This has become a serious concern for the government and public at large. Invariably, we ourselves have created such conditions which make people vulnerable by permitting excessive urbanization in a region of heightened danger. We have allowed multiple constructions which hitherto should have been rejected considering high seismic risk in the region. Identical conditions prevail in all the towns of Assam not to speak of the other states capitals of the region. When the earthquakes of 1897 and 1950 occurred, Assam and the present Northeastern states were sparsely populated. Therefore, death toll was very low, merely 1500 odd people were killed. If the region, which is now densely populated, is to be struck by earthquakes of the same size and magnitude, the effects may be unmanageable. The aspect of the height of buildings has not been adhered to in spite of high seismic risk. Another aspect which afflicts the city like Guwahati, is the hill belt having a fragile ecosystem and landslide in the hill belt sends a danger signal for a catastrophe. To add to these the dangerous areas, the cities are plagued by dense settlements, limited vacant space and narrow streets which have altogether enhanced the degree of vulnerability to earthquake hazards (Singh, 2002).

Furthermore, it may be noted that in Digboi was the only oil refinery in Assam that was in operation during 1950. Subsequently, three more refineries have been established. Nonetheless, about one thousand km long petroleum transporting pipeline from Naharkatia to Barauni has been laid only during last decades. Continuous transportation of highly inflammable products across the entire length of the narrow valley having large concentration of thickly populated villages and towns is a matter of serious concern. It is more surprising that this pipeline crosses many railway bridges, culverts and pools which may be more vulnerable during in any disastrous events where the communication link will be destroyed and paralyzed. Therefore, seeing at these development, it feels a much better standard of preparedness than was necessary in the past. A high degree of societal awareness and support is required to implement hazard mitigation measures.

Earthquakes are natural hazards, but the disasters are man-made events. "Earthquakes don't kill, unsafe buildings do", it has been proved right. The high rise buildings are become vulnerable that turns hazards into disasters. It is the high time to give attention for ensuring structural and operational safety of the buildings not only the structural design, but also the form and configuration of the buildings. The Bureau of Indian Standards (BIS) has laid down the national standards for construction of houses and buildings in seismically vulnerable areas. But it is yet to make mandatory in many of our towns and cities. Moreover, lack of awareness, education and poor enforcement are also some of the reasons because of which the rules are not being followed. This leads to a tendency of using inferior materials and negligence towards construction of buildings with structural design with safety features. As a result, vulnerability increases due to seismically unsafe constructions. Many of the buildings have been seen collapsed due to poor structural design during the recent earthquakes occurred in Assam. Assam and all the Northeastern States, therefore, tend to depict a grim scenario with respect to seismic hazard. Because a major earthquake may adversely affect practically all facets of human life and paralyses the total functioning of society, besides becoming a big economic burden on the country which results in retardation of the pace of sustained economic development and draining out of available resources meant for pursuing development activities (Singh & Tewari, 2002).

Earthquake Disaster Mitigation Measures

Disaster management or mitigation is an integrated process of planning, organizing, coordinating and implementing measures to effectively deal with their impact on people (Deshmukh et.al, 2008).

The basic phases of disaster management are defined as long-term mitigation, preparedness, relief, and reconstruction. Mitigation or risk reduction activities include structural and non-structural measures to reduce the adverse effects of natural hazards (Fazari and Kashim, 2019). With the ever increasing frequency of earthquake disasters in the recent past, particularly after the Uttar-Kashi earthquake of Himalayan region in 1991, Latur earthquake in 1993, Chamoli earthquake in 1999, Gujarat earthquake in 2001, Sikkim earthquake in 2011 and Nepal earthquake in 2015, people in the Northeastern region have become more aware about the natural disasters and its impacts. People have seriously realized the need of analyzing and understanding the relationship between cause and effect, and also visualizing the long-term implications of disasters. Since earthquakes cannot be predicted, there is an urgent need for proper planning for assessment, mitigation, management and rehabilitation (Gaur, 1998). Understanding temporal evolution of some of the natural hazards has, however, led to recognize their space-time pattern, and, to some extent, our own ability to predict the probabilities of their recurrence in future. Accordingly, efforts are being made for designing and implementing protection systems, which are not only cost effective but also maintained and serviced locally.

The need of the day, therefore, is to educate the general public regarding the danger from earthquake hazards and their mitigation strategies. "Prevention is better than Cure" and prevention begins with information. Thus, integrated knowledge about preparedness, rescue, relief, rehabilitation, resettlement, etc., has to be provided to all the stakeholders. This is possible through a long-term program for hazards assessment and mitigation by an integrated approach with the participation of the society and scientific community (Singh and Tewari, 2005).

A systematic scientific mitigation and management strategy for the probable seismic events of varying magnitude needs to be taken up on priority basis. Some of the major strategies and action plans relating to earthquake disaster are discussed below:

Advance Planning:

This is a pre-requisite for each and every type of management strategy. It includes:

1. Identification and demarcation of hazard-prone areas on the basis of historical evidences and current understanding as well as conceptual anticipation.
2. Design and formulation of guidelines for engineering specifications for various kinds of structure, particularly construction of skyscrapers to reduce the damage and minimizing the possibility of collapse of the buildings during the time of earthquake.
3. Scientific assessment of the vulnerability and risk of the existing structure is very much essential. Cost effectiveness should be considered as prime factor for adoption of methods for assessment. In addition, retrofitting of all essential buildings like hospitals, schools and colleges, telephone exchanges, power houses, water supply, heritage buildings and bridges etc. are at utmost need.
4. Design and operational readiness of standard protocols for effective rescue and relief measures, prevention of epidemics, emergency operations of essential services are to be considered as high priority.
5. Regular dissemination of information through carefully designed bulletins, leaflets to evoke a constructive response and avoid panic in the schools, colleges and other important centers.

Vulnerability Mapping and Assessment:

In the last two decades, the term 'vulnerability' has become a critical and widely used concept for hazard research and developing appropriate mitigation strategies at the local, national, and international scale (Friegerio et.al, 2016). The integrated vulnerability mapping and assessment framework is flexible and can easily be applied to urban environments at various geographical scales to review the actual situation. In this sense, the overall vulnerability maps for the urban area have potential for decision-makers in designing vulnerability reduction strategies and hence risk reduction strategies. Further it will also help the planners to suggests specific areas for developmental activities as well as vulnerable to disasters.

Awareness Creation:

There are varied opinions regarding possibility of occurrence of next great earthquake in Northeastern region of the country. whenever the media propagate possibility of occurrence of earthquake in future based on prediction by some scientists, it creates panic among the public. The highly scientific and theoretical seismological studies suggest that such studies could make prediction about the areas of high probability, but about time of occurrence is not possible. The need of the day is, therefore, to educate the general public regarding the danger due to seismic hazards and their mitigation strategies.

Enforcement of Building Construction Regulation:

Building regulation is considered to be an important tool to address the problems of uncontrolled development and increasing disaster risk. In spite of existence of various codes and guidelines (BIS codes, NDMA guidelines, and PWD/CPWD manuals), nothing has proved accountable enough to scale down the magnitude of the vulnerabilities of building stocks (Ghosh et.al, 2017). In the seismic zones, more often these rules are not observed or followed properly. Enforcement of such laws will otherwise go a long way in ensuring safety and longevity of the people and civil constructions.

Formulation of Mitigation Strategy:

Mitigation measures are the actions that eliminate or reduce future risk of hazards or disasters. Seismic hazard mitigation strategy primarily includes structural and non-structural measures. Structural measures include improving the building elements that hold up a building and resist lateral forces from earthquakes, including: foundations, columns, load-bearing walls, floor diaphragms, roof diaphragms and the connections between these structural elements. Nonstructural mitigation measures to restrain, brace, anchor or otherwise improve the seismic resistance of nonstructural building components such as parapets, chimneys, non-load bearing walls, fire sprinkler systems, HVAC systems, suspended ceilings and lights, windows, water heaters, furnaces, air conditioners and emergency generators. In addition, the suitability and adequacy of response and relief infrastructure in the urban areas following an earthquake event also needs to be integrated.

The North-eastern region of the country by virtue of its varied topography and geological set up is vulnerable to earthquakes and in seismic zone V. So, risk is always there if any activity disapproved by the nature results to trigger a disaster. Disasters cannot be stopped to happen, but certainly its effects can be reduced by undertaking various precautions strategy.

Financial and Insurance Institutions:

Disaster Risk Financing and Insurance is one of the key financial tools for funding disasters (NIDM, 2021). Institutions providing construction finance have to be sensitized among the public to make their investment secure. Efforts should be made on insisting well-organized and earthquake resistant construction where necessary mandatory insurance of all buildings above a stipulated value, size or category, both public and private, have to be ensured through suitable legislation.

Training of Stakeholders:

As a matter of fact, when considered the various facets of disaster management, the role of the various stakeholders involved in the entire process cannot be underestimated. This includes the government sector, parastatals, and even the parliament. An effective synergy among the stakeholders has been described as a potent driver for effective disaster management and mitigation (Khan and Rahman, 2007). In context of North-east India, a variety of organization and institution are taking the responsibility to train up personals, officials and institutions etc. on disaster management. Centre for Natural Disaster Management at Assam Administrative Staff College undertakes training programmes at village, panchayat, block, district and state levels aiming at creating awareness amongst the government officials and communities about disasters and possible mitigation measures. Efforts should be made to open disaster management unit at school and college level for better understanding with practical knowledge.

Rehabilitation and Resettlement:

Rehabilitation and resettlement operations are very much integral to disaster response and recovery. It also provides a direct connection between disaster response and long-term development. Rehabilitation involves restoring utmost local services which are of immediate needs. It enables basic services to resume functioning, assist victims' self-help efforts to repair physical damage, restore community facilities, revive economic activities and provide support for the psychological and social well-being of the survivors. Resettlement, on the other hand, is a type of long-term development assistance, which could help people in the affected areas to rebuild their lives and meet their present and future needs. In order to bring back the normal situation preparation should be made for quick constructions of temporary shelf shelters, immediate financial incentives, plan for land pooling, distribution of priority items, viz. cover shelter, food, hygiene, clothes; drug and surgical items and construction; emergency veterinary medicines, vaccines, disinfectants and feeds; and agricultural nutrients etc. Such types of consequential aspect need to be taken up on priority basis to provide living atmosphere to the affected people in which they used to live before the calamity. It is also essential to regain social life, status and self-confidence of the affected people.

Research and Development:

It is an essential part of disaster mitigation, long term management and development. This strategy aims at meticulous scientific preparedness through development of low cost and locally supportable technologies. Efforts should be made for improvement of slope stability, landform classification etc., and other suggestions come from research and practice. Moreover, further research should be taken up for estimating and mapping the pattern of hazards intensities especially in the vulnerable areas using scientific understanding and technology. For the extension of knowledge, it is desirable to have a meaningful interaction between the scientists and technical experts on one hand, and administrators and planners, on the other hand, on regular basis which may help to understand the natural hazard mitigation and management in a better way.

Conclusion

The above discussion on the earthquake scenario in Assam reveals that Assam is prone to earthquakes ranging from small to big events due to its geological uniqueness. Emergence of high magnitude earthquake event may inflict multifaceted socio-economic distress on the society. People should not panic in such situation, and should try to effectively follow the rules by keeping calm. Multidisciplinary awareness programmes should be conducted for creating awareness, preparedness and developing mitigating measures in seismically prone areas. Resilient practices should be encouraged so that affected people can live with the earthquakes. Thus, the devastation caused by this natural calamity can be efficiently mitigated and managed through awareness and effective preparedness.

References

Awasthi, R.K. (2002): 'Geo-hazards in North East India: Experience, problems and mitigation', in: Report on Disaster Management in NE Region: Floods and Earthquakes, 28-29 June 2003, organized by Department of Revenue, Government of Assam and UNICEF, Kolkata, 35-45.

Bahuguna, A. and Sil, A. (2020): 'Comprehensive Seismicity, Seismic Sources and Seismic Hazard Assessment of Assam, North East India', *Journal of Earthquake Engineering*, 24:2, 254-297, DOI: 10.1080/13632469.2018.1453405.

Baruah, S., S. Bora, P.K., Duarah, R., Kalita, A. Biswas, R., Gogoi, N. and Kaya, J.R. (2012): 'Moment Magnitude (MW) and Local Magnitude (ML) Relationship for Earthquakes in Northeast India', *Pure Appl. Geophys*, 16, 1977-1988.

Centre for Natural Disaster Management (CNDM): 'Documentation on past disasters, their impacts, measures taken and vulnerable area in Assam' prepared by Centre for Natural Disaster Management, *Guwahati.*

Deshmukh, R. Rodrigues, L.L.R. and Krishnamurthy, G.R. (2008): 'Earthquake Risk and Knowledge Management'. *Journal of Knowledge Management Practice*, 9 (3).

Fazaril, Al. S. and Kasim, N. (2019): 'Role of Stakeholders in Mitigating Disaster Prevalence: Theoretical Perspective', MATEC Web of Conferences, https://doi.org/10.1051/matecconf/2019 I 2018 266 26603008.

Frigerio, I. De and Amicis, M. (2016): 'Mapping social vulnerability to natural hazards in Italy: A suitable tool for risk mitigation strategies', *Environ. Sci. Policy*, 63, 187-196.

Gaur, V.K. (1998): 'Mitigating disasters in the Himalaya - a basic agenda for development', 8th G. B. Pant Memorial Lecture, delivered on 30th November 1998 at Kulu. G. B. Pant Institute of Himalayan Environment & Development, Kosi-Katarmal, Almora.

Ghosh, C. Chatterjee, R. and Shaw, R. (2017): 'Enforment of building construction regulations in urban centres of India' in Edt. *Disaster Risk Governance in India and Cross Cutting Issues,* published by Springer pp. 315-337.

Jain, S. K. (1998): 'Indian earthquakes: an overview, Indian Concrete Journal, 72, 555-562.

Kayal, J.R. (1987) 'Micro seismicity and source mechanism Study: Shillong Plateau, NE India', *Bull. Seismological Soc.* America, 77, 184-194.

Kayal, J.R. (1996): 'Earthquake source processes in North-East India: A review', *Himalayan Geology,* 17, 53-69.

Kayal, J.R. De, R. and Ram, S. (1992): 'Study of seismicity in North-East Indian (Phase-I): A micro earthquakes survey around the proposed Dam, Ranganadi Hydroelectric Project, Lower Subansiri District, Arunachal Pradesh', *Report of Geol. Survey of India,* 125 (4), 136-139.

Kayal, J, R., Arefiev, S, Baruah, S., Hazarika, D., Gogoi, N., Gautam, J. L., Baruah, S., Dornath, C. and Tatevossian, R. (2012): 'Large and great earthquakes in the Shillong Plateau-Assam valley area of Northeast India region: Pop-up and transeverse tectonics', *Tectonophysics* 532-535, pp 186-192.

Khan, M.R. and Rahman, M.A. (2007): 'Partnership Approach to Disaster Management in Bangladesh: A Critical Policy Assessment'. *Natural Hazards,* 41(2), 359- 378.

Regional Research Laboratory (2004): 'Seismological Bulletin NE India - 2003', Regional Research Laboratory, Jorhat (Assam), India.

Sánchez-Sesma, F. J. Palencia, V. J. and Luzón, F. (2002): 'Estimation of local site effects during earthquakes: An overview', *ISET Journal of Earthquake Technology* 39 (3) 167-193.

Singh, T. (2002): 'Why NE-Region of India falls under high seismic zone, and what may be the preparedness? in: Report on Disaster Management in NE Region: Floods and

Earthquakes, 28-29 June 2003, organized by Department of Revenue, Government of Assam and UNICEF, Kolkatta, pp. 30-34.

Singh, T. and Tewari, P. (2005): Impact of earthquake disasters on New Urbanization Pattern in North Eastern Region of India, *Journal of Environmental Systems*, 32 (1) pp. 69-90.

Walia, V. Virk, H. S. and Bajwa, B. S. (2006): 'Radon precursory signals for some earthquakes of magnitude> 5 occurred in NW Himalaya: An overview', *Pure and Applied Geophysics*, 163(4), 711-721.

Valdiya, K.S. (2002): 'Geodynamic perspective of Arunachal Pradesh: bearing on environmental security and planning for development', In: R.C. Sundriyal, Trilochan Singh, and G.N. Sinha (eds.), Arunachal Pradesh: Environmental Planning and Sustainable Development. HimaVikas Occasional Publication, pp. 1-15. Published by G. B. Pant Institute of Himalayan Environment & Development.

https://www.bgs.ac.uk/discovering-geology/earth-hazards/earthquakes/

https://www.britannica.com/science/earthquake-geology/Tectonics/

https:*//earthquake.usgs.gov.in*

https://ndma.gov.in/Natural-Hazards/Earthquake/

https://onlineasdma.assam.gov.in/kmp/detail/category/earthquake/case-studies/

https://riseq.seismo.gov.in/riseq/earthquake

http://www.mati.gov.in/dmc.html

https://www.k12.wa.us/sites/default/files/public/schfacilities/pdm/pubdocs/mitigation_measures_summary.pdfb

https://reliefweb.int/report/nepal/post-earthquake-relief-rehabilitation-and-reconstruction-measures-undertaken-government

6 Forests

Introduction

Forests are associations of plants,predominantly trees. They constitute one of the major forms of natural landscape. Originally forests covered about 40 per cent of the land area of the earth. But man has been exploiting the forest resources at such a high rate that they now cover only about 17 per cent. However, the forests constitute one of the man's extensively as well as most senselessly used resources of the earth.

Forests are important renewable natural resource. They confer multiple benefits and play an important role in the economic development. Forests help augment natural water supply, preserve wildlife, control floods and prevent soil erosion, besides providing valuable raw materials for a number of industries. They have a moderating influence on climate and provide recreational facilities to human society.

Forest Resources of Assam

Nature has endowed Assam with a rich potential for the development of forest. The forest areas of the state are the meeting grounds of Indo-Tibbetan and Burmese flora and fauna. The luxuriant evergreen and deciduous forests of the state abound in variety of valuable timber species. There is also diffusion of a considerable number of species that must have evolved in the region which has been well demonstrated in the forest types. In certain parts of the state extensive bamboo forests also exist. Assam occupies a significant place in the wildlife map of the country for her rich and rare birds and animal species. In no other parts of the world with comparable climatic, geomorpholgical and edaphic condition, such diversity of flora and fauna exists as they do in the limited biogeographical area of the state of Assam. Of the total geographical area of the state

(78,523 sq. km.), the area under forest is only 22 per cent which is far below the minimum norm of 33.3 per cent prescribed by the national forest policy. Table 6.1 gives the total area under forest in the state during 1974-94.

Table 6.1
Forest area in Assam (in sq. km), 1974-1995

Year	Area	P.C. of total area
1974-75	28,924.00	36.83
1977-78	28,608.00	36.43
1982-83	27,337.00	34.81
1987-88	21,513.20	27.39
1989-09	21,522.60	27.40
1991-92	21,684.52	27.61
1993-94	22,175.85	28.24

Source : Economic Survey, Government of Assam.

From the table it is evident that the total area under forest declined by about 6,748 sq. km. during the period 1974-94. The government, however, planned for further expansion of reserved forest in order to increase the forest area in the state. Table 6.2 shows the reserved forest and proposed reserved forest area since 1989-90.

Table 6.2
Reserved forest and proposed reserved forest area in Assam (in sq. km), 1989-1994

Class of forest	1989-90	1990-91	1991-92	1992-93	1993-94
1. Reserved forest	17427.47	17442.97	17580.54	17567.51	18242.23
2. Propesed reserved forest	4095.13	3909.27	3920.62	3920.62	3933.62
3. Total (excluding unclassed forest)	21522.60	21412.24	21684.52	21488.13	22175.85

Source : Statistical Handbook of Assam 1992, Office of the Principal Chief Conservator of Forest, Assam.

Table 6.2 shows that in the year 1994 Assam had around 18.2 thousand sq.km or reserved forest, while the proposed reserved forest constituted 3.9 thousand sq.km. It has been seen that the total area under forest has increased from 21.5 thousand sq.km in 1989-90 to 22.1 thousand sq.km in 1993-94. But the proposed reserved from area has decreased from 4.0 thousand sq.km in 1989-90 to 3.9 thousand sq.km in 1990-91. From 1991 onward the proposed reserved forest area showed slight increase in area. It may be noted that the area under reserved forest has been increasing in the State. Whereas the area under unclassed forest has been declining since 1993. As a whole, the percentage of forest land is decreasing year by year. As for examble, in 1974 the forest area constituted 36.83 per cent of the State's total area which in 1980-82 decreased to 25.24 per cent. At present it has further decreased to 21.9 per cent (according to NRSA survey). On the other hand, the hilly area of Assam has 19585.14 sq.km.forest area, while the plain region has 11122.43 sq.km. The per capita forest in Assam was only 0.00137 sq.km. (0.13 hectare) in 1991-91.

The following table shows the districtwise distribution of forest area in Assam.

Table 6.3
Districtwise distribution of forest area in Assam (area in hectare)
(Based on satellite survery, 1990)

District	Geographical area	Forest land	%
Barpeta	330730	23411	7.08
Bongaigaon	31335	5354	17.09
Cachar	377927	142953	37.82
Darrang	346530	23215	6.71
Dhemaji	314312	46902	14.92
Dhubri	274550	46525	16.95
Dibrugarh	346763	45275	13.05
Goalpara	253045	22850	9.03
Golaghat	354100	30605	8.69
Hailakandi	132293	50454	38.19
Jorhat	291470	27624	9.46
Kamrup	473380	117260	24.77

Karbi Anglong	1033200	450893	43.64
Karimganj	183900	46328	25.19
Kokrajhar	471650	155879	33.04
Lakhimpur	250328	5576	2.23
Morigaon	121966	39682	32.55
Nalbari	202480	12528	6.19
N.C. Hills	489000	187070	38.55
Nagaon	431905	34369	7.96
Sibsagar	260290	28697	11.01
Sonitpur	525520	107861	20.52
Tinsukia	355626	75076	21.11

Source: Assam's Land Resources - A Satellite-based ARSAC Survey, publised by Assam Science Society, 1993.

The highest concentration of forests in the state occurs in Karbi Anglong (450893 ha i.e. 43.64 per cent) and N.C.Hills (187070 ha, 38.55 per cent) and the lowest concentration occurs in Bongaigaon district (only 5354 ha.). However, there is slight difference in the total forest cover between conventional data and remote sensing data. Conventional data shows 22 per cent forest, but satellite survey shows 21.1 per cent forest cover in the state. The forests in the plain districts are under the management of the State Forest Department, while the District Councils manage the forests in the hill districts.

Forest Types

Forest in Assam can be classified into three types : evergreen, semi-evergreen, deciduous and degraded scrub. The evergreen forest occurs in the tracts of higher rainfall. The evergreen and semi-evergreen forests are found in the districts of Lakhimpur, Dibrugarh, Sibsagar, Golaghat, Jorhat, Karbi Anglong, N.C. Hills, Cachar, Nagaon, Kokrajhar, Dhubri, Darrang, and Sonitpur. On the other hand, deciduous forests are found in the districts of Barpeta, Dhubri, Kokrajhar, Goalpara, Kamrup, Karbi Anglong, Nalbari, Nagaon, Sibsagar, Dibrugarh and Sonitpur. Apart from Barpeta and Dhubri districts all the above districts have degraded scrub.

Evergreen and semi-evergreen forests cover 2455 sq.km of area in the state. Deciduous forest and scrubs cover 14788 sq.km. Moreover, plantation of trees has also increased the amount of the forest area in Assam. Except Cachar, Dhubri, Golaghat, Jorhat, Lakhimpur, Nalbari and N.C. Hills, all other districts have planted forests to increase the forest area.

The semi-evergreen type of forests occur in Nagaon district. In Cachar and Karimganj districts also there are areas of semi-evergreen forest with various kinds of bamboos. The deciduous forest occurs mostly in lower Assam districts of Kamrup and Goalpara. The most important hard wood tree of this forest is Sal (*Shorea robusta*). Assam has also some amount of coniferous trees.

Flora and Fauna

There are different varieties of flora and fauna in the forests of the state. Assam is very rich in biodiversity. As for example, temperate region possesses 10-15 varieties of species in a hectare of land, but same area of tropical and sub-tropical land possesses about 200 varieties of species. The north-eastern region including Assam has 6000-7000 plant species identified so far. Favourable geographical location and ideal climatic conditions have made Assam very rich in biodiversity.

Plant resources of Assam are concentrated mainly in the tropical evergreen and semi-evergreen forests, grasslands, wet deciduous and riverside forests. There are at least 15 species of Angiosperms in Assam. Some of these, for example, *Machilus listeri* are found only in Assam. The important varieties of trees available in the forests of the state are Sal *(Shorea robusta)*, Makai *(Shorea assamica)*, Nageswar (*Mesua ferrea)*, Titachapa *(Michelia champaea*), Sonaru *(Cassia fistula)*, Gandhsarai *(Chinnamomum ceciodaphne)*, Agor *(Aqualaria agalocha)*, Hollong (*Dipterocarpus macrocarpus)*, Simul *(Bombax ceiba)*, Khair *(Acacia catechu)*, Bonsum *(Phoebegoalparensis)*, Amari *(Amora wallichi)*. Various types of bamboo, cane groves, tall grasses and other herbaceous and shruby vegetations are found in the forests of Assam.

Besides the varieties of trees, there are a large number of orchids in the forests of Assam. Depending upon the nature and habit of the orchids, they grow on the branches of trees, on rocks and also on the ground. They are classified as terrestrials, epiphytes and saprophytes. There are more then 100 varieties of orchids available in Assam.

The two hills districts of the state with highest concentration of forest have different varieties of timber-yielding trees. The main tree species found in the North Cachar Hills district are Makari Sal, Poma, Khokan, Bel, Hollock, Sam, Gamari, Bhelkor, Udal, Ahoi, Koroi, Koka, Sida, Ajhar, Amari etc. Besides, some species of bamboo are also found in the forests of this district. In Karbi Anglong district, Sal, Koroi, Jamun, Ajhar, Sam, Gamari, Amari, Poma, Bhelu, Simul, Kanchan etc. are available. The wildlife diversity in the state has been evolved on the basis of the forest and 4000 wetlands, ponds, beels and the Brahmaputra and the Barak river systems.

The mammals of Assam have affinity to the Indo-Chinese group. There are 14 species of monkeys in Assam which constitute 1/6 th of the world. Among these, Hoolock Gibbon, Assamese Macaque, Pigtailed Macaque, Golden Langur, Hanuman Langur, Spectacled monkeys, Capped Langur are the main types. Rare Golden Langur is found only in the western Assam along the Bhutan border while the Phayr's leaf monkey is confined to the forest of Cachar. Pigtailed and Stumptailed Macaque as well as Hoolock Gibbons are confined to the forests on the south of the Brahmaputra only. There are 19 cat families in Assam. India has 27 cat families. Ouite a good number of tigers are found in the region. Manas tiger reserve is a famous Sanctuary.

Among the deer, Sambar and Barking deer are still common and found all throughout the area. Hog deer's present distribution is mainly confined to the grasslands of the Brahmaputra valley. Swamp deer is rare and found mainly in the reserves/ sanctuaries in the Brahmaputra plains. They have become extinct in Cachar. Spotted deer is very rare and is confined to the moist diciduous forest of western Assam, north of the Brahmaputra river particularly in Manas Reserves. Asiatic water buffalo is also found in the reserves of the Brahmaputra valley. Elephant is found in good number in the state though its habitat is under threat. Great Indian one-horned rhinoceros has its last stronghold in the State. A good number of rhinoceros are found in Kaziranga, Manas, Orang, Laokhowa and Pabitora sanctuaries.

Out of the 1200 species of birds in Indian sub-continent, Assam alone possesses about 950 species. Besides the common species, a large number of rare and endangered birds including waterfowls are found, which also include winter vistors to the state from far away places like Siberia, Manchuria, Tibbet, etc.

Crocodiles, snakes, terrapins and monitor lizards with their various species represent the reptiles of the region. Crocodiles are now found mainly in the Brahmaputra river, but they have become extremely rare.

The following are some of the important parks and sanctuaries located in Assam .

Kaziranga National Park

This park is located in Golaghat district in between the Brahmaputra on the north and Karbi hills on the south. It covers an area of 830 sq. km. This park is famous for one-horned rhinos. Table 6.4 shows the number of important animals found in the park.

Table 6.4
Statement of some species of animals in Kaziranga National Park since 1966

Animal	1966	1972	1978	1984	1991	1993
Rhinoceros	366	658	939	1080	1129	1164
Elephant	349	422	773	523	515	511
Wild Buffalo	471	555	610	677	1090	1034
Bison	1	18	23	30	5	-
Swamp Deer	213	516	697	756	635	427
Samber	120	105	215	358	55	34
Hog Deer	1311	4551	6855	987	2911	2048
Wild Bear	155	522	733	1645	555	140
Tiger	20	30	40	52	50	

Source : CCF, Wildlife, Assam

Manas Nationl Park

Manas National Park (Tiger project) is situated at the foothill of Bhutan covering parts of the districts of Barpeta and Kokrajhar. It has a total area of 430 sq.km. The important animals of the national park are tiger, golden langur, capped langur, Assamese macaque , rhesus macaque, panther, golden cat, clouded leopard, fishing cat, jungle cat, Indian civet, bear cat, Himalyan black bear, sloth bear, Indian flying short-nosed fruit bat, Indian elephant, one-horned rhino, sambor, spotted deer, Gangetic dolphin, etc.

Orang National Park

It is situated in the district of Darrang and Sonitpur.The total area of the sanctuary is 72 sq.km. The important animals are rhinoceros, hog deer, wild boar, tiger, elephant, wild buffalo, porcupine, civet, cat, hare, monkey, etc.

Dibru-Chaikhowa Sanctuary

This sanctuary is located in Tinsukia district of Assam. It was declared in 1986. It has an area of 640 sq.km. Important birds and animals are white-winged wood duck, elephant, tiger, sambor, buffalo and wild horse.

Nameri Wildlife Sanctuary

It is situated on the north bank of Bharali river along the Arunachal border of Sonitpur district. It was declared a wildlife sanctuary in 1985. The area of the sanctuary is 137 sq. km. The important animals of this sanctuary are elephant, bison, sambor, barking deer, capped langur and clouded leopared.

Garampani Wildlife Sanctuary

It is situated in Golaghat district. It has an area of only 6.0 sq.km. The important animals are barking deer, sambor, wild bear, hoolock, gibbon, tiger, leopard, elephant and a varity of resident and migratory birds.

Deepar Beel Birds Sanctuary

This wetland sanctuary is located on the south bank of the river Brahmaputra, a few km west of the Guwahati City. The area of the sanctuary is 4 sq. km. Around 100 varieties of birds are found here.

Bornodi Wildlife Sanctuary

It is situated in Darrang district covering 26.2 sq.km of area. It was declared in 1980. Some important animals found in this sanctuary are Pigmy hog, elephant, leopard, barking deer, wild bear, capped langur, etc.

Pabitora Wildlife Sanctuary

It is situated in Morigaon district with an area of 38 sq.km. It is mainly famous for rhinoceros (highest density of rhinoceros in the world), wild buffalo and some colourful native and migratory birds.

Laokhowa Wildlife Sanctuary

This sanctuary is situated in the north -eastern part of Nagaon district. The total area of the sanctuary is 70 sq.km. It was declared a sanctuary in 1979. Rhinoceros, wild buffalo and tiger are the main animals of this sanctuary.

Social Forestry

The social forestry programme was launched in Assam in 1980-81. The programme was designed initially to plant all barren government and community land and also the land where encroachment has already started so that such areas may be made green instead of keeping the same under unauthorised human habitation. Extensive plantation programmes have been taken up on government wasteland, roadside land, community land like village grazing reserves, professional grazing and institutional land.

The social forestry programme was launched in the state keeping in mind the following objectives:

(1) To meet the requirement of leaf-fodders, fuel wood and other forest products including small timber of the local people;
(2) to help the effort of individuals producing fuel wood, fodder and other forest products;
(3) to extend forest to all categories of land without affecting the ownership, and also to bring a sense of responsibility for protection, maintenance and care of forest;
(4) to provide employment and economic upliftment of the present and future generations particularly in the rural sector, and
(5) to ensure maintenance of ecological balance.

Moreover, plantation of avenues in the cities, towns and important rural centres has been taken up as a step for demonstration and to improve the bio-aesthetic impact. Highway parks on the national highways and parks in scheduled caste and scheduled tribe inhabited areas were made for the purpose of recreation. At present, there are as many as 32 such parks in the state. These parks are maintained for the benefit of the people. During the seventh plan period, the schemes undertaken in the state were (a) social forestry – general scheme (b) social forestry under tribal sub-plan, (c) social forestry under scheduled caste component plan scheme, (d) fuel and fodder plantation scheme, and (e) decentralised people's nursery scheme.

The social forestry proposed to plant 38,278 hectares of land during the seventh five year plan and 25,000 hectares of land during the 8th five year plan. The total plantation area under social forestry wing is shown in Table 6.5.

Table 6.5
Plantation area under social forestry in Assam, 1981-1996

Year	Plantation area in hectare
1981-82	3520
1982-83	6030
1983-84	7800
1984-85	2944
1985-86	9266
1986-87	10494
1987-88	5997
1988-89	2733
1989-90	3052
1990-91	5892
1991-92	5417
1992-93	6439
1993-94	5204
1994-95	5656
1995-96	5900

Source : Chief Conservator of Forests (Social Forestry), Assam

Forest Produce and Revenue

There are both minor and major forest products in the state. Among the major forest products, mention may be made of timber and fuel. Every year a huge quantity of these products are extracted from the state's forests to cater to the various needs. Besides the major products, there are varieties of minor products form the forest areas such as gravel or stones, sand, earth and clay, bamboo, cane, thatch, agar, wood, medicinal plants, ivory, rhino–horn, patidoi, honey, fuljaru, etc. Table 6.6 shows the magnitude of the out turn of timber and fuel wood products in the state during the period 1988-92.

Table 6.6
Out-turn of timber and fuel wood in Assam, 1988-1994

Year	Industrial timber (in cubic metre)	Fuel wood (stack cubic metre)
1988-89	7,86,666.28	30,061.45
1989-90	61,369.18	37,026.17
1990-91	65,374.22	35,200.10
1991-92	76,865.31	44,896.29
1992-93	45,545.11	43,094.20
1993-94	47,058.93	29,960.25

Source : Statistical Handbook, Government of Assam.

The annual revenue yield from the minor forest products is substantial. The contributions of forest to the revenue of the state have been given in Table 6.7.

Table 6.7
Revenue earned by the Forest Department, Assam (Rs. in lakh)

Year	Revenue budgeted	Revenue earned
1993-94	2500.00	2260.05
1994-95	2100.00	1692.40
1995-96	2300.00	1726.50

Source: Principal CCF, Assam.

The output of other forest products in 1993-94 is given below:

Table 6.8
Forest products of Assam, 1993-94

1.	Gravel	560327.08M^3
2.	Sand	81208.45 m^3
3.	Earth and clay	31034.50M^3
4.	Bamboo	262705 Nos
5.	Thatches	11003867 bundles
6.	Patidoi	475815 bundles

7. Cane	Nil
8. Tokow leaf	7911359 Nos.

Source: Principal Chief Conservator of Forest, Assam.

Forest-Based Industries

A large proportion of the forest products of the state is processed by a number of forest-based industries located in various parts of the state. There are plywood factories, match factories, timber treatment and seasoning plants, hard board factories, paper mills and saw mills which are based on the forest resources of the state. The first plywood factory in India was set up in Assam after the first World War. These industries are not only utilising the forest resources of the state but also helping in generation of employment opportunities. The number of various forest-based industries and their employment and production pattern have been presented in Table 6.9.

Table 6.9
Employment and production of forest-based industries in Assam, 1984

Factory	Number	Employment	Production
Saw mill	524	3022	262 lakh sq.m.
Plywood factory	59	6243	425 lakh sq.m.
Match factory	4	1050	50.5 lakh gross boxes
Paper mill	3	1867	4821 tonnes

Source : Economic Survey of Assam, 1984.

The saw mills of the state besides meeting the demand for construction materials also cater to the requirements of railway sleepers. The plywood factories manufacture mainly tea chest and commercial as well as decorative plywood. At present, there are three paper mills in Assam namely, the Hindustan Paper Mills at Jagiroad, the Panchgram Paper Mill at Cachar and the Ashok Paper Mill at Jogighopa (which has not been operating for last few years).

Degradation of Forest

A naturalist who had seen the north–eastern states of the country about fifty years back, for instance, most likely would be shuddered if he happens to visit the same area again. He would in most cases see human settlements where he encountered thick forest that time. The depletion of the forest cover is continuing at an alarming rate. It is regretable that the bounty of forest resources bestowed upon the state is not utilised rationally. The reasons for decline in forest cover are manifold. The increasing depletion of this important resource has disrupted the ecological balance leading to catastrophic results in the form of unusual floods and periodic draughts in the state.

However, there is no reliable data available on the extent and the rate of depletion of forest cover annually or over any period of time. The state government has not undertaken any detailed survey in this regard. The report of the working group on development of the north–eastern region during the 7th five year plan published by the North Eastern Council says that among the multiplicity of factors responsible for the rapid depletion of forest cover in the north-east, the two prominent are (i) rackless exploitation of forest and (ii) extensive *jhum* cultivation. An idea about the extent of the damage done to the forest and ecosystem can be obtained from the fact that during 1972-73 to 1980-82 there had been a loss of 157.38 sq.km of forest area in the state of Assam. Recent satellite survey conducted by NRSA shows the deplorable condition of several protected forest areas. For exmple, the Nambor (south) reserve forest has merely 3.6 per cent forest cover, Doyang reserve forest has only 2.5 per cent, Diphu reserve forest has hardly 8.2 per cent and Gohpur reserve forest is almost fully degraded.

Recently illegal forest cutting is also increasing. Assam has 9187 sq.km of forest land under unclassified state forest (both plains and hills) which is not permanently dedicated for the forestry purposes and are subjected to *jhum* cultivation and other non-forestry uses. The area affected by shifting cultivation accounts for about 9 per cent of the state's total area. The two hills districts namely Karbi Anglong and NC Hills have 42.37 per cent and 59.65 per cent of their respective total area under shifting cultivation.

Out of 682 plant species recorded in Red Data Book of Indian plants. 119 plant species of Assam have been considered to be threatened or endangered, or vulnerable. Forest degradation and continuous removal are the evils which result from senseless overcutting. In many areas, specially in the hilly areas, the removal of forests has aggravated soil erosion through gullying or sheet wash. This in turn, results in sediment deposition on the river beds and thus causes flooding in the rainy season.

Encroachment of Forest Land

Encroachment of forest land is one of the burning problems that badly affects resources of the state. Due to rapid growth of population there has been an increasing pressure of man on the limited forest lands. Apart from this, the illegal migrants are also encroacahing upon the forests, especially the reserved forest areas. Besides, some forest land along the state's border are under the occupation of the neighbouring states. It is reported that Nagaland has occupied about 19,478.95 hectares of forest land illegally in the Disoi and Tiru hills reserved forest of Jorhat district. It has also occupied 2950 hectares of forest land of Galaki reserved forest in Sibsagar district.

There are 21 reserved forests in the four districts of northern Assam, namely Darrang, Sonitpur, Lakhimpur, and Dhemaji along the Assam-Arunachal Pradesh border. The total area of reserved forestes there covers 10,24,297 hectares and of this more than 15, 384 hectares of land are reported to have been encroached upon by Arunachal Pradesh. Meghalaya also claims a large part of the foothill areas of Kamrup and Goalpara districts of Assam. Some forest areas along the Meghalaya- Assam border are also being occupied by Meghalaya.

There had been tremendous depletion of forest cover from 1991 to 1995 in the northern Assam mainly because of defective government policy. The state government issued licence for cutting forest of 21 m^3 in 1991-92, 6606 m^3 in 1992-93, 2878 m^3 in 1993-94 and 2283 m^3 in 1994-95. But the actual amount is likely to be more than this official record. In the whole of Assam, out of 22175.85 sq.km forest area (including hill districts), almost 11 per cent is under human encroachment and another 4 per cent comes under the forest villages. Table 6.10 shows the pattern of encroachment during 1991.

Table 6.10
Communitywise number of forest encroachers in Assam, 1991

S.No.	Caste/Community	Number
1.	Scheduled caste	3037

2.	Scheduled tribe	33085
3.	Nepali	6850
4.	Other backward caste	1412
5.	Ex-tea workers	9211
6.	Naga	3562
7.	Others	21692

Source : Principal Chief Conservator of Forest, Assam.

During the 7th plan efforts were, however, made to evict the illegal encroachers from the forests. There was a proposal in eighth plan to remove the illegal settlers and reclaim the forest lands and to raise plantation in such reclaimed lands. But, in practice, it could not be implemented due to some unknown reasons.

Conclusion

Forest can play an important role in the economy of the state. There are a number of industries which have been set up to utilise the forest resources of the state. There is still more scope for setting up a number of other forest-based industries. There should, however, be a proper plan for exploitation and utilisation of the forest resources. Unplanned felling of trees and exploitation of forest resources will lead to severe consequences in the economy and environment of the state.

People's participation in connection with the protection of forest and wildlife is highly essential. In the present day situation, constant people's participation and their active involvement is vital for the success of any such programme. A vital factor in the socio-economic welfare of the tribal and rural people is the development of minor forest products. This will not only help to generate self–employment among them but also contribute substantially to the revenue of the state government.

There is a great demand for fuel wood and fodder in the rural areas, and unless this is met there will be immense pressure on the commercial plantations as well as the reserved forests. So tree plantation in rural areas for fuel is most important.

Government should take firm steps to stop encroachments in the reserved forests and sanctuaries. For this purpose forest protection laws should be enforced and the strength of the protection force should be increased as required. Steps should be taken to make people aware of the need to plant and protect trees and thus to save mankind from disastrous environmental pollution. Keeping in mind the growing tendency of the people in general to destroy forests mercilessly to meet their immediate requirements, relevant awareness campaign should be extended to every nook and corner of the state. Further, some incentive schemes should be introduced to motivate the people for plantation and preservation of trees.

References

Mali, D.D. : *Economic problem and planning in Assam*, Omsons Publications, Guwahati.

Assam Forest at a Glance, published by Chief Conservator of Forest, Assam.

Draft VIII Five Year Plan, Govt. of Assam.

Sarma, K. (1987) : Boundary Problems in N.E. India, Tihu College Magazine, 18th issue.

Economic Survey of Assam, 1998.

Goswami, D.C. (1994) : Assam's Land Resources - A Satellite-based Assam Remote Sensing Application Centre Survey, published by Assam Science Society, Guwahati.

Choudhury, A.U. (1986) : Wild Life in North East India, *Geography of North East India*, NEIGS, Guwahati.

Bio-Diversity and its Conservation, published by Assam Science Technology and Environmental Council, Assam, Guwahati, 1995.

Brikhyaropan aru eyar pratipalan, Social Forestry, Govt. of Assam.

7 Management of Forest Resources and Environment

Introduction

India's North-East including Assam is rich in forest resources. The topographic, edaphic, hydrological and climatic settings of the region favour luxuriant growth of plants of various forms and types. Assam with 78,438 sq km of area alone exhibits six types of forests, namely tropical wet evergreen, tropical semi evergreen, tropical moist deciduous, sub-tropical broad-leaved, sub-tropical pine and riparian forest. These forests with a very wide floral diversity not only provide good habitats for diverse fauna but also contain a wide range of natural resources. The natural forests and forest resources continued to remain almost undisturbed till the occupation of the area by the British in the first half of the nineteenth century. The British for the first time introduced the practice of commercial felling of trees and exploiting other forest resources in the region which in course of time flourished both in dimension and intensity to convert a large proportion of forest area into degraded lands.

In Assam alone, during the last fifty years or so, the forest cover has declined to less than 20 per cent from more than 40 per cent of the State's total area. The spatial shrinkage of forest land and over-exploitation of forest resources have now put the surviving forests almost on the verge of severe degradation except for a few protected pockets. This has led to rapid biodiversity loss (Sagreiya, 2000) degradation on the other. There is, therefore, an immediate need to save the forest from further degradation and manage the left out forests and forest resources sustainably and plan for regeneration of green cover wherever possible and feasible.

Forest Resources

As a part of ecosystem, forest plays a variety of roles. Besides its role in bio-geochemical cycles, forests provide habitats for people to obtain food, fuel, shelter, medicine and construction materials. If of adequate extent, ideally dispersed, scientifically managed and judiciously utilized, forests can be kept perpetually productive and useful conferring many benefits, direct and indirect, on the people (Sagreiya, 2000). Forests, the wildlife they support and the water that flows through them are all renewable resources (Bergman et al., 2008). The biologically very rich forests in Assam provide a wide range of resources from timber to bamboo, cane, thatch, sand, medicinal and other edible plants. Forests also provide facilities for grazing livestock and livelihood to the people in very many ways. In a sense, the lives of the people around the forest are so closely associated with it that it is difficult to draw a line between them. In the hills and plains of Assam, especially the tribal and interior areas, the long tradition of living with the nature and indigenous knowledge and skills evolved by them in course of time have made the local dwellers themselves inseparable part of the forest ecosystem. The symbiotic relation of the people and nature generated vast knowledge on availability and utility of natural resources and spontaneous concern for conservation of nature. The plants and animals identified and narrated by the indigenous people bear the testimony of it. All kinds of folk expressions including belief system, health care system, traditional house types and cane and bamboo works point to people's close association and attachment with forests and forest resources.

In addition to individual/community dependence on forests to earn their livelihoods, the government also earns substantial revenue from the forest products (Table 7.1). It is, however, noteworthy that the amount of revenue earned has gradually declined mainly due to shrinkage of area and vegetation density in the forests of the State.

Table 7.1
Revenue earned by the forest department, Government of Assam

Year	Revenue Earned (Rupees in Lakh)
1993-1994	2260.05
1995-1996	1726.50
1999-2000	1473.00

2000-2001	1476.50
2001-2002	1207.77
2002-2003	1854.50 (estimated)
2003-2004	1916.82
2004-2005	2376.82
2005-2006	4338.26
2016	5497.35
2017	14006.50

Source: Statistical Handbook, Assam, and Office of the Principal Chief Conservator of Forest, Assam

There has been some increase again since 2002 which may perhaps be ascribed to the hike in rate on the products and expansion of activities of sand collection, quarrying, etc.

Deforestation

Clearing forests for fuel, timber and expansion of settlements and economic activities is the most common way through which humans have remarkably changed the face of the earth. There is hardly any need to emphasize that forests play certain key ecological roles as natural repositories of biodiversity, surficial factors affecting hydrological cycle, local and global climate and air quality. They also influenced soil property and the very process of soil formation and prevent land degradation. Although its impact on human health and sustenance is well appreciated, there may be many more unforeseen positive role played silently by the forest cover. Because of all such influences forests are considered to be an integral part of the eventful global natural as well as human history.

Deforestation is not a new phenomenon. It has been going on in varying degree and extent in different parts of the world since the prehistoric time. Compared to many European countries, India including its north-eastern border region came under the grip of deforestation quite late. It was during the colonial period only the country was exposed to the organized process of deforestation mainly to cater to the needs generated by the colonial pattern of industrialization and urbanization. The colonial power could easily visualize that the forest of the region could be an unending source of revenue (Saikia, 2001). At present, like many other nations of the world, India has also qualified itself to be one of the most deforested parts of the globe with only 20% of its total area under forest cover. Interestingly, the very dense forest in the country constitutes only 51,285 sq. km accounting for as negligible as 1.56% of the nation's geographical area. The moderately dense and open forests on the other hand constitute nearly 10% and 8% respectively. These are enough to indicate the present dismal state of the forests in the country. The Brahmaputra valley in particular was till recently one of the country's few areas having very rich forest cover. Right from the banks of the great river Brahmaputra to the foothills of the Himalayas and the Meghalaya plateau, the valley exhibited forest covers of various types and densities. It is, however, important to note that the forest cover of the valley, which was once rightly called the land of jungles, now accounts for not more than that of the national average. The rate and nature of deforestation in the valley have probably surpassed any of the badly degraded areas of the country, especially during the last twenty five years or so. The way, the size and density of green cover are getting reduced during the last few years due to various internal and external factors, it is feared that in near future merely the national parks and the sanctuaries will remain to be connotated as forests in the simplest sense of the term. It is needless to emphasis further that even the so-called protected areas declared and conserved so far may not be able to survive with their fullest biophysical strength as they are today in an increasingly unsustainable situation that would most probably characterize the future ecology of this highly sensitive alluvial plain.

The scenario of dwindling green cover in the Brahmaputra valley and other parts of the State could be reasonably appreciated, if a cursory observation is made in the affected areas - Gohpur, Behali and Balipara, to name a few. The contact zone of Arunachal and Nagaland highlands and the plains of the Brahmaputra valley, which plays a crucial role in regulating water flow from the Himalayas as well as the micro-climatic condition of the concerned areas, sets a hot case of rigorous deforestation and its immediate consequences in the form of sudden flooding, stream channel shifting, gully erosion and significant fall of biophysical productivity of land and above all loss of biodiversity. The sharp decline in tiger population and the ever growing conflict between man and elephant in several areas of the zone, especially in parts of Sonitpur and Biswanath districts, among others, are the best examples of habitat fragmentation and loss caused by indiscriminate felling of trees and human encroachment in the forest areas.

Vanishing Grasslands

As an important component of the earth's variegated green cover, grasslands play a significant role in shaping and sustaining the pattern of global biological diversity. Grasses are the most variable, widespread and useful plant groups that occur in plains, deserts, seas, marshes, high mountains and arctic areas. Their great flexibility and adaptive capacity permit them to sustain in most of the environmental situations.

Besides providing unique habitats for a large variety of mammals, reptiles, birds, insects and amphibians, grasslands fulfill some of the basic human requirements. Grasses produce bulk of the forage for the livestock and house building materials in many of the tropical areas in addition to their universal role in maintaining ecological balance. Interestingly, the lands grazed by livestock constitute about 30% of the earth's land surface. Moreover, certain species of grass are grown as turf for recreational uses, while many others are traditionally used for medicinal purposes. Because of such diverse utilities, grasslands are under constant threat in almost all parts of the world.

Grasses are often affected by the process of cultivation of crops and overgrazing. When the grasses are replaced with crops, the soil become more vulnerable to wind and water actions. This may lead to a desert-like situation, unless proper conservation measures are undertaken in time. The classic example in this regard is the Dust Bowl of 1930s that affected the great plains of the mid-west USA so badly that most of the ill-fated farmers had to migrate either to the Pacific coast or to the eastern industrial areas in search of livelihood. Similarly, overgrazing also eventually exposes the land to subaerial erosion causing thereby sharp decline in the biophysical productivity of soil. The exodus of people from the south of Sahara, called Sahel as ecological refugees to many parts of western Africa is the best example of consequences of land degradation due to overgrazing.

In India, grasslands predominate the high altitude Himalayan slopes, the Terai belt, the semi-arid plains of western and central India, the Deccan plateau and the north-eastern plains and hill slopes. Most of these grasslands have been under constant modification by ever-increasing human activities. It is noteworthy that the proportion of grasslands (permanent pastures) in the country has now come down to only 3.7% of the total land area. Such a low proportion, which is primarily the result of intensive grazing and large scale conversion of grasslands into irrigated agricultural land, probably would not be sufficient to help maintain the natural balance in the concerned areas. This is manifested in the near-extinction of many of the grassland fauna like Cheetah, Wolf, Blackbuck, Chinkara, Bengal Florican, Great Indian Busturd, etc.

The Indian sub-continent, once known for its wild variety of natural vegetations, has now been losing fast its flora diversity, more particularly the valuable grass and herb species. The flat and undulating plains of India's north-east including Assam composed mainly of new and old alluvium are notably rich in grasses. The Brahmaputra riverine tract throughout the length of the valley favours the growth of a large number of important tall grass species. They include mainly Ulukher (Imperata cylindrica), Kahua (S. spontaneum), Ekora (Erianthus ravennae), Nal (Arundo donax), Khagori (Phragmities karka), etc. Similarly, in the contact zone between the Himalayas on the north and the Brahmaputra plain on the south, where Bhabor and Terai elements dominate the ecosystem, generally tall savannah type of grasslands occur. All these grasslands favour the presence and sustenance of important mammals like rhinoceros, tigers, deer and many other rare animals. The role of grasslands does not remain limited to providing good habitats to wildlife alone, its role on regulating surface run-off, checking flood, soil erosion and gulling are also equally significant. Traditionally being the dwellers of a thickly forested territory with gigantic trees like Holong, Holok, Koroi, Dhuna, Simul, etc, the people of Assam generally find it difficult to perceive the open grasslands as integral part of typical forestlands. Although they are closely associated with grasslands and aware of their importance in the ecosystem, yet there is no noticeable evidence of adopting any significant step towards conserving the grasslands except for those inside the protected areas. The State machinery has also been more or less indifferent in this respect and practically allowed encroachment of the professional as well as village grazing reserves at various stages of State's development. The grasslands thus never got due attention and, may be due to lack of vision on the part of the stakeholders, these have now become highly scarce natural resource in most parts of the State. The rapid loss of grasslands, especially in the interfluves and the river banks, has already caused problem of land degradation and even shifting of river courses in many places. It is now universally recognized that accelerated soil erosion, if allowed to continue unabated, may be a kind of slow poisoning to cause even peril to a full-grown human civilization.

The char areas of the Brahmaputra having regular supply of fresh alluvium and biomass during the summer months naturally support luxuriant growth of grasses. Till recently, majority of the char-dwellers earned a substantial part of their income by selling thatches to the people living in the high density flood-free areas. But, with the growth and spatial spread of population, farming and associated infrastructural facilities, these marginal areas have also encountered the problem of acute shortage of good quality grasses needed for forage, and making roofs of the traditional houses. At present, one can easily note the intensity of transformation of grasslands into human habitats and agricultural lands throughout the riverine tracts of the Brahmaputra and its major tributaries. This has resulted in fragmentation and loss of grassland habitats and, in most cases, a clear absence of green cover to check soil erosion. If such a trend continues, certain unforeseen ecological problems may arise to eventually cause enough damage to the local economy and culture. The need of the time is, therefore, to evolve and implement sustainable strategies and action plans for conservation of grasslands, wherever feasible, to protect biodiversity in general and the biophysical productive lands from further degradation in particular.

Human-Animal Conflict

Conflicts of different types and levels between humans and wild animals are now a world-wide phenomenon. The rapid loss and fragmentation of habitats caused by growing human population in all the biomes of the world has led to depriving the wild animals of their required food and natural movement. Even in some of the protected areas, human interference has reached limits no longer tolerable to the natural system. Most of the African and Asian countries especially with rich biodiversity are reported to have suffered from a phenomenon popularly known as human-animal conflict that basically arises out of the extraordinary pressure mounted by human society on the wildlife habitats.

In India, the growing population and associated infrastructures have already reduced green cover to a level much lower than the norms suggested in National Forest Policy. The expansion of settlements to the hill slopes and forest areas, construction of roads, power projects, exploitation of minerals and other natural resources have caused enough harm to the natural environment. According to India's National Economic Survey, 2007-2008, the country's entire road network, which is second largest in the world, measures a total length of 3.34 million kilometres. Roads passing through forests, river valleys, wetlands and other natural areas and even villages not only cause forest loss but also adversely affect landscape continuity and natural movement of wild animals. Death of animals, both wild and domesticated, on roads is a common feature in all parts of the country including Assam. Leaving aside the domestic dogs and cats, an indiscriminate diversity of wild species, such as butterflies, squirrels, lizards, snakes and partridges to more threatened species such as a leopard, deer, tiger, Elephant, loris, langur, macaque meets with their ends on the road (Raman, 2009). It is thus globally agreed that by destroying and segmenting habitats, irrationally aligned roads and railways have substantially added to the growing conflicts between humans and animals. Many examples from the Central Indian Forests, Western Ghats, Himalayan foothills, forest areas of North-East India may be cited in this regard.

Coming to India's north-east, it has been observed that the rapid loss of habitats and for that matter biodiversity has been manifested, among others, in the form of growing conflict between humans and wild animals in different parts of the region. The forest margins in Assam and deforested boundaries between Assam and Arunachal Pradesh, Assam and Nagaland, Assam and Meghalaya are the areas where human-animal conflict has become a regular feature. The destruction of standing crops, valuable trees, houses and kitchen gardens caused by elephants and monkeys has currently reached an alarming dimension at many places. Generally the conflict is found to be maximum in the areas where the scale of deforestation is large. In Assam particularly, human-animal conflict has already attained a critical level. It is amazing to know that during a period of nearly 10 years (1990-2001) more than 450 people in the state were killed by wild elephants. The property damaged by elephants every year in the state has not yet been estimated properly. As a counter action, cases of killing wild elephants by the affected people applying poison and other methods are also not rare in the state. As many as 19 elephants were killed by poisoning during 1998-2000 as per records of the state forest department. Importantly enough, the conflict between people and the monkeys (mostly Assamese Macaque), both in the rural and urban areas of Assam, has currently turned into a kind of hazard needing immediate initiatives for its scientific management. The peasants, who operate small and marginal land holdings particularly in the deforested areas, are the worst sufferers in this regard as major parts of their small produce go to the wild monkeys regularly. Even the golden langurs (Trachypithecus geei) found in the forest of western Assam have started raising home gardens and standing crops indicating signs of impending conflict with the neighbouring human population. Studies conducted in some fringe villages show that golden langurs caused maximum damage to the gamari trees (Gmelina arborea) planted by the villagers (Chetry and Chetry, 2009). All these add further to the poor people's overdependence on the nearby forest and wetland resources.

Some areas located in Assam-Arunachal Pradesh border present certain interesting facts regarding the conflict between people and elephants. The people in the Tipi area in Arunachal Pradesh and lower Bhalukpong in Assam, for instance, have perceived the problem caused by wild elephants to be an accurate one. Human settlements in these two adjacent foothill areas have come up fast ignoring the fact that the elephants have been using these areas for their movement for a long time. The Tipi Area which is identified as one of the most active elephant corridors in between Assam and Arunachal Pradesh is now preferred by the local people for settlement because of its favourable flat topography. As a result, some kind of undeclared competition has been developed between the people and the wild elephants for level space within the broad hilly environment of the area. In the lower Bhalukpong area also, large scale felling of trees and gradation of grasslands during the recent years has made the problem of human-elephant conflict an acute one (Bhagabati and Bhattacharya, 2009). The people living within the Balipara Reserved Forest, south of Bhalukpong, regularly confront raids by elephants whose habitats have been degraded during the recent years.

Around almost all the reserved Forest and wildlife sanctuaries of Assam, the problem of human-animal conflict is quite prevalent and getting intensified in course of time. Initially this problem used to be occasionally arising mainly during the winter when food availability in denuded forests largely declines. But it has now become a regular feature. Forest margins have been converted into war fronts between people and animals. From my personal experience in certain protected areas like Behali Reserved Forest, Gibbon Wildlife Sanctuary and Manas national park, for instance, I can confidently say that the problem of human-animal conflict has already assumed such a dimension that particularly the large animals, which are increasingly being exposed to human habitats, would probably face a state of near extinction in days to come. The villagers living in the forest margins, who do not have any better livelihood options other than traditional farming and small homestead gardening, have already crossed the limit of tolerating raids made by the wild animals, particularly the elephants. The growing population pressure on the forest lands and the declining area and density of natural vegetation have together given rise to a critical situation in almost all parts of the State which will no longer allow a sustainable co-existence of humans and wild animals. In the face of ongoing unhealthy urge for space and wealth on the part of humans not only in the forest fringes but also other naturally rich areas, what kind of a situation is in the making, particularly in terms of biodiversity loss and human health, is really difficult to appreciate. However, strong apprehensions are there about the long-term sustainability of the entire system, if it is allowed to function in the present state and form.

Constituted by various types of lands, water bodies, forest covers and atmospheric elements peculiar to the region, the natural environment of Assam continues to be highly dynamic and ecologically very sensitive needing a befitting approach for their sustainable use and conservation. The regional environment settings covering the highland and lowland sub-systems basically control all kinds of flow including running water, winds and even the movement of wildlife. Further interference with the already degraded nature, particularly the grasslands and other forest habitats, ignoring the natural linkages among the elements and habitats will only aggravate, among others, the problem of conflict between people and wildlife. The exceptionally high rate of shrinkage of wetlands, deforestation and accelerated soil erosion, degradation of hill slopes, foothills, narrow valleys and floodplains have together made the natural environment further fragmented and barren rendering it increasingly incapable to support diverse wildlife. This means more conflicts between people and animals in immediate future. A sincere, whole-hearted and systematic nature-centric massive conservation effort will only be able to mitigate the problem to a certain extent. Otherwise, this important biodiversity hot-spot of the world, as we recognize it, under the changed situation may eventually pave The unwanted way for another kind of hot-spot to be characterized by growing bio-scarcity, poverty, hunger and unforeseen diseases.

Management

As an important component of the environment, forests including grasslands have a far-reaching influence on nature and human culture. Forests form the bases for natural support of life and economy of the people. Management of forests, therefore, is an integral part of overall planning for sustainable development of human society and environment. The objectives of management in this respect may be (i) maintenance of a balance in the environment system, (ii) ensuring long-term productivity of forest resources, (iii) evolving a scientific mechanism for identification and exploitation of resources without disturbing much the biophysical productive capacity of the ecosystem, (iv) conservation of various life forms along with the physical features and soil, and regulation of natural water flow, (v) enhancing scenic beauty, and exploring scope for eco-tourism in places with appropriate geographical context, (vi) developing a culture of regional understanding of nature and working towards coexistence considering the long-term benefit from services rendered by forest ecosystem.

It is worth mentioning in this connection that the process of active participation of people living in the forests/fringe villages in the management programmes has not practically received required consideration in spite of its universally recognized relevance. This has, in addition to the near primitive state of life of the fringe dwellers, contributed to the gradual alienation of them from their traditional attachment with the forest ecosystems. The success of joint forest management (JFM) in certain states of the country is a pointer to at least mutual development of people and the natural environment. Notably, the legacy of the colonial attitude and perception towards forest still continues in different forms, which hinders materialization of ideas concerning coexistence in real sense. Further, it has been observed that whenever there arises any kind of socio-political crisis due basically to weak governance and lack of commitment on the part of stakeholders, the first and foremost damage is generally caused to the forests. This has been perceptibly manifested in the aftermath of almost all socio-political upheavals nurtured and promulgated in the state during the recent past.

It should, however, be remembered that the environmental problems arising out of forest loss do not as such recognise any class or community in their current and potential spread-effect. But the most pathetic fact means that they first victimize the poorer section of the society and render them highly vulnerable. Such a situation shall naturally have a mounting effect on the environmental system including forest and other productive lands. Thus there is sufficient ground to say that if the present state and rate of forest loss and mismanagement of resources continue unabated, The state will definitely face unprecedented environmental problems, which are somehow inherent in its physical frame girdled mostly by steep highlands and heavy precipitation on the one hand and the growing human interferences on the other. A work plan of forest ecosystem management aiming at continuity of well-thought policies and programmes in the broader regional framework of environment, society and economy may go a long way in conservation of forest and sustainable use of forest resources in the State.

References

Bergman, E.F. et al (2008): *Introduction to Geography: people, Places and Environment,* Pearson education, Delhi.

Bhagabati, A.K. (2007): Human Culture and Biodiversity: Some Reflections of their inter-relation, in P. Dwivedi *et al.* (eds.) *Biodiversity and Environmental biotechnology,* Scientific Publishers (India), Jodhpur, pp. 1-12.

Bhagabati, A.K. et al (eds) (2006): Biodiversity of Assam: Status, Strategy and Action Plan for conservation, EBH Publishers (India), Guwahati.

Bhagabati, A.K. and Bhattacharya, P. (2009): Deforestation and Human-Animal conflict in North Bank Landscape (Bhalukpong Area) Of the Brahmaputra Putra Valley, Assam, in A.K. Bhagabati (ed), *Areas of Concern,* Department of Geography, Gauhati University, pp. 59-66.

Bhargav, P. (2007): "Greening India" but Losing Forests, The Hindu, November 2.

Brown, J.H. and Lomolino, M.V. (1998): *Biogeography,* Sinauer Associates, INC. Publishers, Massachusetts.

Chetry, R. and Chetry, D. (2009): The Golden Langur, Gibbon Conservation Centre, Mariani, Assam.

Raman, T.R.S. (2009); Death on the Highway: Crushed by Wheels of Progress, Survey of Environment 2009, The Hindu, pp. 113-118.

Sagreiya, K.P. (2000): *Forest and Forestry,* National Book Trust, India, New Delhi.

Saikia, A. (2005): *Jungles, Resources, Wildlife: A History of Forests in Assam,* WADWT, Guwahati.

8 Biodiversity

Introduction

Assam is a part of a global bio-diversity hot spot (Myers, 1988;1991). It also forms part of two Endemic Bird Areas, viz. Eastern Himalaya and Assam Plains (Collar et al., 1994). Located in the north-eastern part of India (24°09'-27°58' N, 89° 42'-96°01'E), it covers an areas of 78,500 km^2 and is divided into 23 districts. The present state excludes areas such as Naga Hills (present Nagaland state), Khasi Hills, Garo Hills, Jaintia Hills (all in present Meghalaya state), Mizo Hills (present Mizoram state) and NEFA (present Arunachal Pradesh state) which were often referred to in various works on forests and wildlife in the past as part of Assam. Physiographically the major parts are plains with the Brahmaputra valley accounting for more than 60% of the total area of the state. The other plain area is in the south, the valley of the Barak River. The main hilly areas cover North Cachar Hills (N.C.Hills), parts of Cachar and Karbi Anglong (Taher, 1986). The Barail Range, which covers parts of Cachar and N.C. Hills is the highest hill range in Assam with its highest peak rising above 1900 m elevation. The whole of Karbi Anglong and parts of north-western N.C. Hills are part of an Archaean plateau.

The climate of Assam is tropical 'monsoon' type with a hot wet summer and a cool dry winter. Winter rains are also not uncommon. Annual rainfall of the state varies from less than 1000 mm to more than 6000 mm. The temperature generally ranges from 7°C in winter (minimum) to 35°C in summer (maximum).

Habitat Diversity

Assam has widely differing habitats. The tropical wet evergreen forest occurs mainly in the eastern and southern areas covering parts of Tinsukia, Dibrugarh, Sibsagar, Jorhat, N.C. Hills, Cachar and Hailakandi districts. Small areas also occur in the Himalayan foothills, especially in Lakhimpur, Dhemaji and Sonitpur districts. Some important tree species found in such forest areas include Hollong (*Dipterocarpus macrocarpus*), Gurjan (*D. tubinatus*), Mekai *(Shorea assamica)*, Kurta *(Palaquium polyanthum)*, Nahor *(Mesua ferrea)* and Sia-nahor *(Keyea assamica)*. Champion & Seth (1964) classified these forests as 'Assam Valley tropical semi-evergreen forest (2B/C1)' Upper Assam Valley tropical evergreen forest (1B/C2)' and 'Cachar tropical evergreen (1B/C3)'.

The tropical semi-evergreen forest occurs along the *bhabar* tract in the *duars* in Kokrajhar, Bongaigaon, Barpeta, Darrang and Sonitpur districts. In the south bank region, such forests are found in Jorhat, Golaghat, Nagaon, Morigaon, Karbi Anglong and in the low hills of southern Barak valley. Small patches also occur in Sonitpur, Lakhimpur and southern Kamrup. Champion & Seth (1964) classified such habitats as 'Assam Valley tropical semi-evergreen forest (2B/C1)' and 'Cachar semi-evergreen forest (2B/C2)'.

Large part of the lower Brahmaputra valley and parts of Karbi Anglong and N.C. Hills are covered by tropical moist deciduous forest. A speciality of this type of forest is the domination of the Sal tree (*Shorea robusta*). Champion & Seth (1964) had described these forests as 'Eastern hill sal forest 3C/Cla)', Eastern bhabar sal forest (3C/Clb)',and 'Kamrup sal[3C/C2b(ii)]'.

Along the banks of most of the larger hilly rivers and streams occur the riverine or riparian fringing forest. A few large trees with a large number of shrubs, both evergreen and deciduous with grass on the ground characterise this type of forest. The Sissoo (*Dalbergia sissoo*) and Khair *(Acacia catechu)* are the characteristic trees of such type. On the *chars* and *chapories* (sandy islets and tracts) of the Brahmaputra and its larger tributaries, pine like Jhau (*Tamarix dioica)* shrubs occur abundantly.

Throughout the Brahmaputra and the Barak valleys there are a large number of *beels* (ox-bow lakes) and *haors* (wetland complex in Barak Valley) with swamps and marshes in the fringe areas. During summer all such areas remain under flood water. In areas where flood is prolonged the tree growth is less, but in areas where it is for a short period, trees like Hijal *(Barringtonia acutangula)* grows.

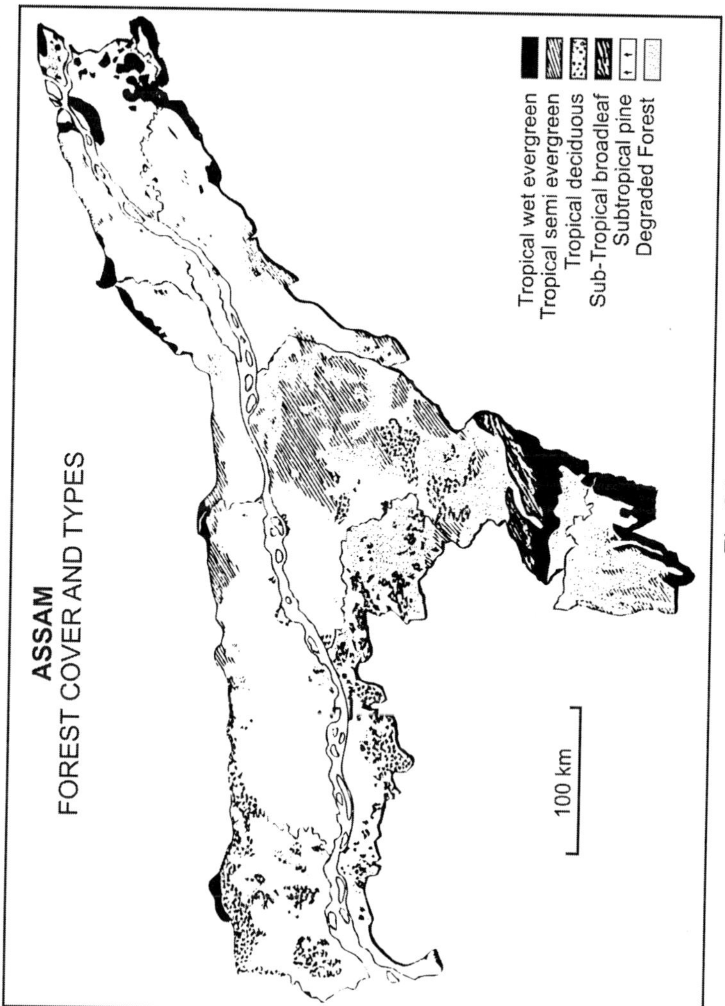
ASSAM
FOREST COVER AND TYPES
100 km
Tropical wet evergreen
Tropical semi evergreen
Tropical deciduous
Sub-Tropical broadleaf
Subtropical pine
Degraded Forest

Fig. 8.1

In the floodplain zone of the Brahmaputra and the *terai* belt along the foot of the Himalayas occur wet savannah grassland with tall elephant grass, reeds and *Alpinia allughas* herbs.Some such grassland pockets are found in Kaziranga, Laokhowa, Burhachapori, Orang, Manas, Kobo chapori and in Dibru-Saikhowa. The main species of grasses found in such grasslands are Khagori or Khag *(Phragmites karka)*, Nal *(Arundo donax)*, Ekra or lkora *(Erianthus ravannae)*, Ulu kher *(Imperata cylindrica)*, Meghela or Bhutang (*Saccharum arundinaccuin*,S. *procerum)*, Kohua or Kash *(S. spontaneum)*. Some trees like Bogori *(Zizyphys spp)*, Koroi *(Albizzia procera)* and Simul or Silk cotton *(Bombax ceiba*) are found scattered in the grasslands (Choudhury, 1997b).

In the higher altitudes of the Barails, both in N.C.Hills and Cachar, a type of evergreen forest occurs with trees 'not very high', which is called the subtropical broadleaf hill forest. The *jhum* cultivation has greatly affected this type of vegetation. In the higher areas of Hamren subdivision of Karbi Anglong, subtropical pine forest occurs with *Pinus kesiya* as the characteristic species.

Species Diversity : Animals

Broadly speaking, Assam forms part of the Indian (zoogeoraphic) Subregion within the Oriental Zoogeograpic Region (now often called the Indomalayan Region). However, if we look into microlevel details, then the areas south of the Brahmaputra River seems to be part of Indochinese Subregion while the Indian Subregion lies towards north of the river. The narrow strip of foothills of the north are part of the Himalayan Subregion. In fact, Assam is located in the transitional zone between the Indian and the Indochinese Subregions where many Indian species have their easternmost limit of distribution (eg., chital or spotted deer *Axis axis*) while many Indochinese species have their westernmost limit of distribution (eg., Tree-shrew *Tupaia belangeri*) (Choudhury, 1997a; Corbet & Hill, 1992).

Mammals : Altogether 230 forms including species and subspecies have been recorded or are likely to be found in Assam (Choudhury, 1997a). The total number of species listed so far is 193, which includes two recently extinct and one true semi-wild (born free, living free) forms. However, occurrence of 18 species (21, including subspecies) is not beyond doubt. Presence of such a diverse mammalian species in a relatively small area is quite significant (c 60% of the total mammalian species found in India). This is one of the highest diversity of mammals in any of the Indian states. Arunachal Pradesh has, however, more species in having 200+ (Choudhury, 1999a).

Large rivers often function as effective barriers in the distribution of mammals. In Assam, the Brahmaputra River, having a minimum width of c1.5km (at places more than 5 km) has played a key role in determining the distribution pattern of many mammalian species. The disrtribution of primates is a glaring example (Choudhury, 1989). The river has been an effective barrier to the movement and dispersal of the Hoolock gibbon, Stumptailed macaque, and Pigtailed macaque. Similarly, the Golden langur is confined to the western areas of the north bank of the Brahmaputra River, where even smaller rivers such as the Manas and the Sankosh have proved to be very effective physical barriers restricting the distribution of the species.

Assam holds the entire known world population of the pigmy hog, about 75% of the world population of the Indian rhinoceros and wild water buffalo, sizeable population of Asian elephant, tiger, and the entire population of the golden langur within India. The endangered and protected species are listed in Table 8.1.

Birds : About 958 species and subspecies of birds have so far been recorded or are likely to occur in Assam. This includes 808 species. Some 280 are winter migrants form the northern latitudes (Choudhury, 1990, 1999c). This richness and diversity in bird species is due to its location in the transition zone between various zoogeographic subregions. This is the highest diversity of birds in any of the Indian states. Arunachal Pradesh comes next with more than 750. Of the 20 orders of birds found in Assam, 12 contain threatened and 11 near-threatened species.

The following orders have more than five threatened and near-threatened species each:

Passeriformes 44, Falconiformes 15, Anseriformes 9, Ciconiiformes 8, and Chaladriiformes 8. The orders of birds that have the highest percentage of threatened and near-threatened species are: Pelecaniformes (43%), Galliformes (37%),

Bucerotiformes (33%), Ciconiiformes (26%), and Anseriformes (25%).

Some globally endangered species such as the Greater Adjutant Stork, White-winged Wood Duck, Swamp Partridge or Swamp Francolin and the Bengal Florican have their largest single concentration in Assam (Choudhury, 1996; Rahmani *et al*, 1991 Saikia, 1998). The endangered and protected species are listed in Table 8.2.

Reptiles : At least 187 species have been recorded or are likely to be found in Assam. This includes two species of crocodiles, 21 turtles and tortoises, 46 lizards and 118 of snakes (including three species of monitors)(A. Choudhury, unpub. data). There is possibility of a few new forms as reptiles are still relatively poorly covered by research in Assam. Like the mammals and birds, presence of such a diverse reptilian species in a relatively small area is quite significant. This is among the high diversity of reptiles in any of the Indian states. Appendix 1 lists all the turtles and tortoises, and some noteworthy other reptiles recorded in Assam.

Threats to Biodiversity

Habitat Destruction : Destruction of natural habitat for commercial felling, encroachment for cultivation and habitation, *jhum* or slash-and-burn cultivation of the hill tribes, and various development activities has become the most serious threat to biodiversity not only in Assam but throughout the world. Construction of roads and railways through forests and massive unplanned bamboo harvesting for two large paper mills (Jagiroad and Panchgram), oil mining and exploration in upper Assam, mini cement plants and coal mining, etc.are some of such activities. Table 5.3 shows the loss of forest cover in different years. It seems that between 1972-75 and 1993 -94, Assam has lost 3, 142 km^2 of dense forest (with 40% crown cover). The percentage of dense forest came down from 23.8% in 1972-75 to 19.8% in 1993-94 (assessment year 1997) (National Remote Sensing Agency; GOI, 1997).

Since the latter half of last century, more than 2200 km^2 of prime habitat, mostly tropical rain forest, have been cleared for tea plantations in Assam. In the hilly areas of the state, the area under *jhum* is more than 2600 km^2. In the Bhairabkunda RF of Darrang district, only 12 out of a total area of 24 km^2 is now left. In the Balipara RF, Sonitpur district, 100 out of 188 km^2 is under encroachment and Forest Villages.There are also extreme cases, in Gohpur RF, in the same district, and Gali RF (Dhemaji district) where almost the entire forests of 133 km^2 and 106 km^2 respecively have been encroached. These illustrate the magnitude of the situation.

Inter-state border problems often make biodiversity the worst sufferer. Almost the entire stretch of Assam-Arunachal Pradesh, Assam-Nagaland, and Assam- Meghalaya border is forested and is excellent natural habitat. Dispute over territory, confusion, break down of law-and-order (not infrequent along Assam-Nagaland border), and ultimately felling and encroachment in cleared and 'disputed' areas have already cost a few thousand square kilometres of prime habitat.The complete loss of once excellent forest comprising Nambor (south block), Doyang, Diphu and Rengma RFs (totalling about 900 km^2 of tropical rain forest on low undulating tract) in Golaghat district of Assam bordering Nagaland is solely due to this factor (Choudhury, 1999b).

Uncontrolled grazing by domestic stock is also a major problem, especially in the grasslands. Since most of the grasslands are confined to a few pockets of protected areas, the pressure is severe. For example, in Dibru-Saikhowa National Park, about 9000 cattle and domestic buffaloes graze (Choudhury, 1998).

The ultimate cause of habitat destruction is, however, the very rapid growth of human population. Every year more than four lakhs of new faces are being added to the already overburdened population of Assam, As a very high percentage of this population lives in rural areas (c. 85%) with farming as the main occupation, the large-scale destruction of natural habitats, i.e., the forest, grassland and wetland seem inevitable.

Poaching : After habitat destruction, poachiong of biological resources, both for commercial purpose as well as for the pot, is the major threat to biodiversity. Species such as the elephant, rhino, tiger, and Himalayan black bear are poached for international trade, while deer, pig, various small animals, birds and reptiles are mostly for domestic consumption. The most pursued species in Assam is the rhinoceros whose horn is in great demand in south-east Asia and Far East for medicinal value while in Yemen for use in dagger-handles. The tuskers are shot mostly for ivory (meat has local demand), tiger for bone and skin, and black bear for biles and gall bladder (all for supposed medcinal value). The poaching figures for the rhino and elephant are listed in Table 8.4. Poaching has played the key role in decimating the elephant population in parts of North Cachar Hills, Assam-Nagaland border and the southern Assam districts of Cachar, Hailakandi and Karimganj. While most were shot with guns and rifles, a few were killed by traditional methods such as pit method and various snares fitted with spear etc.. There are also instances of poaching by electrocution and poisoning.Most poaching goes unreported. Between 1980 and 1998, at least 539 rhinos were poached in Kaziranga alone. In the entire Assam, 875 fell during the same period (as per the official records).

Insurgency and Political Problems : Of late, the biodiversity has been greatly affected by problems arising out of insurgency. In fact, this factor is responsible for virtual elimination of the rhino and tusker in Manas National Park. Besides creating an atmosphere of lawlessness, the insurgency has made available modern sophisticated fire-arms at the doorsteps thus increasing the poaching.

While the insurgents themselves are not directly resorting to poaching (there are reports of connivance with poachers and illegal timber operators by some groups for money to buy arms), the resultant law-and-order situation has helped the poachers and others to carry out their activities with little or no resistance from the authorities. They took full advantage of the unrest in Manas between1989-95.

The problem has also affected to some extent the areas lying towards east and west of Manas. i.e., Barnadi Sanctuary, Sonai-Rupai Sanctuary and the reserved forests around Nameri National Park, and the forests of Kochugaon, Holtugaon and Aie Valley divisions.In Karbi Anglong and North Cachar Hills, besides the local outfits of Karbi and Dimasa tribes, the Nagas from Nagaland and Manipur are also active.

In Laokhowa Sanctuary,almost all the rhinos fell to poachers in the early 1980s due to disturbed situation arising out of AASU (All Assam Students' Union) movement on foreigners.

Others : Problems such as disease (bacterial diseases such as anthrax, rinderpest, etc) occasionally affect wild animals (mostly go unreported), pollution of waterbodies by pesticides like Dieldrin, Thiodan and DDT, occasional poisoning for fishing and by industrial effluents (paper mills, oil installations, etc.), powerlines passing through forest also cause accidental deaths of animals, especially elephants by electrocution, high floods in Kaziranga and other protected areas on the floodplains during monsoon are last but not the least problem for the rhino. While quite a few animals perish every year, but occasional high floods take heavy toll. Between June and September, 1998, as many as 39 rhinos were drowned in Kaziranga while more than a thousand hog deer lost their lives.

Biodiversity Conservation

Measures taken : Legal Protection. Both the habitat and the species are given different types of protection under various Acts and Rules, the notable being The Forest Conservation Act, 1990 and the Wildlife (Protection) Act, 1972. Both provide for stringent protection measures. Most people are aware of the legal status of some major species such as the tiger, elephant and rhino but not in case of others. They are also aware of the status of protected areas, reserved forests, etc. However, it is enforcement which is not adequate in most of the places save for a few well protected areas.

Habitat Protection. Twenty five protected areas in Assam have been notified for biodiversity conservation (Table 8.5). These include five national parks and 20 wildlife sanctuaries. Some noteworthy are Kaziranga, Manas, Nameri, Dibru-Saikhowa, and Orang.However, the protected area network covers a very insignificant 4.99% (3,916 km^2) of the total geographical area of the state. Moreover, in some sanctuaries, the enforcement is inadequate. Again in some key areas such as Dibru-Saikhowa National Park, only about 25 guards are available to patrol an area of 340 km^2.

NGO Activities : Many NGOs (non- governmental organisations) such as the Rhino Foundation for Nature in NE India, WWF-India, Assam Science Sociely, Nature's Beckon, Aaranyak, Green Heart, Megamix, etc., are also supplementing governmental efforts in saving the biodiversity.

Measures Proposed

1. *Creation of new Protected Areas.* With only 3.2% of the habitat under protected area network , there is an urgent need to bring more areas under the system, at least not less than 6%. Some potential areas are, northern and eastern Karbi Anglong, Dhansiri RF, Ripu-Kochugaon- Chirang RFs, Upper Dihing - Dirak-Joypur RFs, Lumding-Kaki RFs, Innerline RF, Katakhal RF, Barail RF and North Cachar RF.
2. *Extension of existing protected areas.* Especially Kaziranga, Garampani, Dibru-Saikhowa, and Barnadi.
3. *Translocation of Rhinos to Laokhowa, Burhachapori, Manas and other areas.* This will be required in near future.
4. *Check on poaching.*
5. *Control of jhum cultivation.*
6. *Check on encroachment and eviction of encroachers.*
7. *Ban on commercial felling in the remaining prime natural habitats.*
8. *Other measures:* Check on grazing,effective protection and management of existing protected areas, conservation education, concrete measures to reduce depredation, expeditious payment of compensation for loss of life and property, large-scale installation of bio-gas, further research and monitoring, and treatment of polluted waterbodies.

Table 8.1
Endangered Mammals of Assam

1. Chinese Pangolin *Manis pentadactyla* ☪
2. Indian Pangolin *Manis crassicaudata* ☪
3. Slow Loris *Nycticebus coucang* ☪
4. Pig-Tailed Macaque *Macaca nemestrina* ☆
5. Assamese Macaque *Macaca assamensis* ☆
6. Stump-Tailed Macaque *Macaca arctoides* ☆

7. Phayre's Leaf Monkey, Spectacled Monkey *Presbytis Phayrei* ☪
8. Capped Langur *Presbytis pileatus* ❑☆☪
9. Golden Langur *Presbytis geei* ❑☪
10. Hoolock Gibbon, White-Browed Gibbon *Hylobates hoolock* ❑☪
11. Dhole, Indian Wild Dog, Red Dog *Cuon alpinus* ☆
12. Himalayan Black Bear, Asiatic Black Bear *Ursus thibetanus* ❑☆
13. Sun Bear, Malayan Bear *Ursus malayanus* ❑☪
14. Sloth Bear *Ursus ursinus* ☆☪
15. Black-Striped Weasel *Mustela strigidorsa* ☆
16. Ratel, Honey Badger *Mellivora capensis* ☪
17. Hog Badger *Arctonyx collaris* ☪
18. Common Otter, Eurasian otter *Lutra lutra* ❑
19. Smooth Indian Otter, Smooth-Coated Otter *Lutrogale perspicillata* ☆
20. Oriental Small-Clawed Otter Clawless, Otter *Aonyx cinerea* ☪
21. Spotted Linsang *Prionodon paradicolor* ❑☪
22. Binturong, Bear-Cat *Arctictis binturong* ☪
23. Leopard Cat *Felis bengalensis* ❑☪
24. Fishing Cat *Felis viverrinus* ☪
25. Golden Cat *Felis temminckii* ❑☪
26. Marbled Cat *Felis marmorata* ❑☪
27. Clouded Leopard *Neofelis nebulosa* ❑☆☪
28. Leopard, Panther *Panthera pardus* ❑☪
29. Tiger, Indian Tiger, Royal Bengal Tiger, Bengal Tiger *Panthera tigris* ❑☆☪
30. Gangetic Dolphin, Susu *Platanista gangetica* ❑☆☪
31. Indian Elephant, Asian Elephant *Elephas maximus* ❑☆☪
32. Indian Rhinoceros, Greater One-horned Rhinoceros *Rhinoceros unicornis* ❑☆☪
33. Javan Rhinoceros, Lesser One-horned Rhinoceros *Rhinoceros sondalicus* Extinct. ❑ *cr.*
34. Sumatran Rhinoceros, Asian Two-horned Rhinoceros *Dicerorhinus sumatrensis* Extinct ❑ *Cr.*
35. Pygmy Hog, Pigmy Hog *Sus salvanius* ❑ *Cr.* ☪
36. Swamp Deer, Barasingha *Cervus duvauceli* ❑☆☪
37. Gaur, Indian Bison *Bos gaurus* ❑✪☪

38. Wild Water Buffalo, Asiatic Wild Buffalo *Bubalus arnee* ✪☪
39. Goral *Naemorhedus goral* ❑☪
40. Serow *Capricornis sumatraensis* ❑☆ (subspecies *jamrachithar,* K *rubidus*✪); ☪
41. Irrawaddy Squirrel, Himalayan Hoary-Bellied Squirrel *Callosciurus pygerythrus* ☆
42. Particoloured Flying Squirrel *Hylopetes alboniger* ✪
43. Crestless Himalayan Porcupine, Chinese Porcupine *Hystrix brachyura* ☆
44. Hispid Hare *Caprolagus hispidus* ❑✪☪

Source: Choudhury (1997 a). ❑ = species included in Appendix I of CITES

✪ = species treated as Endangered by IUCN.
☆ = species treated as Vulnerable by IUCN; Cr. = species treated as Critical by IUCN
☪ = species indcluded in Schedule I of The Wildlife (Protection) Act, 1972 of India.

Table 8.2
Endangered Birds of Assam

Family PELICANIDAE : *Pelicans*
1. Spotted-Billed or Grey Pelican *Pelecanus philippensis* ❑
2. Dalmatian Pelican *Pelecanus crispus* ❑

Family ANHINGUIDE : Darter
3. Darter or Snake-Bird *Anhinga melanogaster* syn.*rufa* ✪

Family ARDEIDAE : Herons, Egrets, Bitterns
4. Great White-Bellied or Imperial Heron *Ardea insignis* syn.*imperialis* ❑

Family CICONIIDAE : Storks
5. Painted Stork *Mycteria leucocephala* ✪
6. Openbill Stork, Asian Openbill *Anastomus oscitans* ✪
7. Oriental White Stork *Ciconia boyciana* ❑☪
8. Greater Adjutant *Leptoptilos dubius* ❑
9. Lesser Adjutant *Leptoptilos javanieus* ❑

Family THRESKIORNITHIDAE: Ibises, Spoonbill
10. White or Black-Headed Ibis *Threskiornis aethiopica* syn. *melanocephalus* ✪

11. Black Ibis*Pseudibis papillosa* ✪
12. Eurasian Spoonbill *Platalea leucorodia* ☪

Family ANATIDAE.Ducks Geese

13. Red-Breasted Goose *Branta ruficollis* ❑
14. Lesser White-Fronted or Dwarf Goose *Anser erythropus* ❑
15. Large Whistling Teal or Fulvous Tree Duck *Dendrocygna bicolor* ☪
16. Marbled Teal *Marmaronetta angustirostris* ❑
17. Baikal Teal *Anas Formosa* ❑
18. Pink-Headed Duck *Rhodonessa caryophyllacea* ❑☪
19. Ferruginous Duck or White- Eyed Pochard *Aythya Nyroca* ❑
20. Baer's Pochard *Aythya basri* ❑
21. Mandarin Duck*Aix galericulata* ✪
22. White-Winged Wood Duck *Cairina scutulata* ❑☪

Familiy ACCIPITRIDAE: Hawks, Eagles, Vultures, etc.

23. Blyth's or Jerdon's Baza*Aviceda jerdoni* ✪☪
24. Black or Black-Crested Baza *Aviceda leuphotes* ☪
25. Nothern Goshawk *Accipiter gentillis* ☪
26. Shikra *Accipiter badius* ☪
27. Crested Goshawk *Accipiter trivirgatus* ☪
28. Eurasian Sparrow- Hawk *Accipiter nisus* ☪
29. Besra Sparrow-Hawk *Accipiter virgatus* ☪
30. Japanese Sparrow-Haw *Accipiter gularis* ☪
31. Imperial Eagle *Aquila heliaca?* ❑
32. Greater Spotted Eagle *Aquila clanga* ❑
33. White-Tailed Sea Eagle *Hailiaeefus albioilla* ✪
34. Pallas's Fish Eagle *Hailiaeefus leucoryphus* ❑
35. Grey-Headed Fish Eagle *lchthyophaga ichthyaetus* ✪
36. Himalayan Grey-Headed or Lesser Fish Eagle *Ichthyophaga nana* Syn. *Humills*
37. King, Black or Red-Headed Vulture *Sarcogyps calvus* ✪
38. Cinereous Vulture *Aegyplus monachus* ✪
39. Long-Billed Vulture *Gyps indicus* ✪
40. White-Backed or White-Rumped Vulture *Gyps bengalensis* ✪
41. Lammergeier or Bearded Vulture *Gypaetus barbatus* ☪
42. Pale or Pallid Harrier *Circus macrourus* ✪
43. Osprey *Pandion Haliaetus* ☪

Family FALCONIDAE : Falcons

44. White-Legged or Pied Falconet *Microhierax melanoleucos* ✪
45. Peregrine Falcon *Falco peregrinus japonensis* ☪
46. Shaheen Falcon *Falco peregrinus peregrinator* ☪
47. Red-Headed or Red-Necked Merlin *Falco chicquera* ✪☪
48. Lesser Kestrel *falco naumanni* ❑

Family PHASIANIDAE: Pheasants. Partridges, Quails, Etc.

49. Swamp Partridge, Swamp Francolin or Kyah *Francolinus gularis* ❑
50. Manipur Bush Quail *Perdicula manipurensis* ❑
51. White-Cheeked Hill Partridge *Arborophila atrogularis* ✪
52. Red-Breasted Hill Partridge *Arborophila mandellii* ❑
53. Bamboo Partridge *Bambusicola fytchii* ☪
54. Blyth's or Grey-Bellied Tragopan *Tragopan blythii* ❑☪
55. Mrs Hume's Barred- Back or Bar-Tailed Pheasant *Syrmaticus humiae* ❑ ☪
56. Peacock-Pheasant *Polyplectron bicalcaratum* ☪
57. Indian or Common Peafowl *Pavo cristatus* ☪
58. Green or Burmese Peafowl *Pavo muticus* ❑

Family GRUIDAE: Cranes

59. Hooded Crane *Grus monacha* ☆☪

Family HELIORNITHIDAE: Finfoots

60. Masked Finfoot *Heliopais personata* ❑

Family OTIDIDAE: Bustards

61. Bengal Florican *Houbaropsis* (Syn. *Eupodotis*) *bengalensis* ❑☪

Family CHARADRIIDAE: Plovers

62. Grey-Headed Lapwing *Vanellus cinereus* ✪
63. Long-Billed Ringed Plover *Charadrius placidus* ✪

Family SCOLOPACIDAE: Curiews, Sandpipers, Snipe

64. Spotted or Nordmann's Greenshank or Armstrong's Sandpiper *Tringa Guttifer* ❑
65. Snipe-Billed Godwit or Asian Dowitcher *Limnodromus semipaimatus* ✪
66. Wood Snipe *Gallinago nemoricola* ❑
67. Spoon-Billed Sandpiper *Eurynorhynchus pygmeus* ❑

Family LARIDAE: Gulls, Terns

68. Black-Bellied Tern *Sterna acuticauda* ❑
69. Indian Skimmer or Scissorbill *Rynchops albicollis* ❑

Family COLUMBIDAE: Pigeons, Doves

70. Purple Wood or Pale-Capped Pigeon *Columba punicea* ❑

Family STRIGIDAE: Owls
Subfamily STRIGINAE : Owls

71. Forest or Spot-Bellied Eagle-Owl *Bubo nipalensis* ✪
72. Tawny Fish Owl *Ketupa flavipes* ✪

Family PODARGIDAE : Frogmouths

73. Hodgson's Frogmouth *Batrachostomus hodgsoni* ☪

Family APODIDAE : Swifts

74. Dark-Backed or Khasi Hills Swift *Apus acuticauda?* ❑

Family ALCEDINIDAE : Kingfishers

75. Blyth's or Great Blue Kingfisher *Alcedo hercules* ❑
76. Brown-Winged Stork-Billed Kingfisher *Haicyon* (Syn. *Pelargopsis) amauroptera* ✪

Family BUCEROTIDAE : Hornbills

77. Brown, Brown-Backed or White-Throated Brown Hornbill *Anorrhinus* (Syn. *Ptilolaemus) tickelli* ✪☪
78. Rufous-Necked Hornbill *Aceros nipalensis* ❑☪
79. Wreathed Hornbill *Aceros* (Syn. *Rhyticeros) undulatus* ☪
80. Indian or Oriental Pied Hornbill *Anthracoceros albirostris* Syn. *malabaricus* ☪
81. Great Pied Hornbill *Buceros bicornis* ☪

Family INDICATORIDAE : Honeyguides

82. Orange-Rumped Honey Guide *Indicator xanthonotus* ✪

Family PITTIDAE : Pittas

83. Blue-Naped Pitta *Pitta nipalensis* ✪

Family DICRURIDAE : Drongos

84. Hair-Crested or Spangled Drongo *Dicrurus hottentottus* ✪

Family STURNIDAE : Starlings, Mynas

85. Spotted -Winged Stare *Saroglossa spiloptera* ✪
86. Collared Myna *Acridotheres albocinctus* ✪

Family MUSCICAPIDAE : Babblers, Fycatchers, Thrushes, Chats
Subfamily TIMALIINAE : Babblers

87. Marsh Spotted Babbler *Pellorneum palustre* ❑
88. Slender-Billed Scimitar-Babbler *Xiphirhynchus superciliaris* ✪
89. Long-Billed Wren-Babbler *Rimator malacoptilus* ✪
90. Long-Tailed or Tawny- Breasted Wren-Babbler *Spelaeornis longicaudatus* ❑
91. Spotted Wren-Babbler *Spelaeornis formosus* ✪
92. Wedge-Billed Wren *Sphenocichla humei* ✪
93. Austen's Spotted or Snowy-Throated Babbler *Stachyris oglei* ❑
94. Jerdon's or Hume's Babbler *Chrysomma Altirostre* ❑
95. Black-Browed or Lesser Red-Headed Parrotbill *Paradoxornis atrosuperciliaris* ✪
96. Greater Red-Headed Parrotbill *Paradoxornis ruficeps* ✪
97. Black-Throated or Black-Breasted Parrotbill *Paradoxornis flavirostris* ❑
98. Slender-Billed Babbler *Turdoides longirostris* ✪
99. Chestnut-Backed Laughing-Thrush *Garrulax nuchalis* ✪
100. Yellow-Throated Laughing Thrush *Garrulax galbanus* ✪
101. Yellow-Breasted or Rufous-Vented Laughing-Thrush *Garrulax delesserti* ✪
102. Spotted-Breasted Laughing -Thrush *Garrulax merulinus* ✪
103. Manipur Streaked or Striped Laughing Thrush *Garrulax virgatus* ✪
104. Rufous-Bellied or Black-Headed Shrike-Babbler *Pteruthius rufiventer* ✪
105. Dusky-Green or Yellow-Throated Tit-Babbler *Alcippe cinerea* ✪
106. Red-Throated Tit-Babbler *Alcippe rufogularis* ✪
109. Grey Sibia *Heterophasia graclills* ✪

Subfamily MUSCICAPINAE: Flycatchers

110. Brown-Breasted Flycatcher *Muscicapa muttui* ✪

Subfamily SYLVIINAE: Warblers

111. Hodgson's Wren-Warbler or Grey-Crowned Prinia *Prinia cinereocapilla* ✪
112. Long-Tailed Grass Warbler or Rufous-Vented Prinia *Prinia burnesii* ❑
113. Bristled Grass Warbler or Bristled Grassbird *Chaetornis striatus* ❑
114. Black- Browed Leaf or Yellow-Vented Warbler *Phylloscopus cantator* ✪
115. Broad- Billed Flycatcher-Warbler *Abroscoupus* (Syn. *Tickellia) hodgsoni* ✪

Subfamily TURDINAE: Thrushes,Chats

116. Rusty-Bellied Shortwing *Bachypteryx hyperythra* ❑
117. Firethroat *Erithacus* (Syn. *Luscinia) pectardens?* ✪
118. Rufous-Bellied Bush Robin *Erithacus* (Syn. *Tarsiger) hyperythrus* ✪
119. Purple Cochoa *Cochoa purpurea* ✪
120. Green Cochoa *Cochoa viridis* ✪
121. Hodgson's Bushchat *Saxicola insignis* ❑
122. Jerdon's Bushchat *Saxicola jerdoni* ✪
123. Pied Ground Thrush *Zoothera wardii* ✪
124. Large Brown or Long-Billed Thrush *Zoothera monticola* ✪
125. Black-Breasted Thrush *Turdus dissimilis* ✪
126. Fea's Thrush *Turdus feae* ❑

Family SITTIDAE: Nuthatches, Creepers

127. Beautiful Nuthatch *Sitta formosa* ❑

Family PLOCEIDAE: Weaver Birds
Subfamily PLOCEINAE: Weaver Birds, Bayas

128. Finn's Baya *Ploceus megarhynchus* ❑

Source : Choudhury (1990. in press), Collar et al. (1994).

❑ species treated as Threatened in *Birds to watch2*

✪ species treated as Near-Threatened in *Birds to watch2*

☆ species treated as Conservation Dependent in *Birds to watch 2*

G species included in Schedule I of The Wild Life (Protection) Act.

Table 8.3
Area under forest in Assam from 1972-75 and 1997 based on visual interpretation of Landsat data

Period/ Year	Total forest area in km^2	% Forest area in km^2	Total dense forest area in km^2	% dense forest area in km^2	Remarks
1972-75	21,055	26.8	18.690	23.8	Source: National Remote Sensing Agency (NRSA)
1980-82	19,796	25.2	15,625	19.9	Source: National Remote Sensing Agency (NRSA)
1991 (1987-89)	24,751	31.5	15,842	20.2	Source: The State of Forest Report, 1991. Based on data of 1987-89
1993 (1989-91)	24,508	31.2	15,998	20.4	Source: The State of Forest Report, 1993. Based on data of 1989-91
1995 (1991-92)	24,061	30.6	15,694	20.0	Source: The State of Forest Report, 1995. Based on data of 1991-92
1997 (1993-94)	23,824	30.3	15,548	19.8	Source: The State of Forest Report. 1997.Based on data of 1993-94
Total loss	927 (since 1987-89)	1.2	3,142 (since 1972-75)	4.0	There appears to be apparent disagreement in the methodology in between NRSA and the State of Forest Report

Table 8.4
Number of known Rhinoceros and Elephants poached in Assam during 1979-1999 (till April)

Year	Number of Rhinos poached	Number of Elephants poached
1979	15	NA
1980	18	NA
1981	38	NA
1982	44	NA
1983	91	NA
1984	45	NA
1985	56	NA
1986	53	NA
1987	43	2
1988	44	14
1989	67	10+
1990	45	11
1991	29	6+
1992	67	4+
1993	70	10+
1994	27	7+
1995	41	11
1996	40	9
1997	29	0
1998	28	4+
1999 (till April)	6	1
Total	**896**	**89**

Source : Rhino--Forest Department of Assam; Elephant - Choudhury (1999, unpub).
The figures of elephants should be treated as minimum known cases. The '0' or NA (not available) should not necessarily be treated as nil.

Table 8.5
Protected areas in Assam

National Park			
S.No.	**Park**	**District**	**Area in sq. km.**
1.	Dibru-Saikhowa	Dibrugrah & Tinsukia	340.00
2.	Kaziranga	Golaghat, Nagaon &	849.00

		Sonitpur	
3.	Manas	Chirang & Baska	500.00
4.	Nameri	Sonitpur	200.00
5.	Orang	Udalguri & Sonitpur	78.81
	Total		**1968.00**

Wildlife Sanctuary

S.No.	Wildlife Sanctuary	District	Area
1.	Bordoibam-Bilmukh	Lakhimpur	11.25
2.	Barnadi	Udalguri & Baksa	26.22
3.	Burhachapori	Sonitpur	44.06
4.	Chakrashila	Kokrajhar	45.57
5.	Dipor Beel	Kamrup Metro	4.14
6.	Garampani	Karbi Anglong	6.05
7.	Gibbon	Jorhat	20.98
8.	Laokhowa	Nagaon	70.1
9.	Pabitora	Morigaon	38.83
10.	Pani-Dihing	Sibsagar	33.98
11.	Sonai-Rupai	Sonitpur	220.0
12.	Nambor-Doigurung	Golaghat	97.15
13.	Nambor	Karbi Anglong	37.00
14.	East Karbi Anglong	Karbi Anglong	221.81
15.	Bhejan-Borajan-Pudumoni	Tinsukia	7.22
16.	Marat longri	Karbi Anglong	451.00
17.	Dihing-Patkai	Dibrugarh & Tinsukia	111.19
18.	Barail	Cachar	326.25
19.	Amchang	Kamrup Metro	78.64
20.	North Karbi Anglong	Karbi Anglong	96.00
	Total		**1947.40**
	Grand Total		**3916.00**

Source: Economic Survey, Assam, 2012-2013, Govt. of Assam.

References

Arnold, E.N. 1989. Towards a phylogeny and biogeography of the lacertilidae relationships wihthin an Old World family of lizard derived from morphology. *Bull. British Mus Nat Hist (Zool.)* 55: 209-257.

Champion, H. & Seth, S.K. 1964. *A revised survey of the forest types of India.* Forest Research Institues & Colleges, Dehra

Dun.

Choudhury, A.U. 1989. Primates of Assam : their distribution, habitat and status. Ph.D. thesis, Gauhati University.

- 1990. *Checklist of the birds of Assam.* Sofia Press & Publishers, Guwahati.

- Choudhury, A.U. 1996 *Survey of the White-winged wood duck and the Bengal florican in Tinsukia dist and adjacent areas of Assam and Arunachal Pradesh.* The Rhino Foundation for nat. in NE India, Guwahati.

- 1997 a. *Checkist of the Mammals of Assam.* Revised second edn. Gibbon Book and ASTEC , Guwahati.

- 1997 b. Survey of grasslands in some parts of central & southern Assam to assess their biodiversity & socio-ecological problems. Final report (BCCP). WWF India NERO, Guwahati.

- 1998. Mammals, birds and reptiles of Dibru-Saikhowa Sanctuary, Assam, India. *Oryx*, 32(3) : 192-200.

- 1999a (in press) *Mammals of Arunachal Pradesh,* WWF-India, New Delhi.

- 1999b. Conservation Strategy For the Indian Rhinoceros and Asian Elephant, Final report /proposals to WWF-India.

- 2000. *Birds of Assam.* Gibbon Books & others, Guwahati.

Collar N.J., Crosby, M.J. & Stattersfield, A.J. 1994. *Birds to watch 2, the world list of ihreatened birds.* Birdlife International, Cambridge.

Corbet. G.B. & Hill, J.E. 1992. *The mammals of the Indomalayan Region : A Systematic review.* Oxford University Press, Oxford.

Das, I 1991, *Colour guide to the turtles and tortoises of the Indian Subcontinent.* R & A Publishing, Portishead.

Das, 1. 1994, The Reptiles of South Asia: Checklist and distributional Summary. *Hamadryad* 19 : 15-40.

- 1996. *Biogeography of the reptiles of South Asia.* Kreiger Publishing Company. Malabar, Florida.

Deoras. P.J. 1970. *Snakes of India,* National Book Trust, New Delhi.

Ditmars. R.L. 1946. *Reptiles of the World.* Macmillans, New York.

Gans, C. 1966. Liste der rezenten amphibien und reptilien.

Uropeltidae. *Tierreich* 84 : 1-29.

Gharpure, K.G. 1937. *Snakes of India*, Popular Book Depot, Bombay.

GOI 1991. *The state of the forest report,* Government of India.

- 1995. *The state of the forest report,* Government of India.

- 1997. *The state of the forest report,* Government of India.

Gyi, K.K. 1970. A revision of colubrid snakes of the subfamily Homalopsinae. Univ. Kansas Publ. *Mus Nat. Hist.* 20:47-223.

Hoge. A.R. and Romano-Hoge. A.R.W.L. 1981. Poisonous snakes of the world. Part I. Checklist of the pit vipers Viperoidea, Crotalinae. *Mem. Inst. Butantan* 42-43: 179-310

King. F.W. and Burke. R.L. 1989. *Crocodilian. tuatara, and turtle species of the world: A taxonomic and geographic reference.* Association of Systematic Collections. Washington.

Kluge, A.G. 1991. Checklist of gekkonold lizards. *Smithsonian Herpetol. Inf. Serv.* (85) : 1-35.

Murthy, T.S.N. 1985. Classification and distribution of the reptiles of India. *The Snake* 17: 48-71.

Myers, N. 1988. Threatened Biotas: "Hotspots" in tropical forests *Environmentalist* 8 (3) : 1-20.

- 1991. The biodiversity challenge : Expanded "Hotspots" analysis. *Environmentalist* 10(4) : 243-256.

Nicholson. E. 1870. *Indian Snakes.* Higgonbotham & Co. Madras.

Rahmani, A, Narayan. G, Rosalind, L., Sankaran, R., & Ganguli-Lachungpa, U. 1991. Status of the Bengal florican in India. *J. Bombay Nat. Hist. Soc.* 88: 349-375.

Saikia. P. 1995. Eco-biology of the Adjutant storks with special reference to the Lesser adjutant. Ph.D. thesis. Gauhati University.

Smith, M.A. 1931. *Fauna of British India, including Ceylon and Burma vol. I Loricata. Testudines.* Taylor & Francis, London.

- 1935. *Fauna of British India, including Ceylon and Burma. vol.II. Sauria.* Taylor & Francis. London.

- 1961. *Fauna of British India. vol. III Serpentes.* Survey of India Reprint, Dehra Dun.

Stidworthy. J. 1969. *Snakes of the World.* Paul Hamlyn, London.

Taher. M. 1986. Physiographic Framework of NE India. *NE Geographer* 18:1-19.

Tikader. B. & Sharma, R. 1992. *Handbook of Indian Lizards,* Zoological Survey of India, Calcutta.

Underwood. G.1967. *A Contribution to the Classification of Snakes.* British Museum (N.H.), London.

Wall, Major F. 1928. *Poisonous Terrestrial Snakes of British India.* Bombay Natural History Society, Bombay.

Waltner. R.C. 1975. Geographical and Altitudinal Distribution of Amphibians and Reptiles in the Himalayas, Parts I-IV. *Cheetal* Vol. 16. Nos 1,2,3,4

Whitaker, R. 1978. *Common Indian Snakes - A Field Guide.* The Macmillan Company of India Limited, Delhi.

Appendix 1
Some noteworthy Reptiles of Assam
(all species of Crocodylia and Testudines, and some others)

Order Crocodylia CROCODILES
CROCODYLIDAE
1. *Crocodylus pelustris* (Lesson, 1831) Marsh Crocodile
GAVIALIIDAE
2. *Gavialis gangeticus* (Gmelin, 1789) Gharial, Gavial

Order Testudines TURTLES & TORTOISES
BATAGURIDAE

3. *Cuora amboinesis* (Daudin, 1801)	Malayan box turtle
a. *C. a. Kamaroma* (Rummier & Fritz, 1991)	
4. *Cyclemis dentata* (Gray, 1831)	Asian leaf turtle
5. *Geoclemys hamiltonii* (Gray, 1831)	Spotted pond turtle
6. *Hardella thurjii* (Gray, 1831)	Crowned river turtle
a. *H. t. thurjii* (Gray, 1831)	
7. *Kachuga dhongoka* (Gray, 1834)	Three-striped roofed turtle
8. *Kachuga kachuga* (Gray, 1831)	Red-crowned roofed turtle
9. *Kachuga smithii* (Gray, 1863)	Brown roofed turtle
a. *K. s. smithii* (Gray, 1863)	
10. *Kachuga sylhetensis* (Jerdon, 1870)	Assam roofed turtle
11. *Kachuga tecta* (Gray, 1831)	Indian roofed turtle
12. *Kachuga tentoria* (Gray, 1834)	Indian tent turtle
a. *K. t. flaviventer* (Gunther, 1864)	

13. *Meianochelys tricarinata* (Blyth, 1856)	Tricarinate hill turtle
14. *Meianochelys trijuga* (Schweigger, 1812)	Indian black turtle
a.*M.t. indopeninsularis* (Annandale, 1913)	
15. *Morenia petersi* (Anderson, 1879)	Indian eyed turtle
16. *Pyxidea mouhotii* (Gray, 1862)	Keeled box turtle
TESTUDINIDAE	
17. *Indotestudo elongata* (Blyth, 1853)	Elongated tortoise
18. *Manouria emys* (Schlegel & Muller, 1840,)	Brown hill tortoise
a.*M.e, emys* (Schlegel & Muller, 1840)	Asian brown tortoise
b.*M.e. phayrei* (Blyth, 1853)	
TRIONYCHIDAE	
19. *Aspideretes gangeticus* (Cuvier, 1825)	Gangetic softshell, Indian softshell turtle
20. *Aspideretes hurum* (Gray, 1831)	Indian peacock softshell turtle
21. *Aspideretes nigricans* (Anderson, 1875)?	Black softshell turtle
22. *Chitra indica* (Gray, 1831)	Narrow-headed softshell turtle
23. *Lissemys punctata* (Lacepede, 1788)	Indian flapshell turtle
a. *L. p. andersoni* Webb, 1980	
Order Lacertilia LIZARDS	
GEKKONIDAE	
24. *Gokko gecko* (Linnaeus, 1758)	Tuckto gecto
a. *G. g. gecko* (Linnaeus, 1758)	Common Indian
a. *G. g. azhari* (Mertens, 1955)	takshak
25. *Gonydactylus khasiensis* (Jerdon, 1870)	Assam rock gecko
a. *G. k. Khasiensis* (Jerdon, 1870)	
26. *Hemidactyius brookii* (Gray, 1845)	Spotted house gecko
27. *Hemidactyius flaviviridis* Ruppell, 1840	Yellow-bellied house gecko
28. *Hemidactyius frenatus* (Dumeril & Bibron, 1836)	Ticticky house gecko
29. *Hemidactyius gamotll* (Dumeril & Bibron, 1836)	Assam greyish-brown gecko
30. *Hemidactyius karenorum* (Theobald, 1968)	Cachar greyish-brown gecko
AGAMIDAE	
31. *Calotes bhutanensis* (Biswas, 1975)	Bhutanese calotes
32. *Calotes emma* (Gray, 1845)	Light-olive Assam
a. *C. e. alticristatus* (Schmidt, 1927)	garden lizard
33. *Calotes maria* (Gray, 1845)	Assam garden lizard
34. *Calotes versicolor* (Daudin, 1802)	Indian graden lizard

a. *C. v. versicolor* (Daudin, 1802)	
35. *Draco blanfordii Boulenger*, 1885	Assam flying dragon
a. *D.b. norvillii* (Alcock, 1895)	
SCINCIDAE	
36. *Lygosoma albopunctata* (Gray, 1846)	Brown dwarf skink
37. *Mabuya carinata* (Schneider, 1801)	Common Indian skink
a. *M. c. carinata* (Schneider, 1801)	
38. *Mabuya multifasciata* (Kuhl, 1820)	Assam olive brown skink
39. *Mabuya quadricarinata* (Boulenger, 1887)	Cachar olive brown skink
LACERTIDAE	
40. *Takydromus haughtonianus* (Jerdon, 1870)	Long headed reddish lizard
ANGUIDAE	
41. *Ophisaurus gracilis* (Gray, 1845)	Indian glass snake
VARANIDAE	
42. *Varanus bengalensis* (Daudln, 1802)	Common monitor , Bengal monitor, Indian monitor
43.*Varanus flavescens* (Hardwicke & Gray, 1827)	Indian yellow monitor
44. *Varanus salvator* (Laurenti, 1768)	Indian water monitor
a. *V. a. Salvator* (Laurenti, 1768)	

Order: SERPENTES SNAKES

Infra order SOCLECOPHIDIA

Family: TYPHLOPIDAE
Subfamily : TYPHLOPINAE

45. *Ramphotyphlops braminus* (Daudin, 1803),	Common worm snake
46. *Typhlops bothriorhynchus* (Gunther, 1864)	Assam worm snake

Family: BOIDAE
Subfamily :PHYTHONINAE

47. *Python molurus* (Linnaeus, 1758)	Rock python
a. P. molurus molurus (Linnaeus, 1758)	
b. P. Molurus bivittatus (Kuhl, 1820)	
48. *Python reticulatus* (Schneider, 1801)	Reticulated python

Infra order: CAENOPHIDIA
Family: COLUBRIDAE
Subfamily: DIPSADINAE

49. *Pareas monticola* (Cantor, 1839)	Assam snail-eater
Subfamily: XENODERMINAE	
50. *Stoliczkaia khasienis* (Jerdon, 1870)	Khasi Hills earth snake
Subfamily: LYCODONTINAE	
51. *Lycodon aulicus* (Linnaeus, 1758)	Common wolf snake
52. *Oligodon amensis* (Shaw, 1802)	Common kukri snake
53. *Trachischium monticola* (Cantor, 1839)	Assam oriental worm snake
Subfamily: SIBYNOPHIINAE	
54. *Sibynophis collaris* (Gray, 1853)	Collared black-headed snake
Subfamily: NATRICIDAE	
55.*Amphiesma stolata* (Linnaeus, 1758)	Striped keelback
56. *Atretium schistosum* (Daudin, 1803)	Olive keelback watersnake
57. *Rhabdophis subminiata* (Schlegel, 1837)	Red-neked keelback
58. *Xenochrophis piscator* (Schneider, 1799)	Checkered keelback watersnake
Subfamily: COLUBRINAE	
59. *Ahaetulla nasutus* (Lacepede, 1789)	Common vine snake
60. *Elaphe frenata* (Gray, 1853)	Assam trinket snake
61. *Elaphe helena* (Daudin, 1803)	Common trinket snake
62. *Ptyas/ Coluber mucosus* (Linnaeus, 1758)	Rat snake or Dhaman
Subfamily: BOIGINAE	
63. *Boiga cynodon* (Bole, 1827)	Bengal cat snake
64. *Boiga quincunclata* (Wall, 1908)	Assam cat snake
Subfamily : HOMALOPSINAE	
65. *Enhydris enhydris* (Schneider, 1799)	Common smooth water snake
Subfamily : DASYPELTINAE	
66. *Elachistodon westermanni* (Reinhardt, 1863)	Indian egg-eater
Family ELAPIDAE	
Subfamily : ELAPINAE	
67. *Bungarus caeruleus* (Schneider, 1801)	Common krait
68. *Bungarus fasciatus* (Schneider, 1801)	Banded krait
69. *Bungarus niger* (Wall, 1908)	Black krait
70. *Callophis macclellandi* (Reinhardt, 1844)	Mac Clelland's coral

	snake
a. C. m. macclellandi (Reinhardt, 1844)	
71. *Callophis malanurus* (Shaw, 1802)	Slender coral snake
72. *Naja naja* (Linneaeus, 1758)	Common or Spectacled cobra
73. *Naja kaouthia* (Lesson, 1831)	Monocellate or Bengal cobra
74. *Ophiophagus hannah* (Cantor, 1836)	King cobra or Hamadryad
Family : VIPERIDAE	
Subfamily: VIPERINAE	
75. *Vipera russellii* (Shaw & Nodder, 1797)	Russell's viper
Subfamily: CROTALINAE	
76. *Trimeresurus albolabris* (Gray, 1842)	Green pit viper

Source : Arnold (1989), A Choudhury (unpub), Das (1994), Deoras (1970), Ditmars (1946), Gharpure (1937), Gyi (1970), Hoge & Romano-Hoge (1981), King & Burke (1989), Kluge (1991), Murthy (1985), Nicholson (1870). Smith (1931,1935,1961), Stidworthy (1969), Underwood (1867). Wall (1928), Waltner (1975), Whitaker (1978).

9

Immigration

Introduction

Human migration is a world-wide phenomenon. Its nature, types and causes vary through time and over space. The study of migration is of particular interest to geographers because of its inherent spatial character. Migration cannot be considered as mere shift of people from one place of residence to another; it is most fundamental to the understanding of continuously changing space-content and space-relationships of an area (Gosal, 1961). The process of population migration is a complex social phenomenon and the factors associated with it are not easy to generalise. Nevertheless, all types of migration are always engendered by a simultaneous operation of 'push' and 'pull' factors, the former operating in the areas of out-migration and the latter in the areas of reception. Whenever migration takes place in whatever form, it generates threefold impact : (i) on the area experiencing in-migration ; (ii) on the area experiencing out-migration and (iii) on the migrants themselves. Once an immigration-current becomes continuous and its magnitude increases substantially, its pervasive impact on demographic, economic and socio-cultural profiles and political life of the receiving areas becomes apparent and conspicuous.

Immigration into Assam vis-a-vis North East India has received attention of scholars from various disciplines in the recent years, because unlike most parts of the country, the problem of immigration into the region is markedly different. Although, all the north-eastern states do reflect the influences of immigration, the magnitude of the problem is negligible in other states as compared to Assam. The process of immigration into undivided Assam which started during the British times has attained such a magnitude during the Post-Independence period that it has surpassed many of the known instances of inter-regional migration of population in the contemporary period. Massive and unchecked influx of population into Assam from Bangladesh, Nepal and other states of India in recent times has not only adversely influenced the demographic, economic and socio-cultural profiles but also has become one of the main reasons of social tension and ethnic unrest in the whole north-eastern region. During the Post-Independence period illegal immigration into Assam has increased alarmingly and has changed the entire migration scenario threatening the identity of the indigenous people. Here, an attempt has been made to explore the histogenesis of the process of immigration into Assam and to examine its impact on the demographic, economic and socio-cultural characteristics and also on the political life of the people of Assam.

Data on Immigration

The study of a complex human phenomenon like migration requires tedious scanning of data to be derived from various sources. The most important problem faced by the researchers is the scarcity of reliable data pertaining to immigration into Assam. Scholars have to work with data of varying level and reliability. In India, the censuses continue to be the most important single source of data for migration analysis. Until 1951 Indian census provided only birth place data. In 1961 census the birth place data were classified as (i) within the district of enumeration; (ii) outside the district, but within the state of enumeration; (iii) outside the state of enumeration, but within India; and (iv) outside migration. The 1961 census also provided data on migration by 'duration of residence' at the place of enumeration, nature of work and age groups. The 1971 census further provided data on marital status of the migrants on the basis of place of 'last residence' and 'duration of residence' at the place of enumeration. In spite of such modifications and improvement of the quality of data, census coverage on immigration into Assam, particularly from outside the country, appears to be inadequate and the data furnished by Indian censuses and other official sources are often not verifiable. Non-availability of 1981 census data (as no census was held in Assam in 1981) has also added to the problem of relevant information on immigration. The 1991 census data on migration are not yet available. Because of such weakness of the data, intellectuals are still debating on the veracity and authenticity of the statistics regarding the actual number of foreign immigrants in Assam provided by various official sources. Moreover, another serious limitation of census data lies in the frequent changes of the administrative boundaries both at state and district levels. Thus, the study of the problems of immigration into Assam holds difficulty because of inadequate and unreliable data. Keeping such limitations in view, data for the study have been taken from Census of India and other relevant statistical reports.

Immigration Streams and Volume

Assam has witnessed a long history of immigration since the beginning of the 19th century. Historically, the whole process of immigration into Assam can broadly be divided into two phases : Pre-Independence Immigration and Post-Independence Immigration. However, the nature, volume and causes of migration during the Pre-Independence period were different from the Post-Independence period.

Pre-Independence Immigration

Historical evidences indicate that a stream of feeble migration into Assam occurred in the later part of Ahom rule. However, the continuous process of immigration into erstwhile Assam started since the Province of Assam was annexed by the British in 1826. The process of immigration, which started initially in a small way just after annexation, attained vigorous momentum as soon as the British consolidated power in the entire region. There were several streams of immigration into Assam during the British rule. Their origin, magnitude and causes, however, differ from one another.

Immediately after annexation of Assam, the British started tea cultivation as early as 1835 near Sadiya. When the tea cultivation proved successful, several England-based Companies took up large-scale tea plantation in Upper Assam after 1835. As labour force was not available locally and the local people of Assam were unwilling to work in the tea gardens, the planters started importing indentured labourers from Chotanagpur region belonging to such tribes like Munda, Santhal, Ho, Oraon and Gond. According to 1901 census, the total number of labourers in the state stood at 6.54 Lakh. With the expansion of tea gardens, the requirements of labourers also went up. Consequently, their immigration was maximum during 1911-31. As per 1931 census there were about 14 Lakh persons falling into the "Coolie" class in Assam. The import of labourers for tea gardens in Assam which continued up to 1931 was practically stopped in 1941. As these labourers did not go back to their original localities, they settled down in the tea gardens and their adjoining vacant lands mainly in old Dibrugarh, Lakhimpur, Sibsagar, Darrang and Cachar districts. As per 1971 census, the total number of tea-tribes population in Assam stood at 22.5 Lakh (15.32 per cent of Assam's total population) and they form a separate class of population giving a new dimension to the broad population base of the state.

The discovery of coal and petroleum in Upper Assam in the later part of 19th century brought a new economic order and a commercial opportunity to Assam. The colonial administrators needed a skilled and educated labour force to man the various departments and to construct roads and railways. They found the local source unable to provide such labourforce and inducted professionals for the service sectors, technicians for mining industry, tradesmen, police and security personels from outside the province. Thus a new trend of induced immigration set in with the coming of various categories of people mainly from Bengal, Bihar, Uttar Pradesh, Rajasthan and also from Nepal.

Tables 9.1 & 9.2 clearly illustrate the alarming trend of immigration of Bengal-born population into Assam during 1911-31.

Table 9.1
Growth of Bengal-born population in Assam, 1911-31

Year	Bengal-born (Total)		Mymensing-born	
	Assam	Assam Valley	Assam	Enumerated as P.C. of Assam Valley
1911	194,000	120,000	37,000	30.8
1921	376,000	301,000	172,000	57.1
1931	576,000	496,000	311,000	62.7

Source : Census of India, 1931, Vol. III, Pt. I.

Table 9.2
Immigration of Bengal-born population into Assam Valley, 1911-31 (in thousand)

District	1911	1921	1931
Goalpara	77 (34)	151 (78)	170 (80)
Kamrup	4 (1)	44 (30)	134 (80)
Darrang	7 (1)	20 (12)	41 (30)
Nagaon	4 (1)	58 (52)	120 (108)
Sibsagar	14	14	12
Lakhimpur	14	14	19 (2)

(Figures within the parentheses indicate persons born in Mymensingh)
Source : Census of India, 1931, Vol.III, Pt.I.

Contemporary to these immigrations, a stream of largescale influx of migrants into Assam started from erstwhile East Bengal (now Bangladesh). When Assam was put under the Bengal Presidency and was brought under a chief Commissioner in 1874, the most populous Sylhet district of East Bengal was tagged with Assam. At that time Assam was a vast virgin land with sparse population. Sir Henry Cotton, the then chief Commissioner of Assam felt that the virgin and fertile cultivable lands be cultivated to step up foodgrain production. Since the turn of the century the process of immigration of land hungry peasants into the Brahmaputra Valley from Sylhet, Mymensingh and Rongpur districts of East Bengal got accelerated. Initially the immigration of peasants from those areas was encouraged by the Jamindars of Lower Assam with the connivance of the British rulers. It virtually opened the floodgates of Muslim immigration into Assam primarily for using the vacant and unexplored lands in the state. The Census Report of 1911 for the first time recorded the immigration of farm labourers from Mymensing, Sylhet and Rongpur districts to Assam who occupied mainly the *Char* lands of Goalpara district. As a result the population of Goalpara district jumped up by 30 per cent in 1901 - 11 as against only 2 per cent in the previous decade. Since then this immigration continued unabated and the immigrants began to spread over the other districts of lower and central Assam, viz. Kamrup, Darrang and Nagaon districts. In 1921, one-fifth of Goalpara district's population and one-seventh of Nagaon district's population were recorded as Bengal-born and a large majority of them were from the district of Mymensing.

The inflow of these immigrants received an increased vigour under political patronage of the Assam Government which was dominated by the Muslim League leaders from 1926-44 with a brief interruption in 1939 gave all encouragements to these new settlers from East Bengal (Davis, 1951). For the settlement of these immigrants the Government appointed Colonisation Officers for the Nagaon district in 1928-29 and for the Barpeta, Mangaldoi and North Lakhimpur Sub-divisions in 1929-30. The probable impact of such immigration was felt seriously by the then Superintendent of Census Mr. C.S. Mullan. In the Census Report of 1931 he made profound remarks regarding the future implications of such unabated immigration which read as "probably the most important event in the province during last twenty five years - an event, moreover, which seems likely to alter permanently the whole future of Assam and to destroy more surely than did the Burmese invaders of 1820 the whole structure of Assamese culture and civilisation - has been the invasion of vast horde of land hungry Bengali immigrants, mostly Muslims, from the districts of East Bengal and in particular from Mymensingh. Wherever there is wasteland thither flock the Mymensighias. In fact, the way in which they have seized upon the vacant area in the Assam Valley seems almost uncanny. Without fuss, without tumult, without undue trouble to district revenue staff, a population which must amount to over half a million have transplanted itself from Bengal to the Assam Valley during the last twenty five years. It looks like a marvel of administrative organisation on the part of the Government but it is nothing of the sort, the only thing I can compare it to is the mass movement of a large body of ants". These hard-working Muslim immigrants, after settling down mainly in the flood plains of the Brahmaputra Valley and cultivable waste lands brought a tremendous change to the agricultural economy of the state by introducing cultivation of jute, vegetables and Ahu rice on a commercial scale. As they become permanent settlers in Assam Valley, with the passage of time, many of the Muslim immigrants have accepted Assamese as mother tongue. However, at one time these immigrants had proved to be of great concern to the indigenous people in certain areas. The inevitable consequence of occupation of charlands, grazing lands and wastelands in the areas settled by the locals, was frequent friction between the immigrants and the local people. The Assam Government could foresee the problem and introduced a "line system" as early as 1920 to impose restrictions on settlement of immigrants in certain areas predominantly inhabited by backward plain tribes. The main objective of the Line System was to isolate the immigrants from the areas settled by the locals and to give land settlement to the immigrants in segregated areas beyond which they were not allowed to settle. But, because of lukewarm response of the Assam Cabinet headed by Sir Sadulla towards the 'Line System', encroachment of reserved forest areas and wastelands in the tribal belt by the immigrants went on unabated. Even during those days the people of Assam were very much apprehensive of a clandestine move in favour of the immigrant muslims for strengthening their hold further in the state.

Besides these immigrants, there was another contemporary stream of migration into erstwhile Assam from Nepal during 1911-31. The majority of the Nepali immigrants were graziers, ex-servicemen, porters and agricultural labourers. Their total number stood at 70,344 in 1921 which increased to 83,306 in 1931 (Table 9.3).

Table 9.3
Nepal-born population in Assam, 1911-31

Year	Persons	% total population
1901	21,347	0.35
1911	47,656	0.67
1921	70,344	0.94
1931	88,300	1.02

Source : Census of India, 1931, Vol. III, Pt. I

The majority of the Nepali graziers occupied the *charlands* and the foothill areas, while many of them settled down in the outskirts of the urban areas. Although the flow of Nepali immigration had almost ceased after 1931, the process again continued after Independence. Most of the Nepali immigrants depended on livestock farming. Many of them also took up agriculture together with cattle rearing as their main occupation. They settled down in all the districts of Assam with high concentration in Darrang, Lakhimpur and Khasi Hill districts.

It is an accepted fact that Assam experienced huge volume of net migration in 1911-21 and 1921-31 as revealed by census reports. Mr. Mc Swiney, the Superintendent of Census, 1911 highlighted the large scale influx into Assam in the census report. According to him there were 8,82,068 immigrants in 1911 constituting 12.5 per cent of Assam's total population. According to Mr. Lloyd, the then Superintendent of Census of Assam, the total number of foreign-born population in Assam stood at 12,90,157 in 1921 constituting 16.1 per cent of the total population. According to 1931 census, there were 14,08,763 persons (15.2 percent) who were born outside the province. But facts and records reveal that during the Second World War also immigration of Muslim farmers continued from East Bengal under the "Grow more Food Programme" initiated by Sir Sadulla. Late Gopinath Bordoloi after forming the Congress Ministry in early part of 1946 initiated eviction of the unauthorised settlers from the reserved land. But owing to the country's tragic partition in 1947 the process was halted and the entire immigration scenario in Assam took a new turn.

Post-Independence Immigration

The country was partitioned into India and Pakistan at the time of Independence. In accordance with the Sylhet referendum, the Muslim majority Sylhet district was carved out from the province of Assam to be included in East Pakistan, while four Hindu majority police stations - Patharkandi, Ratabari, Badarpur and some portion of Karimganj remained in India. Thus, Assam was delimited by a newly drawn boundary at the time of Independence and hence a new phase of mass migration started. Immigration into Assam during Post -Independence period assumed an alarming magnitude under the changed political scenerio in the subcontinent. As an after effect of the country's tragic partition there were communal disturbances in East Pakistan and Assam resulting an unofficial transfer of population. Millions of Hindus finding their lives and properties insecured in the newly formed East Pakistan started migrating to India. Assam and Tripura being adjacent to East Pakistan on the eastern front had to receive a heavy exodus of refugees estimated to be seven to eight lakhs. On the other hand, a considerable number of Muslims had left Assam due to the same reason mainly from Goalpara district. But soon after the Nehru-Liaquat Pact signed on 8th April, 1950, majority of those Muslim immigrants who left during the disturbance came back to Assam. The process of immigration which started as 'forced migration' at the time of partition continued uninterrupted for long.

The "high tide" of immigration during the Post-Independence period was from 1950 to early seventies with the continuous inflows of Hindu refugees followed by Muslim peasant immigrants from East Pakistan (now Bangladesh). The 1951 Census gives the total number of refugees in Assam as 2,74,445, of which 2,72,075 came from East Pakistan. The Cachar district alone, due to its proximity to sylhet district, received 93,177 persons, followed by Goalpara with 44,967, Kamrup 42,871, Darrang 18,883, Lakhimpur 13,965, Sibsagar 7,514, Khasi Hills 5,990 and Garo Hills 5,072. Since then, the number of refugees increased phenomenally. During the period of 1951-58, the number of Hindu refugees who entered Assam stood at 4,87,000 which rose to 6,28,249 according to the Census of 1961. Consequent upon the perennial problem of communal disturbance in East Pakistan, a fresh exodus of Hindu refugees started in 1964. The total number of Hindu refugees who came to Assam since partition of the country rose to 10,68,455 in 1968 (Table 9.4 and 9.5).

The stream of Hindu refugee migration from East Pakistan since partition did not stop altogether. When the pre-partition immigration of the Muslim peasants was induced by 'pull factor', 'push factor' was mainly responsible for the mass migration of the Hindu refugees and other minority communities from East Pakistan during Post-Independence period. Since they become a 'partition liability' for the country, they were rehabilitated in different parts of erstwhile Assam and Tripura. The Hindu refugees who settled down in different parts of the state formed a different occupational category. Most of the refugees belonging to the middle-class Bengali community were urban-minded and quite a large number of them were educated. They preferred to settle down in towns and sub-urban areas. When the educated section was absorbed in different categories of jobs in government offices, banks, railways, tea industries, private companies, the other section took up small trade and business in the towns of Assam. Today, they have become a "small trading community" in the state. Another section of the Hindu refugees belonging to the

Namasudra community of scheduled caste took up cultivation as their occupation and settled down in the countryside. Over the years there has been a substantial increase in number of Hindu refugees in Assam and other North Eastern States.

Table 9.4
Refugee Population in Assam, 1951

District	Refugee Population
Cachar	93,177
Goalpara	44,967
Kamrup	42,871
Nowgong	42,066
Darrang	18,833
Lakhimpur	13,965
Sibsagar	7,514
Khasi Hills	5,990
Garo Hills	5,072
Total	**2,74,445**

Source : Census of India, 1951, Vol. XII, IA

Table 9.5
Refugee Population in Assam, 1961

District	Refugee Population
Kamrup	98,399
Lakhimpur	97,941
Cachar	82,724
Goalpara	82,085
Darrang	82,372
Nowgong	68,287
Sibsagar	57,195
Mikir Hills	18,554
Khasi Hills	15,524
Mizo Hills	14,607
Garo Hills	10,157
Total	**6,28,242**

Source : Census of India, 1961, Vol. III, Pt. IIC, p. 218.

The flow of immigrants from erstwhile East Pakistan assumed alarming proportions during 1950s. According to Mr. R.V.

Vaghaiwalla, the then Superintendent of Census, as many as 13,44,003 immigrants were enumerated in Assam in 1951. In 1961, the number of immigrants enumerated in Assam was 13,53,581. According to Pakyntein, the Superintendent of 1961 Census Operation, 7,74,869 persons enumerated in Assam returned their birth place as Pakistan. He had also expressed doubts about the correctness of the birth place as declared by the migrants. The abnormally high growth rate of Muslim population in Assam during 1951-61 (39% as against 21% in East Pakistan) substantiates the fact that Muslim immigration into Assam continued unabated even after the creation of East Pakistan. As the pre-partition migrants settled down in the districts bordering East Pakistan, the new migrants who came as infiltrators could easily get into these districts crossing the unchecked border and settled down among the old settlers unnoticed. According to Union Home Ministry report (The Assam Tribune, 13 June, 1993) a total number of 11.14 lakh migrants from erstwhile East Pakistan entered into India between 1964-71. With Assam having 300 km open border, one can easily conclude that a major chunk of these infiltrators sneaked into Assam. The magnitude of immigration during 1961-71 was of the order of 15,03,543 persons. Again, according to 1971 Census, 9,03, 429 persons enumerated in Assam returned their place of birth as Pakistan (Table 9.6). Apparently many of these migrants belonged to minority communities including Hindus, Chakmas, Hajongs, Garos, etc.

Table 9.6
Population of Assam by place of birth in 1971

Place of birth	Persons	P.C. of total population
1. Born in the place of enumeration	99,89,676	66.80
2. Born elsewhere in the district of enumeration	28,97,471	19.37
3. Born in other district of Assam	5,47,934	3.66
4. Born in other states of India	5,35,434	3.58
5. Born in Pakistan	9,03,429	6.04
6. Born in other countries beyond India and		

unclassified	82,598	0.55
Total Population	**149,56,542**	**100.00**

Source : Census of India, 1971, Assam, Migration Table, Pt. II D.

Besides the Bengali Muslims and Hindu refugees, there was another simultaneous wave of refugee migration from East Pakistan belonging to minority tribal communities. An appreciable number of tribal refugees, mainly the chakmas and Hajongs, had to abandon their hearth and homes for fear of persecution on the ground of religion and fled from the Chittagong Hill Tract of erstwhile East Pakistan during 1964-71. Although it is not possible to obtain exact number of tribals fled from East Pakistan, thousands of Chakma and Hajong refugees have been rehabilitated in Assam (then Assam includes Mizoram, Meghalaya and Arunachal Pradesh) and Tripura. Initially, it may be mentioned here that about 10,000 Chakma and Hajong refugees came to India in 1964 and they were rehabilitated in erstwhile NEFA, now Arunachal Pradesh. Their number has now increased to about 65,000 and the people of the state have been agitating for their eviction. On the other hand, the Supreme Court of India has directed the State Government to provide adequate protection to the refugees. Thus, the Chakma refugee issue has led to a political cold war between the Centre and the state Government. As a matter of fact, the tribal refugee-flow from Bangladesh to North Eastern States is continuing till 1990s, and their total number in Assam and North Eastern States together will be few lakhs.

Table 9.7
Population composition of Assam according to major linguistic groups, 1951-1971

Language	1951	1961	1971	Growth rate (%) (1951-61)	Growth rate (%) (1961-71)
Assamese	4,946,173	6,757,658	8,904,917	36.62	31.77
Bengali	1,685,472	2,008,809	2,882,039	19.18	43.47
Hindi	325,311	513,979	792,481	58.00	54.18
Nepali	96,028	180,883	349,116	88.36	93.00

Source : Bhuyan, J.C. : "Asomor Jonogathoni, Prabrajan, Asom Andolon aru Samprodayikota", Deuka, Vol. X, 1993.

Although the overall trend of immigration from Nepal was some-what downward during 1940s, the process did not cease altogether. Census reports gave the number of Nepali-speaking population in Assam (including Meghalaya, Mizoram and Nagaland) as 1.25 lakh in 1951 and 2.15 lakh in 1961. According to J.C. Bhuyan, the Joint Director of Census of Assam, 1991, the Nepali speaking population in Assam stood at 96,028 persons in 1951 which rose to 180,883 persons in 1961 and 349,116 persons in 1971 (Table 9.7). Thus, the Nepali population in Assam registered an abnormally high growth rates, 88.36 per cent during 1951-61 and 93.00 per cent during 1961-71. Such an abnormal rise in population cannot be a function of natural growth. Therefore, one can very easily conclude that immigration from Nepal into Assam also assumed an alarming rate during Post-Independence period. There was little or no official resistance to such high magnitude of migration from Nepal. There is no doubt that their number has registered a sharp rise during 1971-91 period also. As the Nepali migrants are expert in grazing and farming activities, they have settled down throughout the state mixing with the old settlers. Many of the Nepalis have adopted Assamese and noted it as their spoken language in the census enumeration.

The pattern of illegal immigration from Bangladesh to Assam has changed during 1970s, especially after the Bangladesh liberation war in 1971. During the Bangladesh war of 1971 there was a heavy exodus of both Hindu and Muslim migrants from Bangladesh to India, particularly to Assam, Tripura and West Bengal as refugees. The creation of Bangladesh in 1971 forced over 10 million people to take refuge in India (Mukherji,1974). Though after liberation of Bangladesh many of these refugees had returned, yet it is safe to conclude that a large number of both Hindu and Muslim refugees did not go back to Bangladesh. Rather they had been absorbed in the host society already settled in Assam. As there was no effective resistance from any corner, frequent crossing of the international border by Bangladesh infiltrators and their onward march to Assam has been continuing unabated since 1971. Some vital information and statistics may be quoted here to substantiate the explosion in Assam arising out of heavy exodus of illegal immigration from Bangladesh after 1971. The report of the General Secretary of the North Eastern Congress-I Co-Ordination Committee placed in its Annual General Conference held in June, 1992 revealed that according

Table 9.8
Growth of Hindu and Muslim population in Assam, 1951-2011

Religion	1951 Population	1971 Population	Growth rate (1951-71)	1991 Population	2011 Population	Growth rate (1971-91)	Growth Rate (1991-2011)
Hindu	58,86,063 (66.65)	10,604,698 (72.51)	80.16%	15,047,293 (67.13)	19,180,759 (61.47)	42.89%	27.47%
Muslim	1,981,859 (24.68)	3,592,124 (24.56)	81.25%	6,373,204 (28.43)	10,679,345 (34.22)	77.42%	67.57%

(Figures in the parentheses are percentages to total population)
Source : Census of India, 1951, 1971, 1991 and 2011.

to Bangladesh Census records there was a reduction of 39 lakh minority population during 1971-81 and 36 lakh minority population between 1981-89 (The Assam Tribune, 6 July, 1992). In other words, according to the Bangladesh Census, 75 lakh population outmigrated from Bangladesh apparently to India during 1971-89, i.e. after the Bangladesh Liberation War. Needless to say that quite a large chunk of these illegal immigrants had sneaked into Assam and other north eastern states bordering Bangladesh. Increasing population pressure on limited land-base and widespread poverty in Bangladesh on the one hand and availability of agricultural land and farm opportunities in Assam on the other, have encouraged the landless Muslim infiltrators to cross over the border illegally and silently . The religion, language, economic and social setup of the district located on both sides of the international border being similar, they could easily mingled with their kith and kin already settled in large areas. Although it is not possible to obtain estimates of illegal migrants as they did not enter through official check gates, the flow was undoubtedly very heavy as because lesser is the obstacle in the way of migration greater is the volume of migrants. As no census was held in Assam in 1981 and migration data of 1991 Census have not been released till now, the actual volume of illegal immigration from Bangladesh since 1971 is not yet correctly known. Another difficulty in estimating the illegal migrants lies in the statistics provided by the census report. Although Indian census reports provide statistics of "foreign - born persons" enumerated in India, the authenticity of such "birth place" data is highly doubtful as the illegal migrants from Bangladesh may tend to hide their actual identity for fear of deportation. However, the census data on religious communities and their growth rates in Assam during 1971-91 released by 1991 Census indirectly throw light on the trend of immigration into Assam (Table 9.8).

The growth rates of Hindu and Muslim population have been taken into consideration because overwhelming majority of the immigrants belonged to these two communities. The rate of growth of population among these major religious communities in Assam since 1951 shows that both Hindus and Muslims had registered equally high growth rates during 1951-71 signifying mass migration from erstwhile East Pakistan. But during the last 20 years (1971-91) the growth rate of Muslim population has remained as high as 77.42 per cent,while the growth rate of Hindu population has dropped to 41.89 per cent. Such a high percentage increase of Muslim population cannot be attributed to the natural increase of indigenous Muslim population alone. In the absence of recent detailed migration data it is difficult to make an estimate of the volume of illegal migrants, but the consistently very high growth rate of the Muslim component transpires a clear indication that unabated infiltration from Bangladesh is still going on.

In addition to the streams of immigration mentioned in the foregoing discussion, there has been different streams of migrants coming into Assam from other parts of the country at different times. The feeble stream of Indo-Aryan migration from Bengal , Bihar, Uttar Pradesh and Rajasthan which started after the annexation of Assam by British became large during Post Independence period. The developmental activities which started in the state immediately after Independence generate a stream of migrants comprising various categories of people from other parts of India mainly due to jobs and trade opportunities. Census of India 1971 enumerated a total population of 5,35,434 persons in Assam as 'born in other states of India'. Though figures for languages and religions are not reliable, yet one can get an estimate about the volume of immigration into Assam from other parts of the country when the distribution of population of Assam according to language spoken is taken into consideration. If we take the people speaking Hindi, Oriya, Nepali, Kannada, Malayalam, Tamil, Telugu, etc. and the 'unclassified group' of people together, they total 628,000 in 1951, 13,10,682 in 1961 and 28,13,047 in 1971 accounting for 7.95 per cent, 11.05 per cent and 19.23 per cent of the total population respectively. These migrants have settled in various parts of the state with their large concentration in the towns. For example, the Marwaris who came to Assam in the early part of the present century have settled in large number throughout the state. Today, this community controls almost the whole of the main business and the wholesale trade of Assam and other North Eastern States. The trend of immigration from other parts of the country thus set in, has been continuing unabated till recently. Recent industrial development, increasing government and commercial activities, modest degree of urbanisation in the state attracted various categories of people to come and settle down permanently. A large segment of the

Table 9.9

Decadal growth of population in Assam and India (in percentage), 1921-2011

	1921-31	1931-41	1941-51	1951-61	1961-71	1971-81	1971-91	1991-2001	2001-2011
Assam	19.91	20.40	19.93	34.98	34.95	-	53.25	18.92	17.07
India	11.0	14.22	13.31	21.51	24.80	24.66	54.72	21.56	17.64

Source : 1991 Census: Demographic Diversity of India, New Delhi and Census of India, 2011.

north Indian migrants work as daily labourers, rickshaw and thela pullers, barbers , coblers, washermen, small traders and transporters. Other migrants from south Indian states and north India came as job-seekers and businessmen.Their number has increased considerably in recent years. Thus, all the categories of people who come under the category of migrants from other parts of country constitute a sizeable proportion and add a new dimension to the population make up of the state.

The above are the main waves and streams of immigration into Assam from various origins since the beginning of the present century. As already stated, different streams of immigrants both from within and outside the country are still coming causing a great stress and strain to the state. The century-long immigration has added a new but sensitive dimension to the population base of the state. The pervasive impact of unabated immigration on demographic, economic, socio-cultural patterns and the state politics is, at present, quite apparent.

Consequences : Among the demographic consequences, the trend of population growth in the state during the post-Independence period has become the most significant and a sensitive issue. During the last 90 years (1901-1991) Assam's population has increased from 3.3 million to 22.4 million, an increase by about 7 times compared to 3.5 times increase for the country as a whole during the same period. The total population which was 3.3 million in 1901 had steadily rose to 8.02 million in 1951 and then increased rapidly to 22.4 million in 1991. Thus, the year 1951 could be considered as the 'year of demographic divide' in the rapidly changing demographic profile of Assam. Immigration has always been a major contributing factor to the exceptionally high growth of population in the state. Since the early decades of the present century the state's population has consistently grown at a higher rate than that of the national average, but the demographic picture got an abrupt turn since 1951.

Although such a change is also noticeable in the national demographic scenario, the abnormal growth rates experienced by Assam cannot be a function of natural increase alone. It can safely be concluded that large scale immigration since the beginning of the present century has been responsible for high population growth in the state. It is worth mentioning here that in 14 districts of Assam the growth rate is between 50-105 per cent during 1971-91. The most important ones are Dhema ji (104.48), North Cachar Hills (96.39), Kokrajhar (75.94), Karbi Anglong (72.79), Kamrup (64.69), Bongaigaon (64.43), Sonitpur (56.94), Lakhimpur (55.91), Dhubri (55.77), Darrang (54.17), Goalpara (52.66), Nowgong (51.17) and Morigaon (51.07). The high population growth in the state either due to natural increase or in-migration has also its impact on other demographic variables such as age and sex, dependency ratios, internal mobility, etc. The age structure has been changing with ever increasing percentage of young population below 15 years of age. It increased from 43 per cent in 1951 to 46.9 per cent in 1971 resulting in high dependency ratio (93.8 per cent). Sex ratio in Assam has shown a continuous decline from 910 in 1921 to 897 in 1971indicating preponderance of male selective migration to the state. Apart from this, internal mobility of the people from rural to urban areas has been increasing day by day as the limited land base in the rural areas cannot absorb more population. The immigrants finding no land for cultivation are getting into the town in search of livelihood. This has resulted in the growth of slums and other associated problems in the towns.

Among the economic consequences, the most important one is related to the capacity of the prevailing economy of the state to absorb huge immigrant population. The simple man-land ratio indicates that the notion of land plenty in Assam as conceived by the immigrants is now a myth only and the ground realities are quite staggering. The arithmetic density of population in the state at present (2011 census) is 398 persons/sq. km as against the country's 368. The physiological density (700p/sq. km) and agricultural density (172p/sq. km) of the state are much higher than those (physiological density : 370, agricultural density : 63) for the country on the average. The population density of Assam which was only 42 persons/sq. km in 1921 increased to 102 in 1951 and then to 286 in 1991. "It was estimated by Mr. Mullan that the plains of Assam could support 500 persons and the Hills only 60 persons per square mile. The optimum population of the state as a whole would be about 12 million. But in 1961 her total population was 11.87 million and the annual growth rate has been placed at 2.3 per cent" (Das, 1970). According to this estimate it is now difficult to say that population of Assam is below the optimum, rather the state is now carrying an additional load of 10 million population. The intensity of population pressure on the agricultural economy of the state can be better understood from the fact that average land holding size in the state has declined from 1.46 hectares in 1970-71 to 1.36 hectares in 1980-81. Assam's per capita land is 0.35 hectare (1991) as against the country's average of 0.39 hectare. Since the reversal of the emerging demographic pattern is not possible now, intensity of population pressure will increase year by year. Whatever economic development is achieved so far, particularly in the agricultural sector, has been neutralized by the abnormal population growth in the state.

The social impact of immigrants on the host population is, of course, determined partly by how different the immigrants are from the population of the receiving area. As the immigrants in Assam are heterogeneous and diverse in caste, religion, language and culture, there may be profound socio-cultural implications. The different migrant groups have brought significant change in the socio-cultural profile of the state, especially in respect of ethnic diversity. However, the impact of the immigrants from Bangladesh are different from those from other parts of the country. Language and religion are two important manifestations of culture, the difference of which could be the source of social tensions. When exodus takes place in large waves and the situation does not permit spontaneous assimilation with the host population, conflict may begin. The inflow of unabated immigrants into Assam has created almost such a situation in the present day Assam. Although the earlier migrants have already assimilated with the local culture, new migrants pose slightly different situation. Assam had already faced a serious language problem in 1960s and the under-current of language conflict still prevails in the immigrant dominated areas. Because of the heavy concentrations of immigrants, ethno-religious composition in some districts of Assam like Dhubri, Barpeta, Goalpara, Nagaon, Hailakandi, etc. has also undergone drastic change.

The most pervasive impact of large scale immigration into Assam is the dominant roles played by the immigrant groups in the state's political millieu in recent times. It is obvious because when influx of migrants adds substantially to the size of a particular group, the concerned group may tend to aspire for political power. It is worth mentioning here that despite the dark implications, there were no genuine efforts on the part of the successive State and Central Governments to check the flow of illegal immigration into the state. On the other hand, immigrants received political patronage from a section of the political leaders and parties just to consolidate their political power taking advantage of the immigrants vote banks. All political parties were fully aware of the far reaching consequences, even then the Government in power practically kept open the flood gate of the illegal immigrants for decades. Thus, Assam became a land of 'EL dorado' for the immigrants, particularly from Bangladesh. It is also true that the indigenous people of the state have seldom welcomed the immigrants with completely open hearts. The Pre-Independence immigrants were seen useful and non-threatening and they were accepted as the population pressure on the state's land resources was minimum at that time. But the Post-Independence migrants are considered by many as a relatively large burden and threat to the local culture and polity. All political parties were aware of the explosive situation, but they showed least concern for the solution of the vexed problem. Ultimately, the All Assam Students Union (AASU) took up the issue and spearheaded the six years long mass movement in 1979 to pressurise the Central Government. In the background of anti-foreigners mass movement the controversial IMDT (Illegal Migration Determination Tribunal) Act, 1983 was adopted by the Parliament. The objective of the Act was to detect and deport foreign infiltrators who entered India after 25th March, 1971 as per the provisions of the India-Mujib Pact. But the objective of the Act remained unfulfilled due to lack of political will of the Government. Unfortunately, certain political circles and minority organisations have mixed up the foreigners issue with Indian minorities living in Assam. It must be remembered that all minorities are not foreigners. Under such paradoxical situation identification of foreign nationals in Assam has become a difficult and complex task.

The historic Assam Accord was signed in 1985 after protracted negotiations between the Central Government and AASU Leaders to find out a solution of the vexed foreigners issue. Incidentally the anti-foreigners movement gave rise to regional politics in Assam. The Asom Gana Parishad came to power in 1985 but the foreigners issue remained unfulfilled and unsolved because of alleged apathetic attitude of the Central Government. Despite the provisions of the Assam Accord that foreigners who entered Assam on or after March 25, 1971 shall continue to be detected and expelled in accordance with law, very little progress has been made in this direction during the last ten years. Although some portions of the border is fenced and roads have been constructed for patrolling, yet immigration seems to be continuing. The Congress Party headed by Late Hiteswar Saikia when regained power in the state in 1991, gave a new twist to the already complex foreigners issue. The Government, despite being aware of the explosive situation, raised a series of contradictory and misleading statements as regard the actual volume of foreign nationals in Assam. The Chief Minister, Late Hiteswer Saikia himself once declared that there were 30 lakhs foreigners in Assam and later on denied his statement when some political leaders of minority community threatened to pull down his government. The suicidal policy of the Government and also of the Election Commission of India to include the names of large number of people with doubtful nationality in the voter lists of Assam, has now brought a radical change in the electoral scenario of the state. Although not all migrant groups seem to aspire for political clout, some groups are exerting political influence in those areas where they are in majority.

Assam abounds in many small and large ethnic groups and communities. The small nationalities are now panicked for the fear of losing their political right and cultural identity. It is also a fact that unabated immigration is the root of unprecedented social tensions in many areas of Assam and other North Eastern States. There can be no two opinions that the 'demographic invasion' brought about by the 'high tide' of immigration has now completely turned the balance against the indigenous population in respect of demography, language, culture, economy and polity. The inarticulate Assamese nationality, of late, realised the dimensions arising out of such a situation. Because of fear psychosis arising out of the fear of being swamped over in their own land consequent upon the outsider's increasing dominance, a section of the 'sons of the soils' have already brought forth the question of self governance.

The sensitive issue of immigration into Assam and for that

matter North East India as a whole, has been accepted as a national problem. The Government of India must have a determined political will to find out a permanent solution of the problem without further delay. All sections of people irrespective of caste, creed, religion and political affiliation should face it unitedly in the interest of peace, harmony and speedy economic development of the state.

References

Bose, A., 1991 : *Demographic Diversity of India*, B.R. Publishing Corporation, Delhi.

Bhuyan, J.C., 1993 : 'Asomor Jonogathoni, Prabrajan, Asom Andolon aru Samprodayikota', *Deuka*, Vol. X (Dec).

Census of India, 1931, Vol. III Pt. I, Assam.

Census of India, 1941, Vol. IX, Assam.

Census of India, 1951, Vol. XII, Assam, Pt. IA

Census of India, 1961, Vol. III, Assam, Pt. IIC.

Census of India, 1971, Series 3, Pt. I & II.

Census of India, 1991, Religion of India, Paper I of 1995, and Religion of Assam at District Level.

Census of India, 2011 : Primary Census Abstract and Religion Data of Assam.

Davis, Kingsley, 1951 : *The Population of India and Pakistan*, Princeton University, New York.

Das, H.P., 1970 : *Geography of Assam*, National Book Trust, New Delhi.

Dutta, N., 1996 : 'Impact of migration on the demographic landscape of Assam' *The Assam Tribune*, March 12, 1996.

Gosal, G.S., 1961 : 'Internal Migration in India - A Regional Analysis, *Indian Geographical Journal*, Vol. 36, p. 106.

Mukherji, P.N., 1975 : 'The Great Migration of 1971', *Economic and Political Weekly*, 9 : 365 - 369, 9 (10) 399-408.

Sharma, H.N., 1988 : 'Pattern of Present Population and Its Impact in Assam', *North Eastern Geographer*, Vol.20, No. 1 & 2.

Taher, M., 1987 : 'The Population Base of Assam', *North Eastern Geographer*, Vol. 19, Vol. 1 & 2.

10 Population

Introduction

Population forms an important component in the whole process of socio-economic development of a region. It is more so because of its dual role as a producer and consumer. Hence, a proper appraisal of its size, growth, composition and quality is considered prerequisite for an effective planning for balanced and sustainable socio-economic development of a region. Such an attempt is highly essential for a socio-economically backward state like Assam which currently witnesses a number of diverse population problems.

With an area of the country's 2.39 per cent, Assam shelters 2.58 per cent of the country's total population. According to 2011 census, it records a population of 31,205,576 and occupies the thirteenth position among the Indian states. In terms of population size, Assam is larger than even a developed country like Australia or Netherlands, and a developing country like Nepal. The arithmetic density of population in the state is also as high as 398 as against 382 persons per sq. km in the country as a whole. But it is discouraging enough to note that Assam accounts for only 1.94 per cent and 1.87 per cent of the country's foodgrain production and state domestic product (at current price) respectively. On the other side, Assam is characterised by a diverse population composition in respect of ethnicity, religion and language having varied socio-cultural practices. This is why Assam is often referred to as "India in miniature". Thus, the prevailing high pressure of population and persistently high population growth in the midst of population diversity and socio-economic backwardness in the state would certainly have a serious pervasive impact upon its demographic and socio-economic structure (Kar, 1995).

Peopling and Ethnic Composition

Assam has a long history of peopling. It has witnessed several waves and streams of migration of people belonging to various racial and ethnolinguistic groups from different directions since prehistoric time. But very little can be deciphered to trace the succession of peopling process in the state. Again, the process of intermingling of different racial and cultural elements over time has made their identity further complicated.

Broadly speaking, Assam has population of four racial traits - (1) Proto- Australoid, (2) Mongoloid, (3) Aryo-Mongoloid and (4) Aryo-Mongolo- Dravidian. But ethnolinguistically, broadly three major groups have been identified in Assam - (1) Austro-Asiatic, (2) Tibeto- Chinese and (3) Indo-European (Taher, 1993). These groups of people had settled in certain parts of the state in segments in different points of time.

Austro-Asiatic : The peopling process in the state is said to have started with the immigration of the Australoids of Austro- Asiatic speaking people from South-East Asia quite a few millennia before Christ. They are now represented by the Mon-Khmer speaking Khasi tribe. Although most of them have settled in Khasi and Jaintia Hill districts of Meghalaya, a small proportion is also found to live in some parts of N.C. Hills and Karbi Anglong districts of Assam. After a long gap, another wave of migration of this Austro-Asiatic stock took place into the state immediately after the British annexation in 1826. Racially, the Australoids are tribal people speaking Mundari dialects, a branch of the Austro-Asiatic linguistic family. They were brought to Assam by the British from central India to work in the tea plantations as labourer. They are known as Munda, Ho, Savara, Oraon, Gond tribes, etc.

Tibeto-Chinese : The Tibeto-Chinese comprised Tibeto-Burman and Siamese-Chinese groups who entered into Assam at different times beginning from the pre-historic period. The Tibeto-Burman people who are racially Mongoloid settled in different parts of the state having migrated from Myanmar, erstwhile Burma. The Mishings of upper Assam and the Bodos of lower Assam form the early settlers of this group. The late settlers (after 1826) of this group of people consisted of the Kachin and Kuki-Chin of Karbi-Anglong and N.C. Hills districts of Assam. On the other hand, the

Siamese-Chinese group of people who came to this part of the country from Shan state of Myanmar during the medieval period, forms a large segment of Tibeto-Chinese group. The Ahoms settled in upper Assam represent this segment of people. The Khamtis, Khamyangs, Aitons, Phakes, and Turungs inhabiting the upper Assam in small groups represent the later-day Tais. Professing Hindu religion they have contributed a lot towards building the present Assamese society and culture.

Indo-European **:** This is another large group of people consisting of both Hindus and Muslims speaking Indo-European languages. Their migration into this part of the country began as early as first millennium B.C. They were originally from Caucasoid racial descent, and subsequently mixed with the later migrants of the same stock and of other racial stocks. The early settlers of this group are the Hindus who migrated from the northern plain of India and settled in the plains of Assam. The Muslims of the same stock started entering Assam from the early thirteenth century. Both these important segments of population who settled in the plains of Assam speak Assamese and Bengali languages. Further, after the British annexation, a large group of tradesmen, technicians and other professionals came to this part of the country from different states of India. They belong to all the major religious groups of the country. The Nepalis also entered into the state during the same period. The Bengali Muslim peasants inhabiting the plains of Assam represent another important segment of immigrants who came to Assam from erstwhile East-Bengal (now Bangladesh) before independence in 1947. Almost during that time a substantially large number of Nepalis entered into Assam. Besides, at the time of India's partition and after that a large number of Hindu refugees and poverty stricken Muslim peasants infiltrated into the state from the erstwhile East Pakistan. Unfortunately, this immigration trend has not yet been completely stopped.

Population Growth

Demographically speaking, Assam is characterized by a very high rate of population growth which is even higher than the country's average growth throughout the present century. At the beginning of the present century, Assam's population was 3.29 million. It increased to 8.03 million in 1951 and 31.21 million in 2011 (Table

10.1) witnessing an average annual exponential growth rate of 1.80 per cent during 1901-1951 and 2.29 per cent during 1951-2011 as against country's corresponding growth rates of 0.83 per cent and 2.03 per cent. As a consequence, the state's population to the country's total increased from 1.38 per cent in 1901 to 2.58 per cent in 2011 through 2.22 per cent in 1951.

When we look at the data on the birth and death rates of the state, it appears that the trend of population growth in the state, as also in the other north-eastern states, is somewhat unnatural as compared to other states of the country. Because the birth and death rates in Assam do not deviate much from that of the national averages. It means that had there been no migration into the state, the growth rates of population in Assam would have been almost similar with that of the average growth rates in the country. The history of peopling in the state as discussed above also clearly reflects this phenomenon. Thus, it can be said that growth of population in Assam has taken place due to both natural increase and migration. It may, however, be mentioned here that the quantum of migration into the state was quite small during the early decades of the twentieth century. Growth rates of population in the state had also remained low during the pre-independence period. Accordingly, two distinct demographic phases with 1951 as a divide may be identified in the demographic history of the state.

During the first phase (1901-1951) of the demographic history, the state's population increased by 144 per cent as against the country's 51 per cent. The decadal variation of the population growth was from 17 per cent to 20.5 per cent in the state, while it was from a decrease of 0.3 per cent to an increase of 14.2 per cent in the country (Table 10.1). Besides small volume of migration the relatively low and uniform population growth in the state is mainly due to the prevalence of comparatively high birth and death rates during the period. Although immigration into Assam had begun since the British annexation in 1826, it became active from the beginning of the present century. The main streams of migrants settled during this period in the then Assam are the peasants from erstwhile East Bengal, tea garden labourers from Central Asia and the graziers from Nepal. Assam's immense economic potential coupled with cheap and plenty of virgin and

Table 10.1
Trend of growth of population in Assam during 1901-2011

Year	Population	Decadal growth (in per cent) Assam Total	Assam Rural	Assam Urban	India Total	India Rural	India Urban	Annual exponential growth Rate (in per cent) Assam	India	Percentage share of State's population to India's total	Poplation density (Persons/ km²)
1901	3289680	-	-	-	-	-	-	-	-	1.38	42
1911	3848617	16.99	17.91	20.55	5.75	6.40	0.35	1.58	0.56	1.52	49
1921	4636980	20.48	20.08	36.80	-0.31	-1.29	8.27	1.88	-0.03	1.84	59
1931	5560371	19.90	19.70	27.58	11.00	9.98	19.12	1.83	1.04	1.99	71
1941	6694790	20.40	20.16	28.30	14.22	11.81	31.97	1.87	1.33	2.10	85
1951	8028856	19.93	18.46	65.73	13.31	8.79	41.43	1.83	1.25	2.22	102
1961	10837329	34.98	30.87	127.57	21.51	20.49	26.41	3.04	1.96	2.46	138
1971	14625152	34.95	32.62	65.01	24.80	21.86	38.23	3.04	2.20	2.66	186
1981*	18041248	23.36	21.92	38.25	24.66	19.37	46.17	2.12	2.20	2.64	230
1991	22414322	24.24	22.56	39.58	23.56	20.00	36.46	2.19	2.14	2.65	286
2001	26655528	18.92	16.51	38.24	21.50	18.10	31.50	1.75	1.97	2.59	340
2011	31205576	17.07	15.47	27.89	17.70	12.30	31.80	1.59	1.64	2.58	398

*Estimated by Census of India

Source: Census of India, Series 1, India, 1971, 1991 and 2011.

fertile lands facilitated such immigrations. India's partition in 1947 had also been responsible for migration of a large number of Hindu refugees from erstwhile East Pakistan to Assam.

On the other hand, after the independence population in Assam began to grow rapidly witnessing an increase of 179 per cent as against country's 134 per cent during the second phase (1951-1991) of the demographic history. This growth of population in the state contributed to a net increase of 14.38 million population. The decadal growth rates during the period 1951-1991 varied between 23.4 per cent and 35.0 per cent in the state, while between 13.30 per cent and 24.8 per cent respectively in the country as a whole (Table 10.1). Except for the other north-eastern states (Arunachal Pradesh, Manipur, Meghalaya, Mizoram, Nagaland and Tripura) and Sikkim, Assam's population growth rate has been the highest among all other states in the country during 1951-1991. The sharp fall of death rates accompanied by only a very slow decline of birth rates have contributed immensely to such increased rate of population growth in the state. Gradual disappearance of severe epidemics and famines coupled with improvement of health and medical facilities, among others, are responsible for the significant decline in death rates. On the other side, the measures to control birth rate could not make any significant headway. A host of inter-related socio-economic factors such as universality of marriage, marriage at lower age, religious belief, illiteracy, poverty, high infant mortality, etc might be responsible for keeping the birth rate high here as elsewhere in the country. Subsequently, however, the gradual expansion of family planning programme accompanied by progress in education contributed to some extent towards bringing down the birth rate in many parts of the country (Gosal, 1993). This is also reflected in the steady fall of crude birth rates in Assam from 38.5 to 27.7 and crude death rates from 17.8 to 9.7 during 1971-1990. It means there had been a marginal decline in the natural growth rate of population from 22.7 to 17.8 during the period. One of the important reasons of decrease in the death as well as birth rates in the state is the decline in infant mortality rates (IMR). As for instance, it witnessed a fall of IMR from 140 in 1971 to 77 in 1990, as against country's 129 and 80 (per thousand live births) respectively. Additionally, rise in life expectancy has also been an important factor of high natural growth of population in the state. However, there exists a significant spatial variation of birth and death rates within the state due to a number of socio-economic factors.

From the above discussion it is clear that migration played a dominant role towards alarmingly high growth of population during 1951-1991 in the state. The decades after independence, in fact, witnessed a new stage of demographic development in the state. The period 1951-1971 experienced an unprecedented volume of immigration when population growth rate in the state shot up to 34.97 per cent and 34.95 per cent respectively during 1951-1961 and 1961-1971 (Table 10.1). The more important streams of migrants during this period consisted of Bengali Hindus and Muslim peasants in large volume from erstwhile East Pakistan. It may be mentioned here that the immigration of Muslims was more spectacular during 1951-1961. The Bengali Hindus from East Pakistan continued to pour into Assam because of the recurrent communal disturbances there. Even after this, immigration from present Bangladesh and Nepal has been continuing unabated till today. On the other hand, immigration of people from other parts of India has also not been so insignificant. After independence, people mainly from Bihar, Uttar Pradesh, Rajasthan, Orissa and Punjab migrated in large number into this state, more particularly to the urban areas for seeking economic opportunities in trade and commerce, industries, construction works and many other occupations. The immigration and inmigration of population having varied socio-economic levels have to a great extent influenced the demographic scenario in the state. For instance, high total fertility rate and child-woman ratio as prevalent among the Muslims, more particularly among the immigrant ones, have been responsible for high rate of population growth among them in the state, as elsewhere in the country. In contrast, it is, however, important to note that as compared to the volume of immigration and inmigration, the size of emigration from this state is very insignificant because of relative absence of mobility of the indigenous people. If the population in the state continues to grow at the present rate it is likely to cross 27.5 million mark by 2001.

It is significant to note that growth rate of population varies spatially within the state depending upon the nature and extent of migration, and a number of socio-cultural factors influencing the natural growth. Before independence, growth rates of population have been considerably high in the districts of middle and lower Brahmaputra valley due to obvious reasons of immigration mainly from East Bengal and Nepal.

The spatial pattern remained more or less same even after the independence. However, the two tribal dominated hill districts, viz. Karbi Anglong and N.C. Hills, had started experiencing high growth rates. During 1951-61 Karbi Anglong recorded the highest growth rate of population in the state (79.21 per cent), closely followed by Dhemaji (68.56 per cent) and Bongaigaon (60.51per cent) as against the state average of 34.97 per cent. But during 1961-71 Dhemaji district had experienced the highest growth rate (103.42 per cent), followed by Karbi Anglong (68.28 per cent) as against the state average of 34.95 per cent (Table 10.2). This is indicative of the fact that the growth of population during 1951-71 in Assam had been quite unprecedented and it was more so in Karbi Anglong and Dhemaji districts.

Satisfactorily, the population growth rate in the state has been checked significantly during the last two decades (1971-1991) with the exceptions of Dhemaji and N.C. Hills districts (Table 10.3). Strikingly, during that period, the population growth in Dhemaji and N.C. Hills districts have been as high as 107.50 per cent and 98.30 per cent respectively as against the state average of 53.26 per cent. From the Table 10.2, it is further evident that throughout the present century population growth rates have been singificantly high in the plain districts of Kokrajhar, Bongaigaon, Darrang, Marigaon and Dhemaji and hill districts of Karbi Anglong and N.C. Hills. On the other hand, Sibsagar, Jorhat and Golaghat, which have witnessed low growth rates of population throughout the century, are the only districts having been least influenced by immigration (Table 10.2). Availability of vacant land in the hilly areas as against a heavy pressure of population on the plains, attracted the migrants to Karbi Anglong and N.C. Hills districts in the recent past, resulting in high growth rates in these districts. Dhemaji district's rapid population growth, particularly after 1951, has been mainly associated with high natural growth rate among the tribal people combined with influx

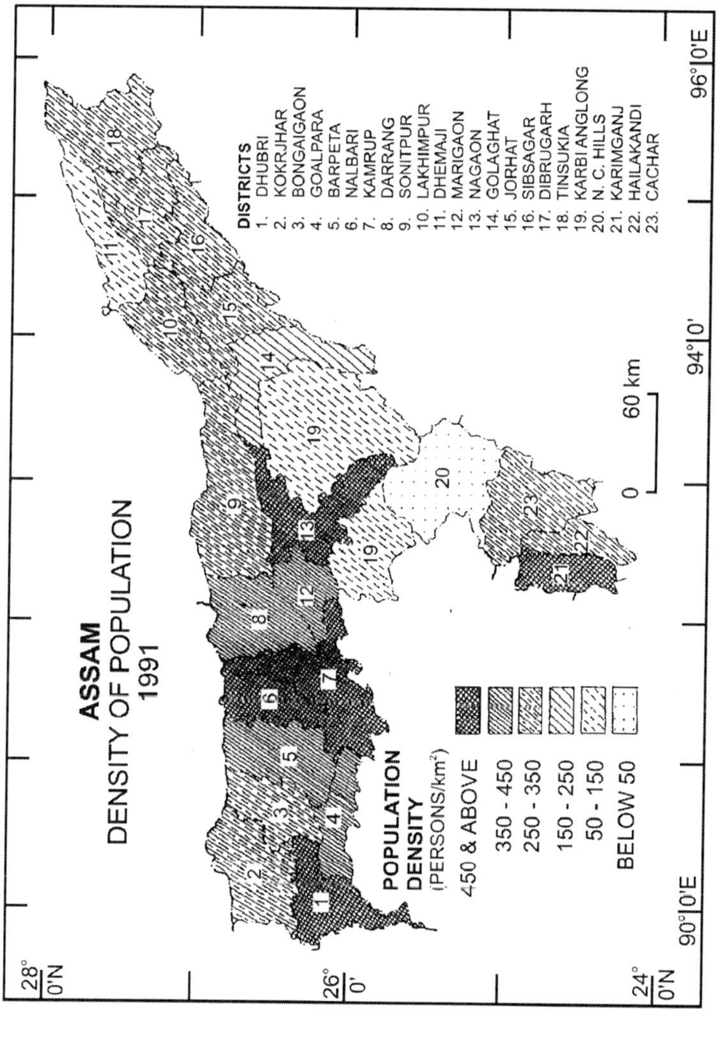
ASSAM
DENSITY OF POPULATION
1991
POPULATION DENSITY (PERSONS/km²)
450 & ABOVE
350 - 450
250 - 350
150 - 250
50 - 150
BELOW 50
DISTRICTS
1. DHUBRI
2. KOKRJHAR
3. BONGAIGAON
4. GOALPARA
5. BARPETA
6. NALBARI
7. KAMRUP
8. DARRANG
9. SONITPUR
10. LAKHIMPUR
11. DHEMAJI
12. MARIGAON
13. NAGAON
14. GOLAGHAT
15. JORHAT
16. SIBSAGAR
17. DIBRUGARH
18. TINSUKIA
19. KARBI ANGLONG
20. N. C. HILLS
21. KARIMGANJ
22. HAILAKANDI
23. CACHAR
0
60 km
90°0'E
94°0'
96°0'E
28° 0'N
26° 0'
24° 0'N

Fig. 10.1

Table 10.2

District-wise decadal variation in population growth (in per cent) in Assam during 1901-2001

District/State	1901-11	1911-21	1921-31	1931-41	1941-51	1951-61	1961-71	1971-91	1991-2001
1. Dhubri	+33.76	+11.42	+6.05	+10.57	+15.21	+27.62	+40.51	+56.57	+22.88
2. Kokrajhar	+16.70	+90.75	+39.68	+26.59	+0.50	+46.34	+54.30	+76.78	+13.13
3. Bongaigaon	+29.94	+26.94	+15.94	+20.85	+4.00	+60.51	+40.29	+64.64	+12.05
4. Goalpara	+27.41	+37.07	+20.50	+ 7.73	+11.33	+37.22	+45.88	+54.12	+23.03
5. Barpeta	+20.02	+34.04	+69.02	+44.06	+18.77	+32.62	+35.81	+43.02	+18.87
6. Nalbari	+13.33	+14.20	+27.92	+39.38	+9.48	+49.62	+42.02	+49.27	+13.03
7. Kamrup	+11.16	+7.06	+9.32	+14.60	+21.87	+37.73	+38.80	+65.72	+26.11
8. Darrang	-0.25	+11.78	+26.00	+35.30	+24.13	+44.75	+43.24	+55.63	+15.82
9. Sonitpur	+24.33	+40.69	+20.50	+19.73	+24.26	+35.82	+27.62	+57.14	+18.06
10. Lakhimpur	+17.30	+22.82	+25.99	+36.98	+21.62	+50.46	+43.39	+56.29	+18.29
11. Dhemaji	+45.30	+63.30	+16.90	+8.41	+-3.35	+68.56	+103.42	+107.50	+19.44
12. Marigaon	+15.50	+31.99	+41.66	+30.87	+24.32	+37.89	+37.51	+50.90	+21.35
13. Nagaon	+15.84	+31.94	+41.35	+10.34	+41.41	+35.91	+38.99	+51.26	+22.26
14. Golaghat	+16.25	+20.14	+18.29	+1.48	+19.52	+26.04	+30.85	+58.12	+14.27
15. Jorhat	+16.62	+17.54	+8.88	+15.12	+15.03	+24.17	+17.47	+33.10	+14.70
16. Sibsagar	+13.41	+20.46	+14.44	+15.64	+15.98	+23.36	+19.47	+38.76	+15.83
17. Dibrugarh	+27.10	+38.40	+22.34	+10.87	+26.23	+31.35	+22.93	+37.78	+13.68
18. Tinsukia	+27.10	+28.32	+27.00	+34.77	+11.09	+35.92	+31.02	+47.03	+19.51
19. Karbi Anglong	N.A	N.A	N.A	N.A	+30.96	+79.21	+68.28	+74.72	+22.72
20. N.C. Hills	-33.12	+5.92	+13.60	+13.75	+6.16	+36.95	+40.00	+98.30	+24.72

21. Karimganj	+12.94	+3.19	++5.91	+9.52	+29.87	+22.96	+25.13	+42.08	+21.87
22. Hailakandi	+16.09	+7.59	+7.08	+10.29	+1748	+27.23	+23.61	+45.94	+20.89
23. Cachar	+12.33	+5.98	+7.60	+13.08	+23.92	+22.60	+23.96	+47.59	+18.88
ASSAM	**+16.99**	**+20.48**	**+19.91**	**+20.40**	**+19.93**	**+34.98**	**+34.95**	**+53.26**	**+19.16**
INDIA	**+5.75**	**-0.31**	**+11.00**	**+14.22**	**+13.351**	**+21.51**	**+24.80**	**+54.38**	**+21.54**

Source: Aspects of Population Profile of Assam, Census of India, Series 4, Assam, 1991 and 2001.

of immigrants from Bangladesh. The growth rate in the district also appears to be significantly high due to its small population base. Thus, it may be concluded that the growth rate is notably high in recent time in most of the sparsely populated districts of the state.

Population Distribution and Density

Distribution and density of population are not uniform throughout Assam. The variation in physical condition, economic advancement and degree of transport accessibility is largely responsible for population distribution in the state. The built-up plain areas are considerably densely populated and the hill areas sparsely populated. Availability of productive land and better transport and communication facilities combined with ongoing process of urbanisation and industrialisation have been greatly associated with high concentration of population in the Brahmaputra and Barak valleys of the state. With an area of 72 per cent and 9 per cent of the state they respectively constitute 84.63 per cent and 11.62 per cent of the state's population. The narrow *bhabar* and *terai* belts of the Brahmaputra valley are thinly populated. Again, due to better transport and communication facilities and high degree of immigration, the districts of the lower and the central Brahmaputra valley are somewhat densely populated as compared to that of the upper Brahmaputra valley (Table 10.4 and Fig. 10.1). According to 2001 census data, the more populated districts in the state include Kamrup (25.2 lakh), Nagaon (23.1 lakh), Sonitpur (16.8 lakh), Barpeta (16.4 lakh), Dhubri (16.3 lakh), Darrang (15 lakh) and Cachar (14.4 lakh). On the other hand, the least populated district is N.C. Hills (1.9 lakh).

The density of population (person per sq km) is very uneven in the state. It has also been increasing overtime as a result of the increase of population throughout the state. During the 1971 census, the population density in the state was 186 as against the country's average of 177. But it increased to as high as 398 in the state as against 382 in the country in 2011. It is further seen that the density of population in Assam is not only higher than the country's average, it is higher than many other states in the country except only West Bengal (1029), Kerala (859), Bihar (1102), Uttar Pradesh (828), Tamil Nadu (555), Punjab (550), Haryana (573) and Jharkhand (414). So far the variation in density of

Table 10.3

Spatial pattern of population growth (in per cent) in Assam during 1951-1971 and 1971-1991

Period	Population Growth (in per cent)								
	30-40	40-50	50-60	60-70	70-80	80-100	100-120	120-140	180-220	220 &above
1951-1971	-	Jorhat Sibsagar	Karimganj, Hailakandi, Cachar	Golaghat, Dibrugarh	Dhubri, Sonitpur, Tinsukia	Barpeta, Kamrup Marigaon, Nagaon, N.C. Hills (State Average: 82.15)	Goalpara, Nalbari, Darrang, Lakhimpur	Kokrajhar, Bongaig-aon	Karbi Anglong	Dhemaji
1971-1991	Jorhat, Sibsa-gar, Dibrug-arh	Barpeta-Nalbari, Tinsukia, ganj, Hailak-andi, Cachar	Dhubri, Goalpara, Darrang, Karim-Lakhim-pur, Marigaon, Nagaon Golaghat (State Average: 53.26)	Bongaig-aon, Kamrup Sonitpur,	Kokra-jhar, Karbi	N.C. Hills Anglong	Dhemaji	-	-	-

Source : Prepared by author using data from Aspects of Population of Assam, Census of India, Series 4, Assam, 1991.

population in Assam is concerned, it is as high as 524 in the Barak valley and 470 in the Brahmaputra valley, and as low as 76 in the Hill region (Table 10.4). On the other hand, among the districts in the state, Kamrup Metro district in the Lower Brahmaputra valley records the highest population density (1313), while Dima Hasao district the lowest (44).

Depending on the variation in population density, the districts of Assam can broadly be categorised into four density levels. Accordingly, the high density (600 and above) districts include Nagaon, Dhubri, Kamrup Metro, Bongaigaon, Marigaon, Karimganj, Nalbari and Barpeta; medium density (300-600) districts Goalpara, Darrang, Golaghat, Sibsagar, Hailakandi, Lakhimpur, Kamrup, Baksa, Udalguri, Cachar, Dibrugarh, Jorhat, Sonitpur and Tinsukia; low density (100-300) districts Kokrajhar, Chirang and Dhemaji; and very low density (below 100) districts Karbi Anglong and Dima Hasao. Thus, the resultant pattern of population in the state reveals that the density is strikingly high in the districts where the impact of immigration is notably high and the Muslims constitute a significant proportion of total population.

Age-Sex Composition

The age-sex composition of a population indicates the demographic and socio-economic well-being of the region. Like India and other developing countries, the birth rate in Assam has still been quite high. This has resulted in a significantly high proportion of children in the state as elsewhere in many other parts of the country. It is discouraging enough to note that while proportion of children in the developing countries is above 35 per cent, it is even below 25 per cent in the developed countries. Strikingly, in 1971, the children below 15 years of age constituted as high as 47 per cent of the total population in Assam. But the adult population (15-60 years of age) constituted only 48 per cent and the old population (60 and above) 5 per cent of the total population in the state. However, as a consequence of the slow decline of birth rate in the contemporary period, the proportion of children below 15 years of age in the state has come down to 40 per cent in 1991. At the same time, the proportion of adult population in age group 15-60 has, on the other hand, increased to 54 per cent. As a result, the dependency ratio (the proportion

Table 10.4
Percentage distribution of population and population density in different geographic regions in Assam, 2011

Region	Area (in sq. km.)	Population	Percentage share of area	Percentage share of population	Population density (persons per sq. km.)
1. **Brahmaputra Valley**	56,194	264,10,562	71.64	**84.63**	**470**
(i) Lower Brahmaputra Valley	20,148	112,52,365	25.69	36.06	558
(ii) Middle Brahmaputra Valley	14,340	74,65,469	18.28	23.92	521
(iii) Upper Brahmaputra Valley	21,706	76,92,728	27.67	24.65	354
2. **Barak Valley**	6,922	36,24,599	8.82	**11.62**	**524**
3. **Hill Region**	15,322	11,70,415	19.54	**3.75**	**76**
Assam	**78,438**	**312,05,576**	**100.00**	**100.00**	**398**

Source: Statistical Hand Book, Assam, 2015, Directorate of Economics and Statistics, Govt. of Assam.

of children and old age population to adult population) of the state's population has declined from 107 per cent in 1971 to 84 per cent in 1991. It may be mentioned here that both children and old age people are generally considered as dependents. Although, there is no significant difference between male and female in this respect, it varies notably between the rural and urban areas. It is found that the dependency ratio is as low as 59 per cent in the urban areas as against 88 per cent in the rural areas in the state. This is indicative of the prevalence of better economic conditions among the urban people in the state.

Sex composition of a population which is indicated in terms of sex ratio (number of females per thousand males) is important to understand the prevailing demographic and social characteristics of a region. The sex ratio of population in Assam has been always lower than the country's average. According to 2011 census, Assam recorded the sex ratio of 958 as against the country's average of 943. The sex ratio in Assam has witnessed a gradual decline from 919 in 1901 to 896 in 1971 and after that there has been a rise. Such a low and imbalanced sex ratio in the state would undoubtedly be the result of the large scale male selective migration and the interplay of a variety of social practices including gender discrimination. It may be noted here that such male selective migration from other states hardly brings any positive impact upon the economic progress of the state. The influence of male selective migration is, however, more prevalent in the urban areas of the state. As a result, urban sex ratio in the state has been as low as 946 as compared to its rural counterpart (960). The sex ratio in Guwahati, the capital city of Assam, is only 933. Depending on a host of diverse factors, the sex ratio in the state varies significantly from one district to another, and also from one social group to another. Among the districts, the sex ratio is found to be the highest in Baksa (974) and the lowest in Dima Hasao (932). On the other hand, the sex ratio in the state is also significantly higher among the scheduled tribes (985) as compared to the scheduled castes (948) and other than scheduled castes and tribes (954).

Rural-Urban Composition

The process of urbanisation in agro-based Assam has been extremely slow. According to 2011 census data, about 86 per cent of the state's population lives in as many as 25,425 villages of different size. The remaining 14 per cent of the population lives in the urban centres numbering 214. On the other hand, as high as 31 per cent of the country's population lives in the urban areas. After independence in 1951 census, the proportion of urban population in Assam was only 4.30 per cent as against India's 17.30 per cent. Assam is, in fact, lagging far behind many of the states of the country in respect of urbanisation. The root causes associated with this are the state's extremely marginal location combined with very slow pace of transport and industrial developments.

According to 2011 census, out of the total number of urban centres in Assam, 174 are small towns having population size of below 20,000. Out of the remaining 40 urban centres, 6 are cities (with population 1 lakh and above), 8 large towns (with population 50 thousand to 1 lakh) and 26 medium towns (with population 20 thousand to 50 thousand). With a population of 9.62 lakh Guwahati turns out to be the largest city in the state. With a high primacy it is gradually emerging as a city of metropolitan character in the entire North-East region. Guwahati city alone constitutes 21.87 per cent of the state's total urban population.

During the period 1971-1991, the urban population growth in Assam has been 93 per cent as compared to 53.26 per cent for the entire population. Although there has been a high growth rate of population in the already existed urban centres during 1971- 1991, the formation of 15 new towns during the period has been primarily responsible for considerably high urban population growth during 1971-91 in the state. Besides, expansion of urban area boundary of the existing towns has also been no less significant in this respect. But it is noticeable that the population growth among the small towns has been very low as compared to that among the large towns and cities. As a matter of fact, during the period 1971-91, the population growth in the large urban centres (with population above 20 thousand) was as high as 138 per cent as against 23.2 per cent in the small towns in the state. The lack of diverse developmental activities in most of the small towns in the state is responsible for such a low population growth. On the other hand, due to the expansion of trade and commerce and diverse employment opportunities in the large urban centres in the state, the migration of people from rural areas and small towns has increased greatly in recent time. This phenomenon is often called concentrated urbanisation. Because large urban centres are growing fast in both population and functions at the cost of small towns. Among the small and medium towns, Sonari, Diphu and Haflong experienced a very high growth of population (above 200 per cent) during the period 1971-91. But, on the other hand, Badarpur and Barpeta towns witnessed unexpectedly a negative growth of population (below zero per cent). All these have resulted in spatially varied topography of urbanisation in the state (Kar and Das, 1998).

The spatial pattern of urbanisation in the state varies from one district to another. According to 1991 census, the number of urban centres is found to be the highest in the districts of Dhubri, Barpeta and Kamrup (7 in each district), and the lowest in Dhemaji district (only 1). In terms of level of urbanisation again, Kamrup district occupies the highest position (32.76 per cent) and Dhemaji district the lowest position (1.86 per cent). The other districts having the urbanisation level above the state average (11 per cent) include N.C. Hills (22.87 per cent), Dibrugarh (17.61 per cent), Tinsukia (16.49 per cent), Jorhat (15.27 per cent) and Dhubri (12.15 per cent). But the districts having urbanisation level significantly below the state average are Nalbari (2.31 per cent), Darrang (4.92 per cent) and Marigaon (5.16 per cent).

The study of occupational structure and characteristics of urban population also reflects the behaviour of different urban centres in the state. Due to the lack of industrialisation most of the urban centres that have emerged in the state are based on administrative and trade and commerce functions. The towns like Duliajan, Digboi, Namrup, etc. are, however, a few industrial towns in the state engaged in manufacturing, construction and repairing works.

Social Composition of the Population

As already mentioned, the state of Assam exhibits a great diversity in respect of population composition. The population of Assam is composed of such a diverse ethnic, cultural and linguistic stocks that the state is often regarded as a living anthropological museum. Its population composition, in fact, hardly differs from that of the country as whole. It may be mentioned here that the Constitution of India has scheduled the tribal people and certain low caste people among the Hindus and Sikhs as scheduled tribes and scheduled castes respectively for their socio-economic upliftment. Accordingly, besides religious and linguistic composition, the population of Assam as also in the country, may broadly be divided into three groups–scheduled castes population, scheduled tribes population and non-scheduled population. There are altogether sixteen scheduled caste communities in Assam. Among them six belong to Assamese Hindus and four Bengali Hindus. But, there is a large number of scheduled tribe communities in the state. Broadly, they can be either hill tribes or plain tribes. The Dimasa and the Karbi are major hill tribes, while Boro, Mishing, Rabha, Tiwa, Sonowal-Kachari and Deuri are major plain tribes in the state .The Muslims and the tribal communities like tea garden labourer and Koch- Rajbongshi are treated under the non-scheduled population group in the state. According to 2011 census, of the total population in the state, scheduled castes constitute 7.15 per cent (India :16.64 per cent), scheduled tribes 12.44 per cent (India: 8.62 per cent) and the remaining non-scheduled population 80.41 per cent (India:74.74 per cent). In 1971 the proportions of scheduled castes and scheduled tribes in the state were 6.24 per cent and 10.98 per cent respectively. It means that both these population groups experienced a growth of 144.51 per cent and 141.77 per cent respectively in Assam during the last four decades (1971-2011).

So far the spatial distribution of population is concerned, it is not uniform both among the scheduled castes and scheduled tribes. While on the one hand, the scheduled castes people are to some extent uniformly distributed, the scheduled tribes people are found in concentration in some areas of the state (Fig. 10.2 & 10.3). In terms of the distribution of scheduled castes, the district having more than 10 per cent scheduled castes in the state include Cachar (14.70 per cent), Karimganj (14.58 per cent), Marigaon (13.78 per cent), Bongaigaon (10.74 per cent) and Nagaon (10.02 per cent). Thus, it is found that the three districts in the Brahmaputra valley and two in the Barak valley together constitute 43.26 per cent of the total scheduled castes population (1,659,412) in the state. The proportion is found to be the lowest in the district of N.C. Hills (2.60 per cent). But strikingly enough, scheduled tribes population constitute as high as 65.54 per cent and 51.56 per cent of the total population respectively in the

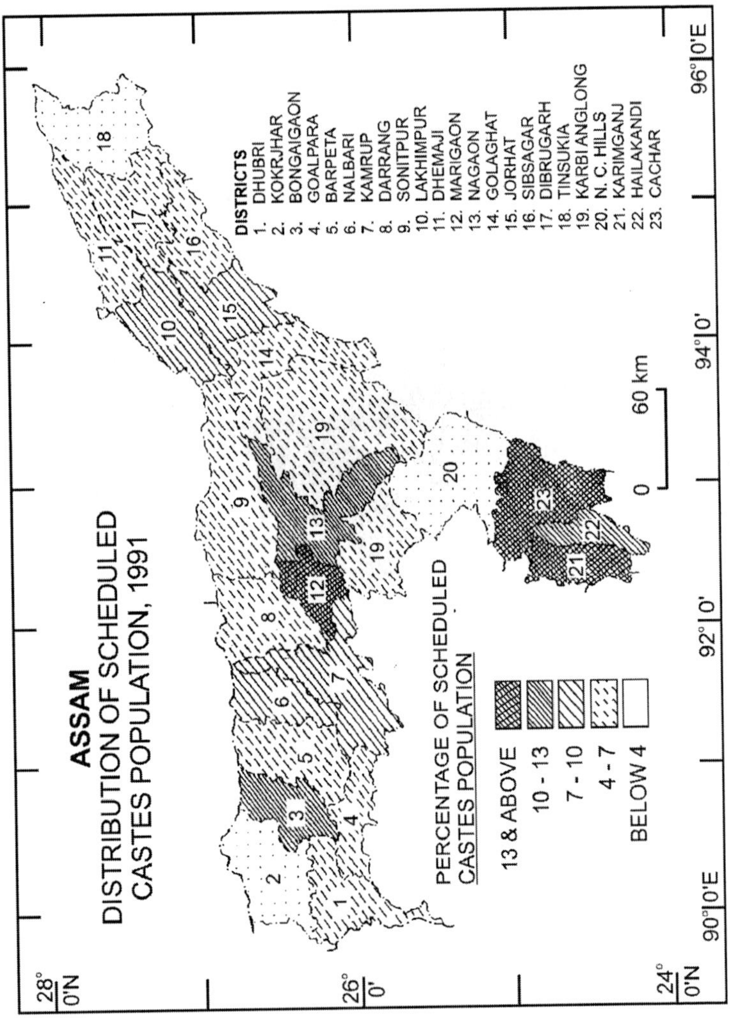
ASSAM
DISTRIBUTION OF SCHEDULED CASTES POPULATION, 1991
PERCENTAGE OF SCHEDULED CASTES POPULATION
13 & ABOVE
10 - 13
7 - 10
4 - 7
BELOW 4
DISTRICTS
1. DHUBRI
2. KOKRJHAR
3. BONGAIGAON
4. GOALPARA
5. BARPETA
6. NALBARI
7. KAMRUP
8. DARRANG
9. SONITPUR
10. LAKHIMPUR
11. DHEMAJI
12. MARIGAON
13. NAGAON
14. GOLAGHAT
15. JORHAT
16. SIBSAGAR
17. DIBRUGARH
18. TINSUKIA
19. KARBI ANGLONG
20. N. C. HILLS
21. KARIMGANJ
22. HAILAKANDI
23. CACHAR
0
60 km
28° 0'N
26° 0'
24° 0'N
90° 0'E
92° 0'
94° 0'
96° 0'E

Fig. 10.2

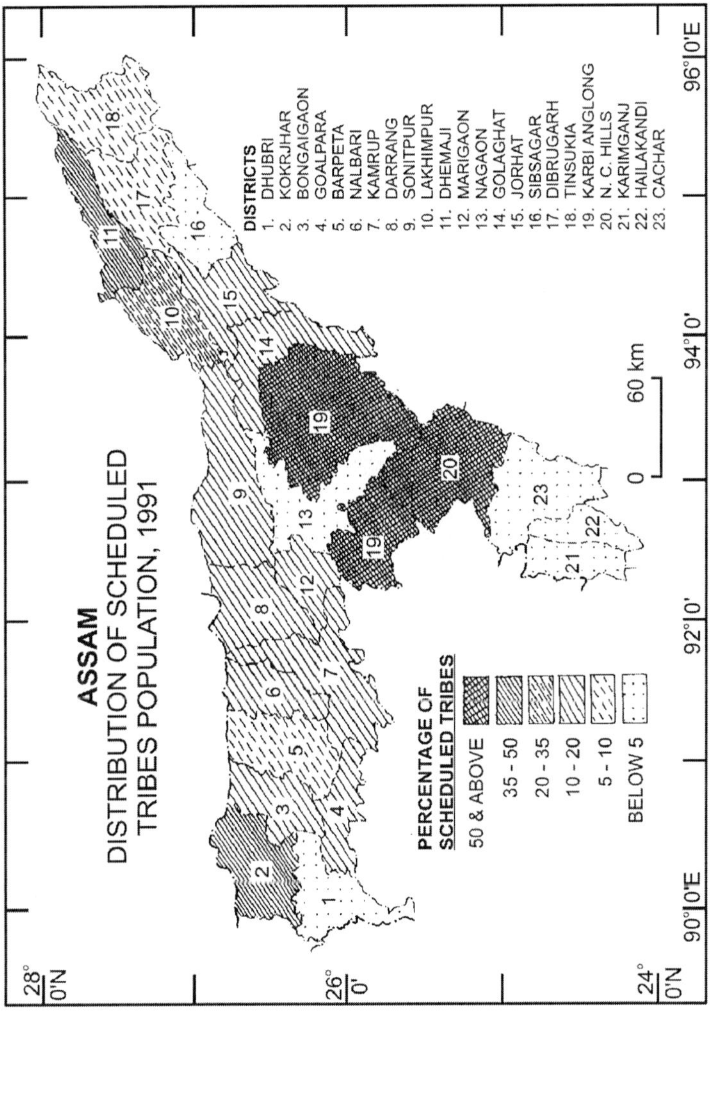
ASSAM
DISTRIBUTION OF SCHEDULED TRIBES POPULATION, 1991
PERCENTAGE OF SCHEDULED TRIBES
50 & ABOVE
35 - 50
20 - 35
10 - 20
5 - 10
BELOW 5
DISTRICTS
1. DHUBRI
2. KOKRJHAR
3. BONGAIGAON
4. GOALPARA
5. BARPETA
6. NALBARI
7. KAMRUP
8. DARRANG
9. SONITPUR
10. LAKHIMPUR
11. DHEMAJI
12. MARIGAON
13. NAGAON
14. GOLAGHAT
15. JORHAT
16. SIBSAGAR
17. DIBRUGARH
18. TINSUKIA
19. KARBI ANGLONG
20. N. C. HILLS
21. KARIMGANJ
22. HAILAKANDI
23. CACHAR
0 60 km
28° 0'N
26° 0'
24° 0'N
90° 0'E
92° 0'
94° 0'
96° 0'E

Fig. 10.3

districts of N.C. Hills and Karbi Anglong. Besides, the proportions of scheduled tribes population are also found to be considerably high in the districts of Dhemaji (43.92 per cent), Kokrajhar (41.15 per cent) and Lakhimpur (23.57 per cent). All these five districts together shelter 40.27 per cent of the state's total scheduled tribes population (2,874,441). The proportion of scheduled tribes population is found to be the lowest in the Barak valley districts of Hailakandi (0.16 per cent) and Karimganj (0.17 per cent). So far the distribution of major tribal communities in the state is concerned, Dimasa-Kacharis are concentrated in N.C.Hills, Karbis in Karbi Anglong, Boros in the north-bank part of the lower and central Brahmaputra valley, Mishings and Deuris in the north-bank part of the upper Brahmaputra valley, Rabhas in the south-bank part of the lower Brahmaputra valley and Tiwas in the south-bank part of the central Brahmaputra valley. It is worth mentioning here that 13.82 per cent and 3.38 per cent of the state's total scheduled castes and scheduled tribes population respectively live in the urban areas. It thus means that the tribal people in the state live mostly in rural areas and hence they are primarily associated with the agro-based economy.

Although Assam is a Hindu dominated state, people of all the major religious groups live here in varied proportions. According to 2011 census data, various religious groups living in the state in terms of their proportions to total population are Hindu (61.46 per cent), Muslim (34.22 per cent), Christian (3.74 per cent), Buddhist (0.18 per cent), Jain (0.08 per cent), Sikh (0.07 per cent) and other minor religious groups including tribal religion (0.85 per cent). In contrast to this, the religious composition of India as a whole is constituted by Hindu (79.80 per cent), Muslim (14.22 per cent), Christian (2.30 per cent), Sikh (1.72 per cent), Buddhist (0.70 per cent), Jain (0.32 per cent), other religions (0.87 per cent), and religion not stated (0.02 per cent). It is seen that the proportion of Muslim population in Assam is significantly higher than that of the country as a whole. Strikingly, Assam occupies the second position, only after Jammu and Kashmir, among the states in the country in terms of the proportion of Muslim population. It means that both the distribution and growth of population among different religious groups are not uniform throughout the country. It is seen in the case of Hindus and Muslims - the two major religious groups in the state - that during the period 1971-1991 the Muslim population (77.33

per cent) experienced a significantly high growth as compared to that among the Hindu population (41.61 per cent). In the case of the country as a whole also, the growth of Muslim population (74.02 per cent) has been higher than Hindu population growth (52.11 per cent). As a result of high population growth rate among the Muslims, the proportion of Muslim population in the state has increased from 24.56 pert cent in 1971 to 28.43 per cent in 1991 and 34.22 in 2011. On the other hand, the proportion of Hindu population has decreased from 72.51 per cent in 1971 to 67.13 per cent in 1991 and 61.46 per cent in 2011. Such a significantly high growth of Muslim population in the state during the last two decades is indicative of the prevalence of high birth rate and immigration among them. Satisfactorily, however, as a consequence of the slow improvement of socio-economic conditions among the Hindus, the natural growth rate of population among them has declined to a great extent in the state during this period.

So far the growth of Muslim population during 1971-91 at district level is concerned, it was the highest in Dhemaji (682.62 per cent) and the lowest in Hailakandi (55.56 per cent). The other districts witnessing high growth of Muslim population in the state are N.C. Hills (409.92 per cent), Sonitpur (131.72 per cent), Karbi Anglong (111.42 per cent), Darrang (108.10 per cent), Lakhimpur (106.65 per cent) and Golaghat (104.83 per cent). It is thus noted that the growth of Muslim population in all the districts is higher than that of the Hindu population growth in the state as a whole. In respect of Hindu population, the growth is found to be the highest in N.C. Hills district (87.70 per cent) and the lowest in the undivided Nagaon district (32.27 per cent). It is worth mentioning that except Karbi Anglong and undivided Lakhimpur districts, the growth of Hindu population during the period 1971-91 has been below 45 per cent in all other districts.

In respect of the spatial distribution also, both the Hindu and the Muslim population groups exhibit varied patterns. In the case of Muslim population, the districts like Dhubri (70.45 per cent), Barpeta (56.07 per cent), Hailakandi (54.79 per cent) and Goalpara (50.18 per cent) have occupied dominant positions. About 36 per cent of the state's total Muslim population (6,373,204) live in these four districts. Other districts having considerably high proportions of Muslim in the state include Karimganj (49.17 per cent), Nagaon (47.18 per cent), Marigaon (45.31 per cent), Cachar (34.49 per cent), Bongaigaon (32.74 per cent) and Darrang (31.97 per cent). The district having the lowest proportion of Muslim in the state is Dhemaji (1.48 per cent). But out of 23 districts in Assam, as many as 19 are Hindu dominant where the proportion of Hindu is above 50 per cent. But among these, the districts having significantly high proportion of Hindus are Dhemaji (93.87 per cent), Jorhat (93.58 per cent), Dibrugarh (91.30 per cent), Tinsukia (90.18 per cent), Sibsagar (89.25 per cent), Golaghat (86.11 per cent), Karbi Anglong (84.82 per cent) and Sonitpur (80.19 per cent). The proportion of Hindus is found to be the lowest in Dhubri (28.73 per cent). On the other hand, the districts having considerable proportions of Christian population in the state are N.C. Hills (24.47 per cent), Karbi Anglong (12.48 per cent), Kokrajhar (9.81 per cent) and Goalpara (7.89 per cent). It may be mentioned here that a large proportions of tea garden and ex -tea garden labourers and a considerable proportion of scheduled tribes, particularly the hill-tribes, profess Christianity. Nowadays, however, many tribal people more particularly in the plains follow Hindu-way of life and social practices. The Hindus, being the most numerous in most parts of the state, occupy the highest socio-economic rank in the society. The Hindu population of the state consists of the followers of the Vaishnavaites, Saktas and Saivaites. The role of the Vaishnavism has been found to be very significant in removing the feeling of caste-creed differences across the state.

Language is the most important medium of human interaction. The language data of any area reflect the socio-cultural diversity of its population. It is worth mentioning that Census of India has been providing basic information on linguistic characteristics of the country's population.The language spoken by any individual is ascertained from his mother tongue. As per the Census of India, there are altogether 114 languages in the country each having a numerical strength of above ten thousand. Out of these, 18 are called scheduled languages as they are included in the Eighth Schedule of the Indian Constitution and the rest numbering 96 are non-scheduled languages. The Eighth Schedule languages are Hindi, Bengali, Telugu, Marathi, Tamil, Urdu, Punjabi, Malayalam, Kannada, Oriya, Gujarati, Rajasthani, Assamese, Nepali, Konkani, Manipuri, Kashmiri, and Sanskrit. On the other hand, the minor languages which have the numerical strength of below ten thousand each are clubbed together to call them 'total of other languages'. Among the major languages, while the scheduled ones have 85 mother tongues, the non-scheduled have 131 mother tongues.

So far the linguistic composition is concerned, India's population is broadly composed of five linguistic families- (1) Indo-European (75.30 per cent), (2) Dravidian (22.53 per cent), (3) Austro-Asiatic (1.13 per cent), (4) Tibeto-Burmese (0.97 per cent) and (5) Semito-Hamitic (0.01 per cent). It means that the Indo-European (primarily the Indo-Aryan) and the Dravidian are the two major linguistic families in the country. On the other hand, Indo-European (86.95 per cent) and Tibeto- Burmese (11.23 per cent) form the major language families in the state of Assam. In the country as whole, in terms of numerical strength of speakers Hindi is the largest language (39.85 per cent of the total population), followed by the Bengali (8.22 per cent), Telugu (7.80 per cent), Marathi (7.38 per cent), Tamil (6.26 per cent), Urdu (5.13 per cent), and Sanskrit the smallest (0.01 per cent). The proportion of the Assamese speaking population in the country is only 1.55 per cent. The proportion of the scheduled language speakers in the country is 96.29 per cent. It may be mentioned here that the population of the non-scheduled languages is largely concentrated in the north-eastern states of the country.

So far the linguistic composition of Assam's population is concerned, unlike many other states of the country it has two major scheduled languages– Assamese and Bengali. They respectively constitute 57.81 per cent and 21.67 per cent of the total population of the state as per the 1991 census data. Although Assamese is the state language, it is largely confined to only the Brahmaputra valley. On the other hand, Bengali is the major language in the Barak valley. About four-fifth of its population speak Bengali. In addition, Bengali speaking population is also found in considerable number in other districts of the state. Further, being inhabited by a number of tribal groups, Assam also has a significantly high proportion of non-scheduled language speakers (12.53 per cent). Among the non-scheduled languages, Bodo/Boro, Mising, Karbi, Santali, Garo and Rabha are important in terms of their number of speakers (above 1 lakh population each). The tribal language speakers like Boro live largely in the lower and middle Brahmaputra valley, Mishing in the upper Brahmaputra valley and the Karbi in Karbi Anglong hill-district. The Nepali speaking people, on the other hand, live mostly in the districts of Sonitpur, Nagaon and Tinsukia. The tea-gardens and the urban areas are the major concentrations of Hindi speaking population in the state. The state altogether has 87 languages including 69 non-scheduled ones. This is indicative of the prevailing high degree of linguistic diversity in the state.

It may be mentioned here that among some of the major scheduled and non-scheduled languages in the state, the proportion of speakers of the Assamese language has declined from 60.88 per cent in 1971 to 57.81 per cent in 1991; Bengali increased from 19.71 per cent to 21.67 per cent; Hindi decreased from 5.42 per cent to 4.62 per cent; Nepali decreased from 2.39 per cent to 1.93 per cent; Bodo/Boro increased from 3.65 per cent to 5.28 per cent; Mising increased from 1.21 per cent to 1.70 per cent; Karbi increased from 1.31 per cent to 1.58 per cent; and Rabha increased from 0.22 per cent to 0.50 per cent. At national level also, the proportion of Assamese speaking population has decreased from 1.63 per cent in 1971 to 1.55 per cent in 1991.

The probable causes behind such changing pattern of the linguistic composition of the state's population may be understood when the growth rates of population of each language is compared with that of the average growth of state's population during 1971-91 (53.26 per cent). Accordingly, it has been noted that Assamese (45.52 per cent), Hindi (30.66 per cent) and Nepali (23.89 per cent) language speakers have witnessed significantly low growth rate even below the state average as compared to Bengali (68.51 per cent), Karbi (85.34 per cent), Mising (115.30 per cent), Bodo/Boro (121.95 per cent) and Rabha (246.99 per cent). The increasing proportion of the tribal language speakers accompanied by high growth rates among them during 1971-91 are indicative of the prevailing high natural growth rate among them, and also return of many tribal people who included themselves as Assamese in the 1971 census. This process is known as retribalization. As a consequence of this, the proportion of Assamese speakers in the state has declined significantly witnessing a much lower gowth rate among them during 1971-91. Moreover, the natural growth rate of Bengali speaking population in the state would be mainly due to the prevalence of high natural growth rate among them and partly due to migration. Besides, inclusion of the Chakma, Hajong and Rajbongshi under Bengali language by the Census of India might be an added factor towards high growth rate among them. On the other hand, significantly low growth rate as evidenced particularly by Nepali speakers might be the result of low natural growth and also return of many Nepali people as Assamese during the census. But the low population growth among the Hindi speaking population is associated with low natural growth and emigration. In any case, the growth rate of the non-scheduled language speaking population in the state (99.03 per cent) during 1971-91 was much higher than that among the scheduled language speakers (48.37 per cent). On the whole, it is found that Assam presents a diverse linguistic composition and each language maintains its identity and distinctiveness in the broad framework of Assamese society.

Literacy Pattern

Literacy and educational attainment are considered to be the basic ingredients of socio-economic development of any society. As a matter of fact, the quality of human population of any society can be judged through its educational attainment. But unfortunately, Assam is lagging much behind in respect of literacy and education than many other parts of the country. According to 2011 census, the literacy rate in Assam is found to be 72.19 per cent as against country's 73.00 per cent. Although the literacy rate in Assam appears to be slightly lower than that of the national average, it is notably lower than the states like Kerala (94.00 per cent), Mizoram (91.30 per cent), Goa (88.70 per cent), Maharashtra (82.30 per cent), Himachal Pradesh (82.80 per cent), Tamil Nadu (80.10 per cent), Nagaland (79.60 per cent), Gujarat (78.00 per cent), Tripura (87.20 per cent) and so on. Its position at state level is as low as eighteenth in the country and sixth in the North-East.

In fact, like many other parts of the country the progress of literacy in Assam was slow prior to independence. In the beginning of the twentieth century, about 96 per cent of the state's population were illiterate. There was hardly a literate woman per one hundred females in the state at that time. The illiteracy rate came down to only 87 per cent in 1941. There was no productive effort as such in the country for spread of education.

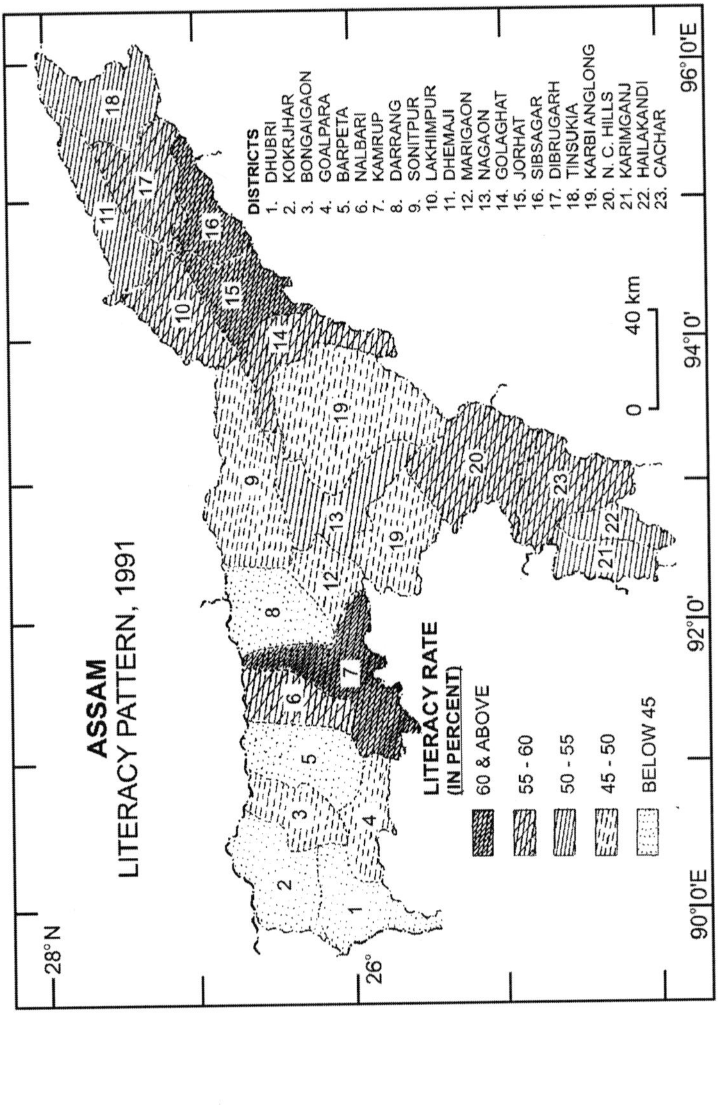
ASSAM
LITERACY PATTERN, 1991
LITERACY RATE
(IN PERCENT)
60 & ABOVE
55 - 60
50 - 55
45 - 50
BELOW 45
DISTRICTS
1. DHUBRI
2. KOKRJHAR
3. BONGAIGAON
4. GOALPARA
5. BARPETA
6. NALBARI
7. KAMRUP
8. DARRANG
9. SONITPUR
10. LAKHIMPUR
11. DHEMAJI
12. MARIGAON
13. NAGAON
14. GOLAGHAT
15. JORHAT
16. SIBSAGAR
17. DIBRUGARH
18. TINSUKIA
19. KARBI ANGLONG
20. N. C. HILLS
21. KARIMGANJ
22. HAILAKANDI
23. CACHAR
0 40 km
90°0'E
92°0'
94°0'
96°0'E
28° N
26°

Fig. 10.4

But increase in literacy rate had attained a momentum after independence of the country. After independence, during a period of four decades (1951-1991), the litetracy rate increased from 21.73 per cent to 52.89 per cent (Table 10.5). Satisfactorily, as a result of the rise in literacy rate, the prevailing male-female differential in this respect has also come down to a great extent. During the period 1951-1991 the female literacy rate in the state witnessed a significant rise from 9.48 per cent to 43.03 per cent. It is thus clear that the literacy and educational development is essentially a post independence phenomenon. But even after fifty years of independence, about two-fifth of the state's population is illiterate. The literacy rate, however, varies widely across the state. The variation in literacy is even found to be more significant in respect of rural-urban, male-female and different social components in the state. Such variations in literacy rate in the state are associated with cumulative effects of physical, economic and social constraints and the historical legacy.

Table 10.5
Progress of literacy in Assam, 1901-2011

Year	Literacy rate* (in per cent)		
	Person	Male	Female
1901	4.17	7.45	0.59
1911	5.59	9.97	0.77
1921	7.11	12.22	1.40
1931	8.57	14.40	1.87
1941	13.14	20.76	4.38
1951	21.73	32.25	9.48
1961	32.98	44.28	19.63
1971	34.60	44.31	23.52
1991	52.89	61.87	43.03
2001	63.30	71.30	54.60
2011	72.19	77.80	66.30

Source : Calculated on the basis of data from Census of India, 1961 & 1971, Assam, Part 1, General Report and 1991, India, Paper 2 of 1992, Final Population Totals and PCA, India, 2001 and 2011.

*Effective Literacy Rate (excluding 0-4 years for 1901-1971 and excluding 0-6 years in 1991). There was no Census in Assam in 1981. Assam included Meghalaya and Mizoram till 1961.

Spatial Variation in Literacy : When looked at the district level, it is found according to 1991 census data that the literacy rates in the state vary from 38.31 per cent in Dhubri to 65.51 per cent in Jorhat (Table 10.6). In fact, literacy rate is found to be significantly high in the urbanized and socio-economically advanced districts like Kamrup, Jorhat, Sibsagar and Cachar. On the other hand, it is quite low in the socio-economically backward districts like Dhubri, Kokrajhar, Barpeta, Darrang and Karbi Anglong where either immigrant Muslims or tribal people constitute a considerable proportion of total population. But the impact of Christianity has, however, made the situation in the tribal-dominted N.C. Hills district somewhat better in this respect (Fig. 10.4).

Table 10.6
Districtwise literacy rates in Assam, 1991

Year	Literacy rate* (in per cent)		
	Total	Male	Female
1. Dhubri	38.31	47.32	28.75
2. Kokrajhar	40.57	49.57	30.92
3. Bongaigaon	49.06	58.67	38.72
4. Goalpara	46.81	55.47	37.58
5. Barpeta	43.24	52.61	33.20
6. Nalbari	55.99	66.95	44.19
7. Kamrup	65.04	73.67	55.01
8. Darrang	42.00	50.80	32.53
9. Sonitpur	48.14	56.70	38.60
10. Lakhimpur	58.96	68.28	48.85
11. Dhemaji	53.84	65.43	41.12
12. Marigaon	47.99	56.17	39.19
13. Nagaon	54.74	62.49	46.30
14. Golaghat	58.54	66.50	49.75
15. Jorhat	65.51	73.29	56.88
15. Sibsagar	64.46	71.91	56.14
17. Dibrugarh	58.32	66.72	48.89
18. Tinsukia	50.28	59.27	39.99
19. Karbi Anglong	45.57	55.55	34.35
20. North Cachar Hills	57.76	66.39	47.34
21. Karimganj	54.71	64.05	44.76
22. Hailakandi	53.07	64.08	41.04
23. Cachar	59.18	68.79	48.76

ASSAM	**52.89**	**61.87**	**43.03**
INDIA (Excl. J & K)	**52.21**	**64.13**	**39.29**

Source : Primary Census Abstract, General Population, Scheduled Castes and Scheduled Tribes, Series-4, Assam, Part II -B, Census of India, 1991.

The spatial variation in literacy rates in the state becomes further clear when it is viewed separately for its rural and urban areas. Strikingly, the urban literacy rate in the state is as high as 79.39 per cent as against 49.32 per cent for its rural counterpart. Among the districts, the rural-urban differential in literacy rates is found to be the highest in Kokrajhar (urban : 77.14 per cent ; rural : 37.90 per cent), and the lowest in Jorhat (urban : 79.95 per cent; rural: 62.78 per cent). The differential in this respect is generally found to be low in the areas where the overall literacy rate is already high. So far the spatial variation in rural literacy rate is concerned, it is significantly higher than the state's average in the districts of Sibsagar, Jorhat, Lakhimpur, Kamrup, Golaghat, Cachar and Nalbari. The rural literacy rate is, however, below 40 per cent in the districts of Dhubri and Kokrajhar. Among a host of factors, poverty combined with lack of adequate educational facilities and consciousness towards education are primarily responsible for the prevalence of low literacy in most of the rural areas of the state.

Male-Female Disparity in Literacy : As elsewhere in the country there exists a significantly high male-female disparity in literacy in Assam . In fact, the disparities between male and female educational development are rooted, among other factors, in the persistence of some attributes of the spurious modernisation of the colonial era. The prevalence of strong societal prejudices against women's mobility and education has been responsible for the slow progress of literacy in the state. It may be mentioned here that the relative deprivation of women in the field of education results in deprivation in many other attributes (Raza and Agarwal, 1986). According to 2011 census data, the female literacy rate in the state stands at 66.30 per cent as against the male literacy rate of 77.80 per cent. It is, however, satisfactory to note that the female literacy rate in the state is slightly higher than that of the corresponding national average (64.60 per cent). Besides, it is encouraging to note that the sex disparity in literacy in the state has been gradually decreasing since the beginning of the twentieth century. This is indicative of increasing spread of female education in various parts of the state. Nevertheless, sex differential in literacy is still much higher in the rural areas (male: 75.40 per cent; female : 63.00 per cent) than that of the urban areas of the state (male: 91.80 per cent; female : 84.90 per cent). It is significant to note that the male-female disparity in literacy is much low in its neighbouring Christian-dominated tribal states, namely Meghalaya, Mizoram and Nagaland.

So far as the spatial variation in male-female differential in literacy as per 1991 is concerned, it is strikingly high in the districts of Dhemaji, Hailakandi and Nalbari, and considerably low in the districts of Sibsagar, Nagaon and Golaghat (Table 10.6). The same is also true with respect to the rural and urban areas of the state. The lowest and the highest male-female disparity in urban literacy are recorded respectively by Sibsagar district (male: 85.25 per cent; female: 80.71 per cent) and Barpeta district (male: 82.81 per cent: female : 66.07 per cent).

Literacy among Different Social Groups

So far as the literacy rate among the scheduled castes, scheduled tribes and other population in the state is concerned, it is satisfactory to note that unlike many other parts of the country it is quite uniform. According to 1991 census data, the scheduled castes and scheduled tribes record literacy rates of 53.94 per cent and 49.15 per cent respectively as against 53.39 per cent for the non-scheduled population group. The corresponding percentages for the country as a whole, however, are 37.41 per cent, 29.60 per cent and 57.69 per cent. Although the levels of literacy in Assam among different social groups are relatively better than the country's ones, there exists a significantly high rural-urban and male-female differentials in literacy among them in the state. The rural-urban differential in literacy rate in the state is, however, found to be quite low among the scheduled castes (urban: 65.71 per cent; rural: 51.95 per cent) as compared to the scheduled tribes (urban : 78.18 per cent; rural: 48.09 per cent). On the other hand, scheduled castes record the highest sex disparity in literacy (male: 63.88 per cent; female: 42.99 per cent), followed by scheduled tribes (male: 58.93 per cent ; female : 38.98 per cent) and non-scheduled population (male: 62.13 per cent; female: 43.69 per cent). Besides, the literacy rate is also found to be significantly low among the immigrant Muslim population and tea garden and ex-tea garden labour population in the state. Thus, all these spatial characteristics and social strands make the literacy topography in the state varied and complex. (Kar and Shrma, 1994). It may, however, be kept in mind that only literacy rate cannot reflect the educational development scenario of any area in true perspective. For instance, of the total literates in Assam according to 1991 census, as high as 50.29 per cent have the educational attainment of only up to below primary and primary levels. Besides, there are some more literates who do not have any formal education. On the other hand, the proportions of matriculates and graduate and above are found to be only 13.48 per cent and 3.76 per cent respectively. It, therefore, necessitates the analyses of educational attainment of the literates. Because the quality of human population in an area depends greatly upon the educational levels of its people.

Occupational Pattern

The proportion of population engaged in economic activities (work participation rate) and the occupational structure of the workers indicate to a great extent the overall economic development scenario of the state. According to 2011 census data, the total work participation rate in the state is found to be 38.40 per cent (main workers : 27.80 per cent; marginal workers: 10.60 per cent) as against the country's average of 39.80 per cent. It is, however, satisfactory to note that the work participation rate has witnessed a significant rise during the last two decades. As per the 1971 census data, the work participation rates in the state and the country were 27.96 per cent and 32.93 per cent respectively. But there is still a significant gender disparity in work participation. Both in the state and the country, the female work participation rates (Assam - 22.50 per cent; India - 25.50 per cent) are notably lower than their male counterparts (Assam - 53.60 per cent; India - 53.30 per cent). Within the state again, the work participation is considerably higher in the rural areas (38.70 per cent) than that of the urban areas (36.40 per cent). Further, it is found that the gender differential in work participation in urban areas of the state is noticeably higher than rural counterparts. On the whole, the low work participation rate in the state, as in other parts of the country, is largely associated with lack of development of the non-agricultural sector. Besides, the considerably high proportion of non-working population in the age group 0-15 is also no less important in keeping the work participation low in the state.

So far the spatial variation in work participation rate in the state is concerned, the participation rate is found to be high in the tribal dominated and economically backward districts like Dhemaji (44.97 per cent), Lakhimpur (42.86 per cent), Karbi Anglong (42.78 per cent) and N.C. Hills (42.05 per cent), and the tea garden dominant districts like Sibsagar (41.94 per cent), Tinsukia (40.68 per cent), Golaghat (40.98 per cent), Sonitpur (40.23 per cent), Darrang (40.19 per cent) and Dibrugarh (39.52 per cent). But the work participation rate is found to be quite low in the Muslim dominated districts like Karimganj (28.90 per cent), Dhubri (29.45 per cent) and Barpeta (30.03 per cent). The female work participation rate is also found to be the lowest in these districts (Table 10.7). Thus, it may be said that the economic conditions and the social prejudices play an important role in determining the female work participation in different parts of the state.

Table 10.7
Districtwise work participation rates in Assam, 1991

District	Total work participation rate (in per cent)		
	Person	Male	Female
1. Dhubri	29.45	49.60	8.23
2. Kokrajhar	35.53	50.74	19.37
3. Bongaigaon	35.09	49.52	19.78
4. Goalpara	33.68	48.92	17.58
5. Barpeta	30.03	47.92	10.96
6. Nalbari	33.13	46.02	19.36
7. Kamrup	32.58	48.96	13.94
8. Darrang	40.19	50.61	29.07
9. Sonitpur	40.22	51.52	27.85
10. Lakhimpur	42.85	47.53	37.84
11. Dhemaji	44.97	48.09	41.59
12. Marigaon	34.87	51.24	17.49
13. Nagaon	35.46	50.67	19.07
14. Golaghat	40.97	49.61	31.59

15. Jorhat	38.35	47.97	27.81
16. Sibsagar	41.93	49.64	33.43
17. Dibrugarh	39.52	48.92	29.14
18. Tinsukia	40.68	50.31	29.87
19. Karbi Anglong	42.78	51.08	33.63
20. North Cachar Hills	42.05	52.36	30.03
21. Karimganj	28.90	48.10	8.68
22. Hailakandi	31.49	49.63	11.97
23. Cachar	31.84	49.33	13.07
ASSAM	**36.09**	**49.45**	**21.61**
INDIA (Excl. J & K)	**37.50**	**51.61**	**22.27**

Source : Primary Census Abstract, General Population, Scheduled Castes and Scheduled Tribes, Series-4, Assam, Part II -B, Census of India, 1991.

Table 10.8
Districtwise work participation in different occupational sectors in Assam, 1991

District	Sectorwise percentage of workers to total main workers		
	Primary	Secondary	Tertiary
1. Dhubri	75.87	5.86	18.27
2. Kokrajhar	84.03	2.95	13.02
3. Bongaigaon	75.04	4.61	20.35
4. Goalpara	75.20	4.88	19.92
5. Barpeta	76.33	5.28	18.39
6. Nalbari	71.66	6.44	21.90
7. Kamrup	45.65	11.94	42.41
8. Darrang	83.97	3.13	12.90
9. Sonitpur	77.81	3.97	18.22
10. Lakhimpur	80.54	3.26	16.20
11. Dhemaji	85.66	2.17	12.17
12. Marigaon	82.23	4.66	13.11
13. Nagaon	74.42	5.63	19.95
14. Golaghat	81.08	3.94	14.98
15. Jorhat	69.49	6.27	24.24
16. Sibsagar	78.79	3.54	17.67
17. Dibrugarh	70.75	6.25	23.00
18. Tinsukia	73.83	7.95	18.22
19. Karbi Anglong	85.00	3.50	11.22
20. North Cachar Hills	64.33	6.25	29.42

21. Karimganj	68.99	7.45	23.56
22. Hailakandi	77.28	5.08	17.64
23. Cachar	69.43	5.94	24.63
ASSAM	**73.99**	**5.56**	**20.45**
INDIA (Excl. J & K)	**67.53**	**11.97**	**20.50**

Source : Primary Census Abstract, General Population, Scheduled Castes and Scheduled Tribes, Series-4, Assam, Part II -B, Census of India, 1991.

The predominance of the agrarian economy of the state is also reflected in its occupational structure of the working population. It is found that as high as 63 per cent of the total main workers in the state is engaged in agriculture. On the other hand, only 6 per cent of the total main workers of the urban areas in the state is engaged in agriculture. The proportions of main workers engaged broadly in the primary, secondary and tertiary sectors in the state as a whole are found to be 73.99 per cent, 5.56 per cent and 20.45 per cent respectively. The pattern in this respect does not vary much also in the case of the districts (Table 10.8). In the urban areas of the state, however, about 90 per cent of the main workers are engaged in secondary and tertiary sectors. Unlike many other districts of the state, the proportions of main workers engaged in tea plantation and to a certain extent in livestock farming, fishing and lumbering are considerably high in the districts of Dibrugarh (30.74 per cent), Tinsukia (30.72 per cent) and Sibsagar (24.57 per cent). On the other side, the proportion of agricultural workers (cultivators and agricultural labourers) is found to be significantly high (above 75 per cent) in the districts of Dhemaji, Karbi Anglong , Kokrajhar, Marigaon and Barpeta. In the case of Kamrup district, the most urbanised district in the state, the proportions of main workers engaged in manufacturing, and trade and commerce are found to be 8.80 per cent and 13.26 per cent respectively. It thus reveals that Assam is still long way to go in the field of industrialisation. It is highly discouraging to note that the prevailing traditional agricultural system and slow progress of the non-agricultural sector have not been able to generate job opportunities in accordance with the rising population in the state. This situation has resulted in increasing unemployment even among the educated youth in the state. Besides, there has been marked increase of rural to urban migration.

Table 10. 9
Selected Indicators of Socio-Economic well-being

District	Percentage of population in age group 0-6	Sex Ratio	Adjusted-Crude Birth Rate	Infant Mortality Rate	Mean Female Age at Marriage	Male Literacy Rate	Female Literacy Rate	Percentage of Urban Population	Percentage of main workers	Percentage of Non-primary workers	Composite z-Score	Rank of Socio-Economic Well-being
1. Dhubri	23.3	950	38.2	128	16.9	47.3	28.8	12.2	27.2	24.1	-10.25	23
2. Kokrajhar	20.1	940	39.7	78	18.2	49.6	30.9	6.3	33.9	16.0	-3.92	19
3. Bongaigaon	21.3	940	29.3	122	17.6	58.7	38.7	9.1	30.3	25.0	-3.31	17
4. Goalpara	22.7	947	35.2	106	17.5	55.5	37.6	7.8	29.1	24.8	-5.28	20
5. Barpeta	21.2	939	43.2	101	17.8	52.6	33.2	7.0	26.4	23.7	-7.25	22
6. Nalbari	18.6	936	47.0	96	18.1	67.0	44.2	2.3	26.5	28.3	-2.58	16
7. Kamrup	17.4	879	28.3	77	18.4	73.7	55.0	32.8	28.7	54.4	10.50	2
8. Darrang	20.2	938	42.0	111	18.0	50.8	32.5	4.9	31.2	16.0	-6.91	21
9. Sonitpur	19.2	912	34.0	77	18.5	56.7	38.6	7.3	35.3	22.2	0.06	9
10. Lakhimpur	20.1	931	34.1	112	18.2	68.3	48.9	6.5	32.1	19.5	-0.24	10
11. Dhemaji	22.6	927	25.9	114	18.5	65.4	41.1	1.9	33.6	14.3	-2.27	14
12. Marigaon	21.9	941	30.5	88	18.0	56.2	39.2	5.2	28.9	17.8	-3.54	18
13. Nagaon	20.7	929	33.9	97	17.9	62.5	46.3	10.9	29.3	25.6	-0.96	11
14. Golaghat	18.9	920	32.1	61	18.8	66.5	49.8	5.9	35.4	18.9	4.35	6
15. Jorhat	16.4	912	31.5	47	19.0	73.3	56.9	15.3	34.4	30.5	10.79	1
16. Sibsagar	16.9	907	31.1	75	19.0	71.9	56.1	7.2	36.5	21.2	7.03	4
17. Dibrugarh	17.7	905	24.4	51	19.1	66.7	48.9	17.6	34.7	29.3	9.30	3
18. Tinsukia	19.4	891	37.5	73	18.6	59.3	40.0	16.5	36.5	26.2	1.49	8

19. Karbi Anglong	21.3	906	33.1	86	18.8	55.6	34.4	10.6	38.0	15.0	-1.42	13
20. N.C. Hills	19.3	857	29.7	108	19.2	66.4	47.3	22.9	38.3	35.7	5.20	5
21. Karimganj	19.5	946	48.5	111	18.2	64.1	44.8	7.3	26.9	31.0	-2.56	15
22. Hailakandi	20.3	929	32.1	99	18.2	64.1	41.0	7.6	29.1	22.7	-1.33	12
23. Cachar	18.4	931	32.6	97	18.5	68.8	48.8	9.8	29.7	30.6	3.40	7
Mean	**19.9**	**922.3**	**34.5**	**92.0**	**18.3**	**61.8**	**42.7**	**10.2**	**31.8**	**24.9**		
Standard Deviation	**1.8**	**22.9**	**6.1**	**21.3**	**0.6**	**7.5**	**7.8**	**6.9**	**3.7**	**8.4**		

Source: Census of India, 1991

Table 10.10

Data about a few Population Attributes at District Level in Assam, 2011

District	Area (in km²)	Population	Population Density (Persons/ km²)	Decadal Growth Rate of Population (2001-11)	Sex Ratio (Females per thousand Males)	Percentage of 0-6 Population	Literacy Rate (in per cent)		
							Person	Male	Female
1. Kokrajhar	3296	887142	269	5.21	959	15.43	65.22	71.89	58.27
2. Dhubri	2176	1949258	896	24.44	953	18.89	58.33	63.10	53.33
3. Goalpara	1824	1008183	553	22.64	964	17.03	67.37	71.46	63.13
4. Barpeta	2282	1693622	742	21.43	953	17.00	63.81	69.28	58.05
5. Morigaon	1551	957423	617	23.34	967	17.11	68.03	71.90	64.04
6. Nagaon	3973	2823768	711	22.00	962	16.29	72.37	76.51	68.06
7. Sonitpur	5204	1924110	370	15.55	956	14.44	67.34	73.65	60.72
8. Lakhimpur	2277	1042137	458	17.22	968	15.04	77.19	83.52	70.67
9. Dhemaji	3237	686133	212	19.97	953	15.19	72.70	79.84	65.21
10. Tinsukia	3790	1327929	350	15.47	952	13.69	69.65	77.19	61.73
11. Dibrugarh	3381	1326335	392	11.92	961	12.31	76.05	82.82	69.00
12. Sivasagar	2668	1151050	431	9.44	954	12.11	80.41	85.84	74.71
13. Jorhat	2851	1092256	383	9.31	962	11.41	82.14	87.63	76.45
14. Golaghat	3502	1066888	305	12.75	964	12.63	77.43	83.56	71.09
15. Karbi Anglong	10434	956313	92	17.58	951	15.86	69.25	76.14	62.00
16. Dima Hasao	4888	214102	44	13.84	932	15.14	77.53	83.29	71.32
17. Cachar	3786	1736617	459	20.19	959	14.78	79.34	84.77	73.68
18. Karimganj	1809	1228686	679	21.90	963	17.25	78.22	84.12	72.09

19. Hailakandi	1327	659296	497	21.45	951	16.88	74.33	80.74	67.59
20. Bongaigaon	1093	738804	676	20.59	966	15.75	69.74	74.86	61.84
21. Chirang	1923	482162	251	11.34	969	15.18	63.55	70.24	56.65
22. Kamrup	3105	1517542	489	15.69	949	13.18	75.55	81.30	69.47
23. Kamrup Metro	955	1253938	1313	18.34	936	10.00	88.71	92.13	85.07
24. Nalbari	1052	771639	733	11.99	949	12.35	78.23	84.36	72.57
25. Baksa	2457	950075	387	10.74	974	12.93	69.25	77.03	61.27
26. Darrang	1585	928500	586	22.19	954	16.84	63.08	67.87	58.04
27. Udalguri	2012	831668	413	9.61	973	13.62	65.41	72.57	58.05
ASSAM	**78438**	**31205576**	**398**	**17.07**	**958**	**14.86**	**72.19**	**77.85**	**66.27**
INDIA	**3,160,000**	**1210854977**	**382**	**17.70**	**943**	**13.60**	**73.00**	**80.90**	**64.60**

Source: Computed from Census of India, Primary Census Abstract, India and Assam, 2011; and Statistical Hand-Book, Assam, 2015, Directorate of Economics and Statistics, Assam, Guwahati.

Table 10.11
Socio-Economic Data of a few Population Attributes of Assam at District Level, 2011

Districts	Percentage of Urban Population	Total Work Participation Rate (in per cent)			Percentage of Main Workers under different occupational categories			
		Persons	Male	Female	Cultivators	Agricultural Labourers	Household Industry	Other Workers
1. Kokrajhar	6.19	38.45	52.18	24.13	50.59	11.66	2.89	34.86
2. Dhubri	10.45	34.36	52.60	15.23	35.32	21.27	3.17	40.24
3. Goalpara	13.69	35.96	52.04	19.21	40.88	14.08	2.89	42.15
4. Barpeta	8.70	33.17	51.36	14.09	42.04	12.95	3.75	41.26
5. Morigaon	7.65	36.67	53.01	19.79	51.66	15.33	2.52	30.49
6. Nagaon	13.08	34.70	53.27	15.40	39.10	14.64	2.74	43.52
7. Sonitpur	9.03	40.05	54.59	24.82	35.16	9.13	2.47	53.24
8. Lakhimpur	8.76	41.26	53.29	28.83	52.49	5.69	1.66	40.16
9. Dhemaji	7.03	46.17	53.11	38.89	72.45	2.37	1.54	23.64
10. Tinsukia	19.94	41.95	53.90	29.42	25.07	4.74	1.57	68.62
11. Dibrugarh	18.38	42.26	54.40	29.63	21.98	5.56	1.66	70.80
12. Sivasagar	9.56	42.10	55.17	28.59	22.02	4.39	1.92	71.67
13. Jorhat	20.19	45.65	57.60	33.22	29.13	5.95	4.26	60.66
14. Golaghat	9.16	44.98	56.91	32.62	40.35	6.22	2.34	51.09
15. Karbi Anglong	11.81	40.10	50.22	29.44	62.88	7.90	1.95	27.27
16. Dima Hasao	29.19	39.71	50.83	27.79	51.79	2.68	1.09	44.44
17. Cachar	18.17	35.01	53.00	16.26	22.13	7.89	0.24	69.74
18. Karimganj	8.93	32.49	51.10	13.17	27.30	12.48	2.79	57.43
19. Hailakandi	7.30	32.49	50.55	13.52	33.98	10.72	1.97	53.33

20. Bongaigaon	14.86	34.62	52.56	15.50	34.83	13.00	2.48	49.69
21. Chirang	7.33	40.21	52.58	27.45	48.03	11.43	3.05	37.49
22. Kamrup	9.38	41.44	55.04	27.12	35.54	10.00	7.00	47.46
23. Kamrup Metro	82.70	39.15	58.00	19.01	5.00	2.00	1.93	91.07
24. Nalbari	10.72	34.68	53.08	15.29	26.09	9.96	4.36	59.59
25. Baksa	1.29	42.81	55.39	29.90	44.71	15.97	3.75	35.57
26. Darrang	5.98	34.98	51.94	17.21	44.51	19.22	1.91	34.36
27. Udalguri	4.52	41.49	54.71	27.88	41.51	14.26	2.47	41.76
ASSAM	**14.09**	**38.36**	**53.59**	**22.45**	**36.13**	**10.34**	**2.79**	**50.74**
INDIA	**31.14**	**39.80**	**53.30**	**25.50**	**26.46**	**23.77**	**3.40**	**46.37**

Source: Computed from Census of India, Primary Census Abstract, India and Assam, 2011.

Socio-economic Well-being

From the foregoing discussion it is clear that the population characteristics of Assam do not portray a satisfactory picture as compared to many other states of the country. The level of socio-economic well-being is, in fact, the result of the cumulative performance of a number of related population attributes. The population attributes like crude birth rate, infant mortality rate, literacy rate, urbanisation level, work participation in non-primary sectors, etc. are indicative of the overall development process in an area. Again the attributes like literacy and educational attainment may also act as factors of development. The way education raises the quality of human population, the same way it also helps to enhance the quality of any production or service. The importance of female education as a crucial determinant of fertility amongst women in different parts of the country is well known. This is also found to be true in the case of death rate, infant mortality rate, life expectancy at birth and female age at marriage. That is why female education is often treated as an effective instrument in the reduction of birth rate and improvement of physical well-being of human life.

So far the UNDP's (United Nations Development Programme) Human Development Report is concerned, the determination of human development index at country level is based on only three indicators-GDP per capita, life expectancy at birth and adult literacy rate including gross enrolment ratio. Here, in the present study the relevant data available at district level pertaining to demographic (proportion of population in age group 0-6, sex ratio, adjusted crude birth rate, infant mortality rate and female mean age at marriage), social (male literacy rate and female literacy rate) and economic (work participation rate and work participation in non-primary sectors) aspects have been taken into consideration (Table 10.9) for determination of level of socio-economic well-being. The technique applied here for calculation of composite value is the one well known as composite Z-score [1]. It may be mentioned here that while calculating the composite score the algebraic signs of the individual Z-values of the negative indicators (population in age group 0-6, adjusted crude birth rate, infant mortality rate) have been reversed in order to make them compatible with that of the positive indicators.

The resultant composite scores clearly reveal that the level of socio-economic well-being in the state is quite uneven (Fig. 10.5). On the whole, the south-bank districts of the upper Brahmaputra valley in the state witness a high level of socio-economic well-being, followed by the hill-districts and the Barak valley districts. In fact, the socio-economic well-being of human population in various parts of the state is the result of the interplay of a host of factors ranging from historical legacy through population composition and social practices to infrastructure and economic development characteristics. Accordingly, the level of socio-economic well-being is found to be the highest in the considerably socially and economically developed district of Jorhat (10.79) in the upper Brahmaputra valley and the lowest in the Muslim population dominated and economically backward district of Dhubri (-10.25). On the other hand, the most urbanised and considerably industrially and economically advanced district of Kamrup in the lower Brahmaputra valley occupies the second position (10.50) in this respect. Unfortunately, because of overall economic backwardness and lack of any positive historical legacy, certain districts having considerably high proportion of Muslim and tribal population in the plains like Barpeta, Darrang, Marigaon, Bongaigaon, Kokrajhar, Nalbari, Karimganj and Dhemaji occupy a discouragingly low position. But among the two hill-districts, the N.C. Hills district which has a considerably high proportion of urban population and Christian tribal population witnesses encouragingly high socio-economic well-being (5.20) as compared to the economically backward tribal dominated Karbi Anglong (-1.42). This pattern of socio-economic well-being also reflects the prevailing unevenness in the overall development scenario in different parts of the state. Such an attempt is therefore proved to be useful in identifying problem areas and determining necessary planning inputs for a balanced human resource development across the state. While doing so more emphasis needs to be given to uplift the weaker sections of the society in respect of education, health and family welfare. The role of NGOs in this respect is vital not only for cost effective implementation of various social development programmes but also for ensuring active participation of people towards success of such programmes.

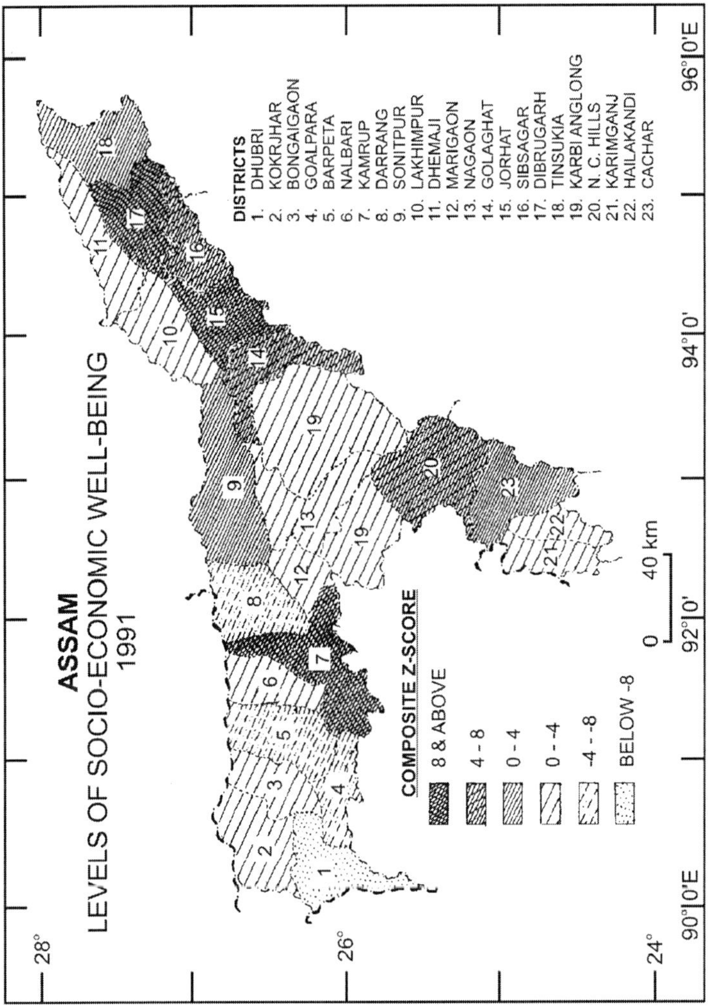
ASSAM
LEVELS OF SOCIO-ECONOMIC WELL-BEING
1991
COMPOSITE Z-SCORE
8 & ABOVE
4 - 8
0 - 4
0 - -4
-4 - -8
BELOW -8
DISTRICTS
1. DHUBRI
2. KOKRJHAR
3. BONGAIGAON
4. GOALPARA
5. BARPETA
6. NALBARI
7. KAMRUP
8. DARRANG
9. SONITPUR
10. LAKHIMPUR
11. DHEMAJI
12. MARIGAON
13. NAGAON
14. GOLAGHAT
15. JORHAT
16. SIBSAGAR
17. DIBRUGARH
18. TINSUKIA
19. KARBI ANGLONG
20. N. C. HILLS
21. KARIMGANJ
22. HAILAKANDI
23. CACHAR
0
40 km
28°
26°
24°
90°0'E
92°0'
94°0'
96°0'E

Fig. 10.5

Note

1. Composite-Z-Score, ΣZ_j-ΣX_{ij}-Xmi/d_i

Where, Xij is the value of ith attribute for jth area and x_i and *d_i respectively are the mean and standard deviation of the ith attribute for all areas.

References

Census of India (1971, 1981 and 1991) : *General Population Tables, Series 1,* India, Paper 1.

Census of India (1901) : Vol. IV, Assam, Part I, Report by B.C. Allen.

Census of India (1931) : Vol. III, Assam, Part I, Report by C.S. Mullan.

Census of India (1951) : Vol. XIII, Assam, Manipur and Tripura, Part I-A, Report by R.B. Vaghaiwalla.

Census of India (1991) : *Primary Census Abstract, General Population, Scheduled Castes and Scheduled Tribes,* Assam, Series - 4, Part II-B.

Census of India (1991) : Series 1, India, Paper-2, *Final Population Totals - Brief Analysis of Primary Census Abstract.*

Census of India (1991) : Series -4, Assam, Part IV-B (ii), *Religion,* (Table C-9).

Census of India (1991) : Series -4, Assam, Part IV, *Language.*

Census of India (2001 and 2011) : PCA, India.

Census of India (2011) : Religion, India.

Census of India (1991): *District Level Estimates of Fertility and Child Mortality and their Inter Relations with other Variables,* Occasional Paper No. 1 of 1997.

Directorate of Economics and Statistics (1980, 1996 and 2015) : *Statistical Hand Book, Assam.*

Gosal, G.S. (1993): 'India and the Demographic Transition Model', in Ahmad, A. (ed.) *Social Structure and Regional Development : A Social Geography Perspective,* Rawat Publications, Jaipur and New Delhi, pp. 105-119.

Kar, B.K. and Sharma, H.N. (1994): 'Women Literacy in Assam', *North Eastern Geographer, Vol. 25, pp. 18-39.*

Kar, B.K . (1995): 'Population Growth Trend in Assam and its Socio- Economic Implications', *Geographical Review of India,* Vol. 57, No. 4, pp. 321-35.

Kar, B.K. and Das, M.M. (1998): ' Spatio-Temporal Pattern and Characteristics of Small Towns in Assam', in Datta Ray, B. (eds) *Urban Development in North East India,* Reliance Publishing House, New Delhi, pp. 118-131

North Eastern Council (1995): *Basic Statistics of North Eastern Region,* Shillong.

Raza, M. and Agarwal, Y. (1986): 'An Analysis of Inequities in the Levels of Literacy in India' in Malhotra, P.L. (ed.) *School Education in India: Present Status and Future Needs,* NCERT, New Delhi, pp. 97-120.

Taher. M. (1993) : 'The Peopling of Assam and Contemporary Social Structure', in Ahmad, A. (ed.) *Social Structure and Regional Development : A Social Geography Perspective,* Rawat Publications, Jaipur and New Delhi, pp. 201-18.

11 Urbanization

The urban centres play an important role in socio-economic development in any area as centres of dissemination of administration, trade and commerce, industry, culture, education, technology and various important services and facilities. That is why the urbanization level is considered as one of the important indicators of socio-economic development. It is, however, unfortunate that Assam is lagging much behind in respect of urban development compared to the states like Maharashtra, Gujarat and Tamil Nadu. According to 2011 census, urban population constituted only 14.09 per cent of the total population of Assam as against country's 31.4 per cent. On the other hand, the proportions of urban population in Maharashtra, Gujarat and Tamil Nadu were as high as 45.20 per cent, 42.60 per cent and 48.40 per cent respectively. It is, however, noteworthy that the number of towns in Assam had increased from 12 in 1901 to 214 in 2011 through 87 in 1991.

Historical Background

The history of urbanization in Assam dates back to the pre-historic period. Information on urbanization, especially in the pre-British period, is not easily available. Assam has always been an agrarian state. A large proportion of its population engaged in the primary sector generally live in the villages. The emergence of non-agricultural activities like trade and commerce, services, transport and communication, industry, etc. has been altogether very slow until the recent past. Some traditional industries which developed in the long past have been still functioning.

The process of urbanization in Assam may be discussed according to the following characteristic time-periods:

1. Urbanization during the ancient period
2. Urbanization during the medieval period

3. Urbanization during the British period
4. Urbanization during the post-independence period.

Tezpur and Pragjyotishpura have been identified as ancient cities of Assam. Situated on the north-bank of the river Brahmaputra, Tezpur was the administrative capital of the Salastambha rulers (Barpujari, 1992). On the other hand, Pragjyotishpura was the capital of Koch kings, and later residence of Barphukan during the Ahom rule. Recent archaeological evidences proved that there were two more ancient capital cities- one in Baitabari, the capital city of Barman ruler, in the Phulbari region of Garo Hills district of present Meghalaya and another in Dhubrani region of Barpathar in Golaghat district (Choudhury, 1996). Both these cities date back to 6-8 century A.D.

During the medieval period , few capital cities built by the Ahom rulers emerged in places like Gargaon, Rangpur, Sibsagar and Jorhat. Other cities in Assam came into being during this period were Dimapur (capital of the Kacharis), Maibong (new capital of the Kachari Kingdom) and Sadiya (capital of Kundillya Kingdom). Situated mostly on the river-bank and characterized by palaces, high walls, temples, pavilion, tank, etc. these towns were basically rural in character.

During the British period, with the process of consolidation of British rule a new set of towns emerged in different parts of the state in association with the development of rail and motor transport, exploitation of forest and mineral resources, expansion of modern education, promotion of tea cultivation and trade and commercial activities. The medieval town had also undergone change both spatially and functionally. Tinsukia and Lumding may be cited as examples of trade-cum-rail town. British gave much priority towards expansion of transport and communication in the state with a view to exploit its rich natural resources. Accordingly, the Calcutta port was connected by feeder network. Obviously, the urban centres at that time were mainly acting as centres of syphoning out the surplus resources from the state. The British encouraged foreign capital investment especially in production. The British administration had also been responsible for large-scale migration of tea labourers and tradesmen into the state. Unfortunately, no adequate effort had been made to modernize agriculture. Tea industries and Digboi oil refinery were the only worth mentioning manufacturing units in the state during the British period. The rural-urban interaction in the state was almost absent. The towns that emerged were mostly the centres of civil administration.

It is, however, worth mentioning that the British admnistrators played a positive role towards urban development planning in the state. The steps taken in this respect were planning the lay-out of the urban centres, reserved land for public uses, development of public building in modern line, improvement of road network and introduction of Municipal/town improvement acts. The total effect in respect of urban growth in the state during the long British period was hardly significant. The number of towns in the state increased from 12 in 1901 to only 23 in 1941. The proportion of urban population had also increased from 2.3 per cent to 3.10 per cent during 1901-1941.

During the post independence period, both the number of urban centres and proportion of urban population in the state have witnessed a considerable increase. In fact, at the time of partition a large volume of migrants from within and outside the country had settled in the urban and semi-urban, and also rural areas of the state. There has been again the pressure of rural to urban migration within the state. In addition, due to development initiatives under the different five year plans, decentralization of administration and slow process of economic diversification and industrialization the number of urban centres and proportion of urban population in the state reached 87 and 11.08 per cent respectively in 1991. These figures further increased to 214 and 14.10 per cent in 2011.

Urban Growth : A Size-Class Analysis

The Census of India categorizes the urban centres into six groups on the basis of population size. This grouping helps understand the urban development pattern in any area. The urban centres with population of 1 lakh and above are grouped as class I, 50 thousand to 1 lakh as class II, 20 thousand to 50 thousand as class III, 10 thousand to 20 thousand as class IV, 5 thousand to 10 thousand as class V, and below 5 thousand as class VI. It may be mentioned here that the Census of India recognizes the places with a minimum population of 5 thousand, population density of 400 persons per km^2, 75 per cent of male workers in the non-agricultural activities and municipal committee or town committee as towns or urban centres. Besides, a place without satisfying the first three criteria may be recognized as town through the state government notification. In 1901, the urban population size in the state was only 77 thousand. Being distributed in 12 towns this consituted only 2.3 per cent of the state's total population. Out of 12 towns, Guwahati (11,661) and Dibrugarh (11,227) were the only Class IV towns in the state. The class V towns were Barpeta, Goalpara, Karimganj, Sibsagar, Silchar and Tezpur, while the Class VI were Nagaon, Dhubri, Jorhat and Golaghat. The towns under the category of class V together constituted about half of the total urban population of the state in 1901.

In 1921, the number of towns in the state increased to 22 as a result of the emergence of 10 new towns during 1901-1921. The important towns among them were Tinsukia, Lumding, Doomdooma, North Lakhimpur, Palasbari and Gouripur. The growth rate of urban population became 3.68 per cent per annum during that period. In 1941, the proportion of urban population in the state became only 3.1 per cent, with two-third being concentrated in 2 class III and 6 class IV towns. The urban population in the state, however, witnessed a significant increase during 1941-1951 (6.9 per cent per annum). Accordingly, the average size of population per town had increased from 8,669 to 14,071 during the period 1941-1951. Such a rapid growth of urban population was mainly because of the influx of migrant population at the time of partition. This put tremendous pressure on the existing infrastructures, services and job availabilities in the selected towns. Among the towns, Lumding, Hailakandi, Karimganj, Nagaon, Silchar, Dhubri, Mangaldoi and Makum witnessed a very high growth rate of population during that period. It is worth mentioning that the urban population growth rate in the state became as high as 12.6 per cent per annum during 1951-1961 as a result of the emergence of as many as 29 new urban centres and growth of existing ones. Moreover, Guwahati became the single class I city in the state after crossing the one lakh mark by 1961. The size of urban population in the state had further increased from 7.9 lakh in 1961 to 13.2 lakh in 1971 experiencing an annual growth rate of 6.68 per cent. The number of urban centres had increased to 71, and out of these 47 belonged to class V and VI categories. With the emergence of three more Class I urban centres called cities (Dibrugarh, Silchar

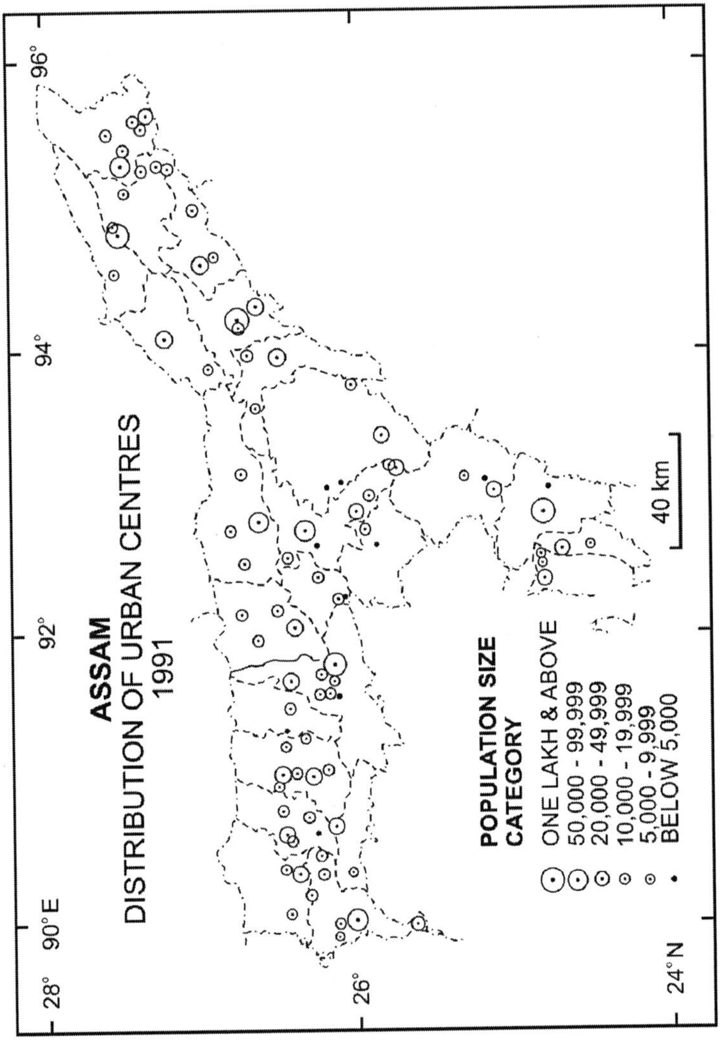
ASSAM
DISTRIBUTION OF URBAN CENTRES
1991
POPULATION SIZE
CATEGORY
ONE LAKH & ABOVE
50,000 - 99,999
20,000 - 49,999
10,000 - 19,999
5,000 - 9,999
BELOW 5,000
40 km
28°
90° E
92°
94°
96°
26°
24° N

Fig. 11.1

and Jorhat), the size of urban population became 24.7 lakh in 1991 being distributed in 87 urban centres (Fig. 11.1). Thus, Guwahati city together with these newly emerged cities have constituted about 38 per cent of the total urban population in the state. It is worth mentioning that due to the promotion of the small and medium urban centres to the respective higher orders, the proportion of urban population in the Class V and VI towns had decreased from 25 per cent in 1961 to 14.5 per cent in 1971 and 6.3 per cent in 1991. It means that there has been high concentration of urban population in the Class I and II urban centres. Some of the urban centres like Bongaigaon, Dhubri, Goalpara, Jorhat, Makum, Namrup, North Lakhimpur, Rangia, Sonari, Guwahati and Haflong witnessed considerably a high growth of population during the last two decades (1971-1991) due to functional diversities combined with inclusion of new areas within the urban boundaries. For instance, the population of Guwahati city was increased by more than 4 times during 1971-1991. This is mainly attributed to large-scale migration due to shifting of capital of Assam from Shillong to Guwahati and also increasing functional diversities, and physical expansion of urban boundary encompassing many new rural areas in the urban fringe.

During 1901-1991, the urban population in the state was increased by about 30 times, while the number of urban centres was increased by 7 times only. It may be mentioned that although the small towns constitute a very small proportion of population, their role is quite significant in providing direct services to the rural areas. However, a proportionate distribution of population in the different size category of urban centres is necessary towards bringing a balanced urban and regional development in the state. The proportion of urban population in the state has become 14.10 per cent in 2011. The Class I urban centres constitute 37.60 per cent of the total urban population. The population of Guwahati city has increased from 5.84 lakh in 1991 to 9.62 lakh in 2011.

Pattern of Urbanization at District Level

After about five decades of independence, Assam still remains as one of the least urbanized states in the country. On the other hand, Assam is well endowed with mineral and forest resources, fertile soil, high potential for hydel power development and extensive cultivation of tea and jute. It is discouraging enough to note that

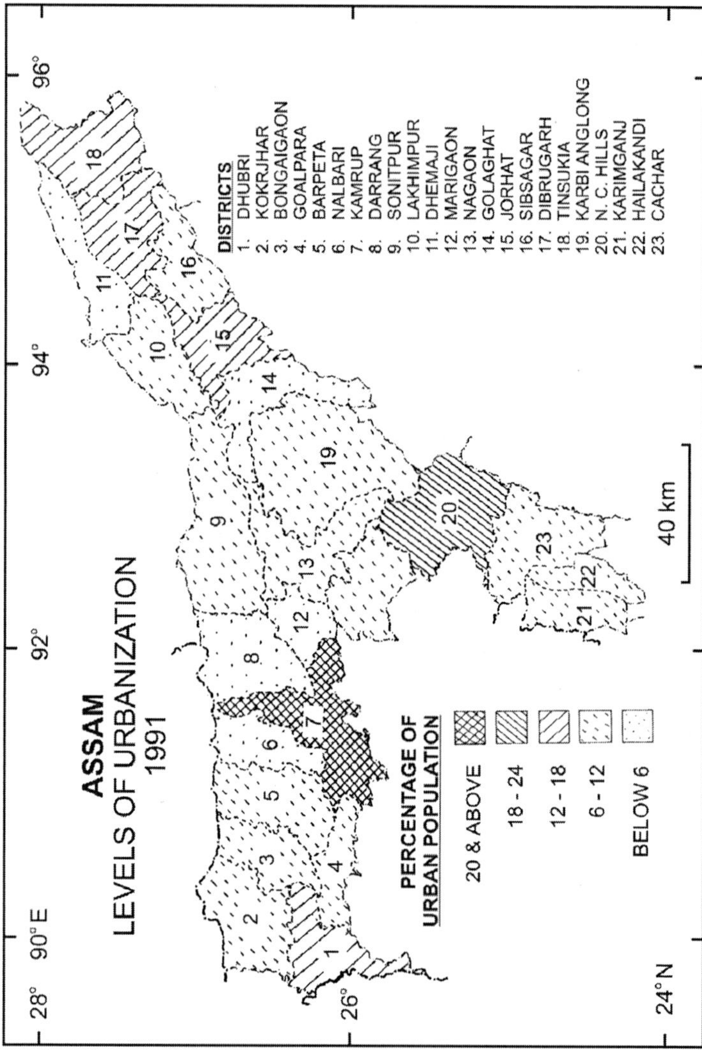
ASSAM
LEVELS OF URBANIZATION
1991
PERCENTAGE OF URBAN POPULATION
20 & ABOVE
18 - 24
12 - 18
6 - 12
BELOW 6
DISTRICTS
1. DHUBRI
2. KOKRJHAR
3. BONGAIGAON
4. GOALPARA
5. BARPETA
6. NALBARI
7. KAMRUP
8. DARRANG
9. SONITPUR
10. LAKHIMPUR
11. DHEMAJI
12. MARIGAON
13. NAGAON
14. GOLAGHAT
15. JORHAT
16. SIBSAGAR
17. DIBRUGARH
18. TINSUKIA
19. KARBI ANGLONG
20. N. C. HILLS
21. KARIMGANJ
22. HAILAKANDI
23. CACHAR
40 km
28°
26°
24°N
90°E
92°
94°
96°

Fig. 11.2

only Kamrup district in the state has got the higher proportion of urban population (32.4 per cent) than that of the country's average (25.8 per cent). Guwahati, the state capital of Assam and the headquarters of Kamrup district, alone accounts for 23.4 per cent of the total urban population of Assam. The least urbanized district (Dhemaji) has the urban population of only 1.88 per cent.

So far the level of urbanization is concerned, it is quite uneven across the state. As already mentioned, it varies from as low as 1.88 per cent in Dhemaji district to as high as 32.4 per cent in Kamrup district. Out of all the districts, only Kamrup, N.C. Hills, Dibrugarh, Tinsukia and Jorhat have the urbanization level above the state average (11.08 per cent). On the other hand, urbanization level is between 5 per cent and 11 per cent in as many as 14 districts. Dhemaji, Nalbari and Darrang are among the least urbanized districts in the state with urbanization level below 5 per cent (Fig. 11.2).

So far districtwise distribution of urban centers is concerned, Dhubri, Barpeta and Kamrup each has seven urban centres, while Nagaon and Karbi Anglong six urban centres each, Sibsagar, Dibrugarh and Tinsukia five each, Bongaigaon and Darrang four each, and the remaining districts have got 2 to 3 each, except Dhemaji district with only one urban centre.

Regional Pattern of Urbanization

Spatial variation in the levels of urbanization reflects a mixed scenario of regional economy in the state. For better understanding of the prevailing urban characteristics as many as six micro urban regions may be identified in Assam. They are (1) North-Bank Plain, (2) Lower Brahmaputra Valley, (3) Central Brahmaputra Valley, (4) Upper Brahmaputra Valley, (5) Barak Valley and (6) Hill Zone.

1. The North-Bank Plain Region

The districts of Darrang, Sonitpur, Lakhimpur and Dhemaji form the North-Bank Plain region. It accommodates 17.6 per cent of the state's total population. The population density is moderate (274 persons per km^2). The ubanization level is as low as 5.7 per cent as against the state average of 11.08 per cent in 1991 census. There are 10 urban centres in the region. Agriculturally, it is one of the advanced regions in the state. The share of its tea production is about 23 per cent of the state. So far industrial growth is concerned, the major industrial units and SSI units constitute about 10.32 per cent and 16.46 per cent of the state's respective total. The overall population growth in the region during 1971-1991 (60.42 per cent) was slightly higher than the state average (53 per cent). There is not a single Class I urban center in the region so far. Tezpur, with a population of 54,999 is the largest urban centre in the region belonging to category II. The road density in the region (0.36 km/km^2) is lower than the state average (0.40 km/km^2).

2. The Lower Brahmaputra Valley Region

This region comprises the districts of Dhubri, Kokrajhar, Bongaigaon, Goalpara, Barpeta, Nalbari and Kamrup. It constitutes 35.77 per cent of the state's total population. The density of population is considerably high here (395 persons per km^2). The growth of population is moderate. The level of urbanization here (13.8 per cent) is slightly higher than the state average. There are 32 urban centres in the region. The region is agriculturally productive. The share of its cereal production is quite high. Industrially also, the region is quite developed. 28.14 per cent and 40.95 per cent of the state's major industrial units and SSI units are located in this region. Guwahati, the capital city of Assam is the only Class I urban centre in the entire region. The other important urban centres include Dhubri, Bongaigaon and Goalpara. Significantly, about half of the state's total urban population lives in this region. The transport system is also considerably developed. The road density is found to be slightly higher than the state average.

3. The Central Brahmaputra Valley Region

The districts of Marigaon and Nagaon form the Central Brahmaputra Valley Urban region. This constitutes about 11.36 per cent of the state's total population. The population density is as high as 457 persons per km^2. The population growth rate was also significantly high during the last three decades. The proportion of urban population in the region is found to be only 9.50 per cent. There are 8 urban centres in the region. With a population of 93,324 Nagaon is the largest urban centre of the region. The area is agriculturally advanced. The road density of the region (0.57 km/km^2) is also significantly higher than the state average.

4. The Upper Brahmaputra Valley Region

This urban region is comprised of Golaghat, Jorhat, Sibsagar, Dibrugarh and Tinkukia districts. About 20.48 per cent of the state's total population lives in this region. The population density is moderate (282 persons/km^2). The population growth during 1971-1991 in the region was also moderate. The level of urbanization in the region (12.30 per cent) is slightly higher than the state average. There are 19 urban centres in the region. It is worth mentioning that this is the only urban region where there are two Class I urban centers– Dibrugarh and Jorhat. Tinsukia is emerging as an important trade and commercial centre and transport node in the region. Again, Digboi, the first industrial town of the entire north-east region is also located in this region. Being rich in mineral (coal and petroleum) and forest resource-base, this urban region alone contains as high as 45.80 per cent of the major industrial units and 24.14 per cent of the SSI units of the state. It is also the major tea producing area of the state. It is worth mentioning that more than 60 per cent of the state's tea is produced from this region. It thus reflects that the region has a high potential for further economic development in future.

5. The Barak Valley Region

This urban region consists of Karimganj, Hailakandi and Cachar districts. It shelters about 11.77 per cent of the state's total population. The density of population is found to be 360 persons/km^2. The level of urbanization is also quite low (8.55 per cent). There are 7 urban centers in this region. Unfortunately, this region has not witnessed any significant progress in agricultural front. Its share of tea production in the state is only about 10 per cent. Silchar is the only Class I urban centre of this region.

6. The Hill Zone

This urban region is comprised of Karbi Anglong and North Cachar Hills districts. Only 3.6 per cent of the state's total population lives in this region. The density of population here is extremely low (53 persons/km^2). The proportion of urban population in this region (12.93 per cent) is found to be slightly higher than the state average. There are 9 urban centres in the region. Diphu and Haflong, both belonging to category III, are the major urban centres in this region. This region is lagging behind both agriculturally and industrially. The proportions of major and SS industrial units in the region are only 2.2 per cent and 2.1 per cent of the state's total respectively. This region is, however, rich in forest resources and limestones. There is immense potentiality of horticulture development in the region in near future.

Problems of Urbanization

Although the pace of urbanization has been quite slow in the state, the urban problems like shortage of housing, lack of adequate safe drinking water, lack of efficient transport system, lack of adequate educational and health services, lack of proper drainage and sewerage system, lack of adequate employment opportunity, traffic congestion, lack of proper garbage management and shrinking open and public spaces, etc are not so uncommon in most of the urban centers. As a matter of fact, a sizeable amount of financial support and infrastructures are essential for minimum maintenance of the 87 urban centres in the state with a population of about 2.5 million. The maintenance of the urban roads, drainage and garbage disposal becomes crucial in most of the large urban centres particularly during the summer monsoon season. Because of lack of proper planning and management, the living condition in the urban areas gets extremely deplorable.

Out of so many problems, congestion in public places and, traffic and drainage problems deserve immediate attention. During the British period towns were too small both in respect of area and population. There was land reserved for different public utilities in those days. But, growth of population combined with diversification and intensification of urban functions over the years without proper planning have resulted in overcrowding in the public places and congestion of traffic in the urban areas. Besides, lack of maintenance of drainage , and sometime absence of proper drainage, results in stagnation of drain-water and makes the urban areas unhealthy.

The shortage of shelter and housing is another growing problem of the urban areas. As per the estimate of the National Building Organization (NBO), there was shortage of about 3.12 lakh houses in the urban areas of Assam in 1990. Among others, rapidly increasing cost of land and building materials is the important factor of growing shortage of houses. There are poor people constituting about one-third of the total urban population who want both work place and shelter in close proximity as far as possible. Besides, improvement of existing road network and construction of new roads are another issue. It may be mentioned here that of the total road length in the urban areas of Assam in 1991-92 only 28.11 per cent was surfaced road.

Urban problems are interdependent in nature and can be solved only with sound and integrated framework of action plan. Creation of physical and socio-economic infrastructure requires heavy investment. Hence, any step for solution of the problems has to be made in phased manner on priority basis. The approach in any such case needs be holistic one. In order to reduce the expenditure a cost effective technology is to be explored especially in the areas of housing, drinking water and sanitation. The success of any such effort depends largely on the people's participation.

Urban Development Prospect

The above discussion reveals that in spite of having a rich natural and agricultural resource potential, the state of Assam still occupies a low position in respect of urban development. The process of industrialization in the state has been altogether very slow. Only 3.78 per cent of the total workers in the state are engaged in the secondary sector. The primary sector still dominates the state's economy. In the agricultural front also, only 34.66 per cent of the total area is brought under cultivation and out of that only 27.90 per cent area is under double and triple cropping. The yield of food grains per hectare of land and per capita food production in the state are much lower than that of the national average. The large-scale tea plantation also could not make any significant headway in the process of urbanization in the state. The per capita income in the state has still been quite low. The migration and its associated implications have also been greatly responsible for low pace of overall development in the state. In spite of these constraints there is prospect and potentiality for urban and economic development in the state.

Under the present circumstances, there is need of formulation of a long-term economic development plan for the entire state. The long-term investment plan should be prepared by integrating various functional sectors as well as rural and urban areas in a regional framework where in urban centres would act as the engines of the growth process. The existing agriculture and allied sector is to be modernized so as to have surplus production. The manufacturing activities are to be promoted in consideration with resource availability and feasibilty. The trade and commercial activities are to be strengthened in the urban areas establishing adequate linkage with the rural settlements. The trade linkages need be established with the neighbouring countries also. Entrepreneurship skills need be developed through innovative training and motivation. In all such endeavour the urban development is to be perceived in the context of overall economic development programme of the state.

References

Barpujari, H.K. (ed.) (1992) : *The Comprehensive History of Assam*, Vol. 4, Publication Board, Assam, Guwahati.

Census of India (1991 and 2011): *Primary Census Abstract, Assam.*

Census of India (1971): *Town Directory, Series 3,* Part IV, Assam.

Census of India: *Housing Report and Tables,* Series 3, Part IV, Assam.

Choudhury, N. (1996): 'Recent Archaeological Excavations in Assam', *The Assam Tribune, August 29, Guwahati.*

Govt. of Assam (1987, 1994 and 2015): *Statistical Hand Book of Assam,* Directorate of Economics and Statistics, Guwahati.

Govt. of Assam (1995): *Urban Development Strategy, Assam* (Interim Report), Directorate of Town and Country Planning, Assam, Guwahati.

Govt. of Assam (1987): *Report of the National Commission on Urabanization,* Ministry of Urban Development, New Delhi.

Govt. of Assam (1983) : "Planning of Urban Development", Task Force on Housing and Urban Development, Planning Commission, New Delhi.

North Eastern Council (1987) : *Basic Statistics of North Eastern Region, Shillong.*

Ramachandran, R. (1989) : *Urbanization and Urban Systems in India,* Oxford University Press, New Delhi.

12 Agriculture

Introduction

The thickly populated areas with a very high proportion of rural population cultivating individually owned land, generally exhibit a kind of farming which may be called small scale agriculture. It is characterised by small size of landholding and technologically a low level of farming practice. The small size of landholding that too fragmented into several scattered plots creates a situation quite detrimental to adoption of modern agricultural practice. Such characteristics along with other adverse socio-economic factors limit the size of agricultural output resulting in a disability of the sector to keep pace with the increasing demand for crops (Bhagabati, 1990).

In India, there are large number of farmers operating very small to moderately small and medium-size landholdings. On rough estimate (save the intensive irrigated areas) the minimum subsistence land for a family under the present condition can be considered to be four hectares (Singh and Dhillon, 1984). According to the agricultural census, 1990-91, nearly 89 per cent of the operational holdings of the country is of below 4 ha size. The figure is well indicative of the dominance of smallholders' farming in the country. It is more so in the case of Assam where the average size of operational holdings is 1.28 ha and the percentage of holding below 4 ha size is nearly 95 (1990-91). As a matter of fact, in spite of various efforts made by the government towards improving the condition of small farming through land reforms and a number of agricultural modernisation programmes, Assam still continues to be dominated largely by smallholders, less productive subsistence type of farming, Only in the case of a small fraction of relatively rich farmers, the subsistence mode of farming has been in transition from predominantly traditional to a modernised one. However, any development in the sector undoubtedly bears special significance on the overall economic development of the state.

In fact, broadly two types of farming have been in practice in Assam: small-scale peasant and large-scale tea plantation farming. Between the two, the peasant farming is more important in the sense that it alone contributes slightly less than half of the State's Domestic Product. Moreover, while it covers 65 per cent of the State's total working force, the tea plantation shares less than 10 per cent. Besides the peasant farming is more problem-ridden and enigmatic, than the tea plantation (Das, 1984). As the mode of production, management and the nature of problems of plantation farming are quite different from those of the wide space-cum-people covering peasant farming, the former is excluded from the purview of the present discussion. The study is primarily concerned with the pattern of cropping, performance of the agricultural sector in terms of yield and output and the factors influencing it. Special emphasis has been given to rice, the most dominant crop in the state in terms of both hectareage strength and productivity.

Agro-Climatic Zones

Based on rainfall and temperature, terrain and soil characteristics, Assam has been broadly classified into six agro-climatic zones:

1. ***Upper Brahmaputra Valley (North)*** : comprising the districts of Dhemaji, Lakhimpur, Sonitpur and Darrang. Main crops grown are rice, rape and mustard and sugarcane.
2. ***Central Brahmaputra Valley*** : comprising the districts of Nagaon and Morigaon. Rice, jute, rape and mustard, pulses are mainly cultivated here.
3. ***Upper Brahmaputra Valley (South)*** : comprising the districts of Tinsukia, Dibrugarh, Sibsagar, Jorhat and Golaghat. Main crops include rice, rape and mustard and sugarcane.
4. ***Lower Brahmaputra Valley*** : comprising Kamrup, Goalpara, Dhubri, Kokrajhar, Bongaigaon, Barpeta and Nalbari district. Main crops are rice, jute, rape and mustard, potato, wheat and pulses.
5. ***Hills*** : comprising the districts of Karbi Anglong and North Cachar Hills. Main crops are rice, maize and sugarcane.

6. ***Barak Valley*** : comprising the districts of Cachar, Hailakandi, and Karimganj. Main crops include rice, sugarcane and potato.

Each of these zones has its own particularities. However, there is a great deal of uniformity in the agricultural practice in the plain zones. During the summer months when monthly average rainfall remains above 300 mm, the main crops cultivated are rice and jute. During the winter months on the other hand, the rainfall remains very low (monthly average being below 10 mm) and *rabi* crops like rape and mustard, wheat, potato and various kinds of vegetables are cultivated. In the hills, the dominant agricultural practice is *Jhumming* (shifting cultivation). During the recent period, the *jhum* cycle in most part has declined to below five years resulting in degradation of land and natural vegetation. In the valley floors, however, wet cultivation is performed wherever possible. There is thus a basic variation in the farming practice over the agro-climatic zones in the State.

Agricultural Land Use

Agricultural land use constitutes a dominant feature of the diverse uses of land for productive purposes. It includes not only the land used for growing crops and fodder, but also that for raising livestocks. Cropland, as defined by Eastal (1976) in the context of the US, is an assessment of potential as well as of current use. It includes harvested areas, land that is fallow, idle or in cover crops. It also includes land in the crop rotation currently used for pastures.

The present pattern of agricultural land use in Assam is an outcome of long-continued human settlement and culture. Within the state, diverse ecological setting of the river valleys, hills and ridges and other microphysiographic units combined with varying socio-economic factors have produced significant spatial variation in the pattern of agricultural land utilisation (Bhagabati, 1990). However, in Assam agricultural land use basically means the cultivation of soil for growing crops only, leaving insignificant areas for grasslands, horticulture, pisciculture and dairy farming unlike the developed agricultural regions of the world.

In Assam, there has been a slow increase in area under different crops. The gross cropped area in the state increased

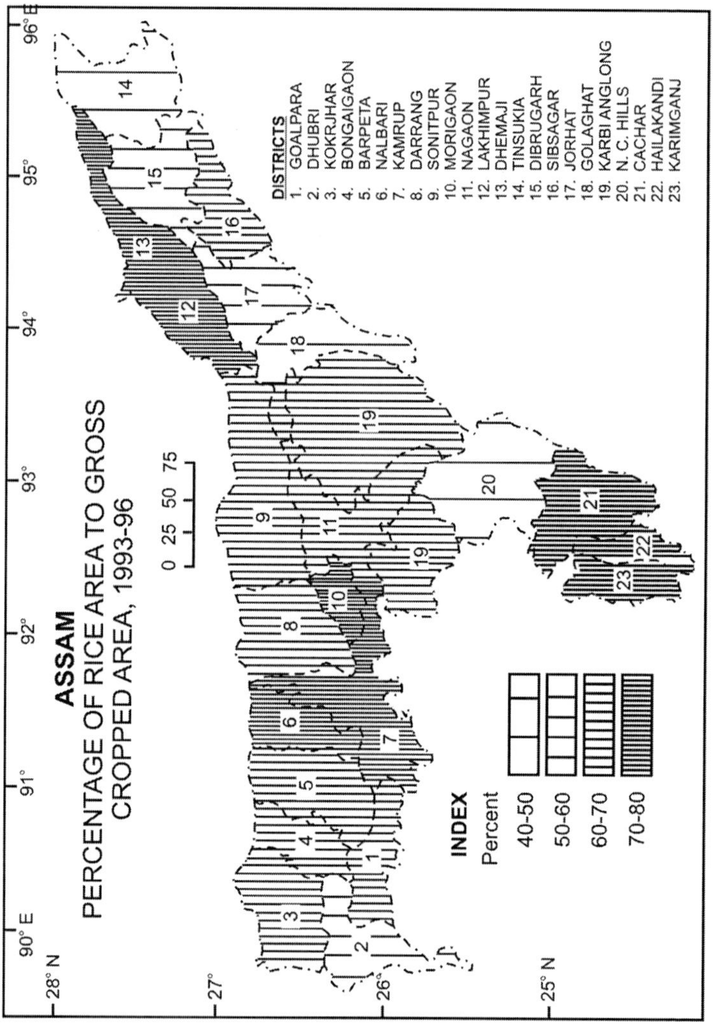
ASSAM
PERCENTAGE OF RICE AREA TO GROSS
CROPPED AREA, 1993-96
0 25 50 75
INDEX
Percent
40-50
50-60
60-70
70-80
DISTRICTS
1. GOALPARA
2. DHUBRI
3. KOKRJHAR
4. BONGAIGAON
5. BARPETA
6. NALBARI
7. KAMRUP
8. DARRANG
9. SONITPUR
10. MORIGAON
11. NAGAON
12. LAKHIMPUR
13. DHEMAJI
14. TINSUKIA
15. DIBRUGARH
16. SIBSAGAR
17. JORHAT
18. GOLAGHAT
19. KARBI ANGLONG
20. N. C. HILLS
21. CACHAR
22. HAILAKANDI
23. KARIMGANJ
90° E
91°
92°
93°
94°
95°
96° E
28° N
27°
26°
25° N

Fig. 12.1

from 22.03 lakh ha in 1951-52 to 31.76 lakh ha in 1975-76 and recently to 39.26 lakh ha (1993-96). About 71 per cent of the total cropped area has been occupied by the foodgrains.

Area under Major Crops

Rice : The cropland of Assam is generally dominated by the cultivation of foodgrains, viz. rice, wheat, maize, gram and small millets. Among these, rice as the staple food crop continues to dominate the state's agriculture. In 1993-96 (triennium average), the area under rice was 25.33 lakh ha accounting for 64.52 per cent of the total cropped area and 91.43 per cent of the total area under foodgrains (Table 12.1and Fig. 12.1). It is note worthy that rice area increased from 15.4 lakh ha in 1951-52 to 25.3 lakh ha in 1995-96 registering a growth of 64.28 per cent during the period. Since the introduction of high yielding varieties of rice in the last part of the 1960s, the area under rice in the state noticeably increased. Rice cultivation again received a boost after the great flood of 1988, when the farmers of the state made all efforts to bring more and more flood-free area under it.

Table 12.1
Rice area as P.C. of the total cropped area in Assam, 1993-96
(area in hectares)

District	Total cropped area	Rice area	P.C. of rice area to total cropped area
1. Goalpara	98461	68503	69.57
2. Dhubari	226550	127605	56.32
3. Kokrajhar	142605	94947	66.58
4. Bongaigaon	153243	95596	62.38
5. Barpeta	312965	201482	64.37
6. Nalbari	196883	144199	73.24
7. Kamrup	223534	157906	70.64
8. Darrang	279094	171133	61.31
9. Sonitpur	222459	146940	66.05
10. Morigaon	125746	92630	73.66
11. Nagaon	380889	253474	66.54
12. Lakhimpur	157337	113955	72.42
13. Dhemaji	92379	69598	75.33
14. Tinsukia	128250	64119	49.99
15. Dibrugarh	159702	87579	54.83

16. Sibsagar	163947	105187	64.15
17. Jorhat	163838	97256	59.36
18. Golaghat	175230	81326	46.41
19. Karbi Anglong	181269	122351	67.49
20. N.C. Hills	34655	14908	43.01
21. Cachar	139275	100466	72.13
22. Hailakandi	59679	42199	70.70
23. Karimganj	108153	79843	73.82
Assam	**3926143**	**2533202**	**64.82**

Source: Directorate of Agriculture, Govt. of Assam.

Considering the growing and harvesting period, rice in Assam is categorised into three main groups: autumn rice (*ahu*), winter rice *(sali)* and summer rice *(boro)*. An important variety of winter rice locally known as *bao* is cultivated in the low-lying areas where floods and waterlogged conditions make cultivation of other rice difficult. *Bao* has the special ability to grow keeping pace with rising water during flood. It is interesting to note that among the three groups of rice a trend of taking over others area has recently emerged which may be attributed to the growing awareness among the farmers to select the varieties best suited to a particular ecological set-up.

Winter Rice *(Sali)* : Nearly 70 per cent of the state's rice area is under *sali* rice. It is transplanted in prepared soil from nursery beds in July-August and harvested in November-December. Its area increased from 1.22 million ha in 1951-52 to 1.80 million ha in 1990-91 and thereafter recorded a slight decrease to 1.76 million ha in 1995-96. During the period 1951-96, *sali* rice area increased by 44.26 per cent. Among the districts of Assam, Sibsagar recorded the highest proportion of area under it (61.96 per cent of the district's gross cropped area). It is followed by Hailakandi (61.04 per cent), Cachar (60.81 per cent), Karbi Anglong (60.71 per cent), Dhemaji (58.60 per cent). Karimganj (56.88 per cent), Lakhimpur (55.61 per cent), Nalbari, (51.81 per cent) and Jorhat (50.87 per cent). The remaining districts have *sail* rice area below 50 per cent of their respective gross cropped area. Dhubri and North Cachar Hills exhibit the lowest proportion (around 24 per cent).

Autumn Rice *(Ahu)* : *Ahu* rice occupies 24.0 per cent of the state's total rice area. It is cultivated by both transplantation and broadcast methods. *Ahu* is usually sown in February-March and harvested in

July-August. From 0.32 million ha in 1951-52 its area increased to 0.63 million ha in 1995-96, the percentage of increase being 96.87. The cultivation of this group of rice is mainly concentrated in the lower Brahmaputra valley districts where density of population is relatively high. In each of the districts of Dhubri, Goalpara, Kokrajhar, Bongaigaon, Barpeta, Nalbari, and Darrang, *ahu* rice accounts for more than 20.0 per cent of the gross cropped area. The upper Brahmaputra valley districts on the other hand have below 10.0 per cent of gross cropped area under it.

Summer Rice *(Boro)* : *Boro* rice occupies about 6.0 per cent of the state's total rice area. The share of this rice in the total cropped area of almost all the districts is substantially lower than that of *sali* and *ahu*. But interestingly, its area has increased manifold since 1951-52. From a mere 0.03 lakh ha in 1951-52, the area under

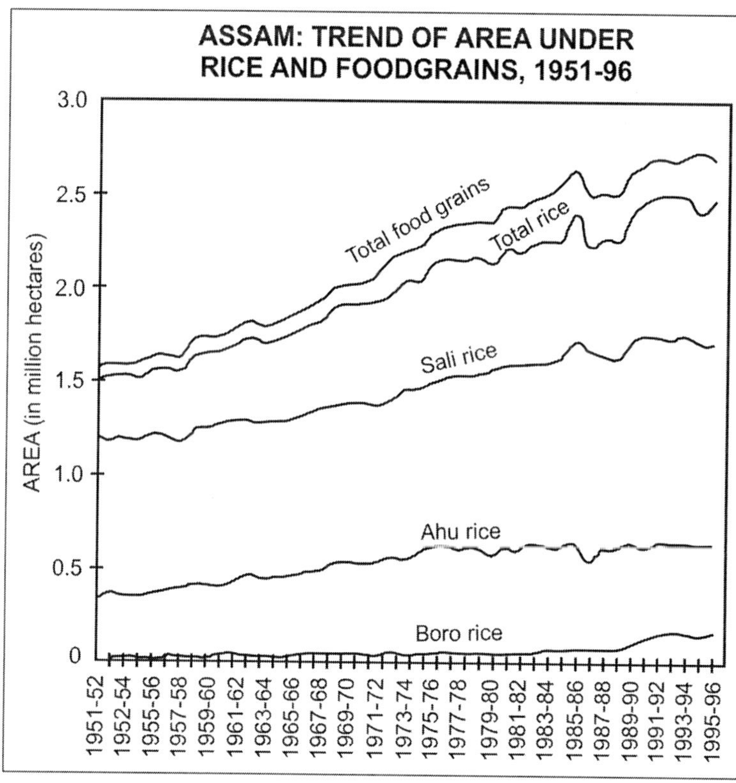

Fig. 12.2

boro rice increased to 0.14 ha in 1995-96. This rice is grown in the low-lying marshy lands during the winter and harvested just before the arrival of the first wave of flood in June-July.

A look into the district level pattern shows that *boro* rice occupies less than 1.0 per cent of the gross cropped area in each of the districts Nalbari, Sibsagar, Jorhat, Golaghat, Dibrugarh, Tinsukia, Lakhimpur, Karbi Anglong and N.C Hills. Its position is slightly better in the districts of Goalpara. Dhubri, Kokrajhar, Bongaigon, Barpeta, Darrang, Sonitpur, Dhemaji, Cachar and Hailakandi (1-5 per cent). Compared to the above districts, Karimganj (6.02 per cent), Kamrup (6.73 per cent) and Nagaon (10.52 per cent), and pariticularly Marigaon (20.98 per cent) show a much better picture in respect of *boro* rice cultivation. In Morigaon district, the presence of a large number of wetlands provides suitable land for the cultivation of this rice.

The trend of increase in *boro* rice area in the State (Fig 9.2) is really noteworthy. It show a sharp increase after 1989. The main reason behind this is the fact that *boro* rice has gained increasing popularity among the farmers of the chronically flood-affected areas where *sali* rice generally gets damged by floods.

Wheat : Although the arrival of wheat *(Triticum aestivium)* with some significance in the crop-landscape of Assam dates back only to the mid-sixties of the last century, presently it has been able to occupy a notable position in the crop distribution pattern of the state. The indigenous people of the region even at present do not consider wheat to be their staple food and generally they devote not more than a small portion of their land to this crop. But very recently a growing tendency among the immigrants as well as indigenous peasants to cultivate wheat has been observed in the lower and the central Brahmaputra valley regions.

In 1951-52, wheat occupied only 0.11 per cent of the state's total cropped area. Its area increased to 0.72 per cent in 1970-71 after the introduction of HYV wheat in the state under the influence of green revolution that took place in western India. With 0.85 lakh ha of area in 1995-96, wheat has now spread to all the districts of Assam. The lower Brahmaputra valley comprising the districts of Dhubri, Goalpara, Kokrajhar, Bongaigaon, Barpeta, Nalbari and Kamrup is the main area where wheat has been successfully grown.

Pulses : Occupying 4.32 per cent of the state's total cropped area in 1995-96, next to rice pulses are the important food crops of Assam. Of the total area under pulses, *rabi* pulses such as lentil *(mosur),* blackgram *(matimah)* green gram *(moog)* and pea *(matar)* cover more than 90 per cent. The central and lower Brahmaputra valley districts, particularly Goalpara, Barpeta and Kamrup produce pulses successfully.

Oil Seeds : With 3.3 lakh ha of land under cultivation, oil seeds account for nearly 10 per cent of the state's total cropped area. The important oil seeds grown in Assam include rape and mustard, sesamum, linseed and castor. Among these, rape and mustard alone occupies more than 90 per cent of the area under the oil seeds. In all the districts of the Brahmaputra valley, rape and mustard is cultivated. In the Barak valley, its cultivation is not significant (below 2 per cent).

Jute : Among the cash crops such as jute, sugarcane, mesta, cotton, etc., jute is the most important in terms of hectareage strength. The history of commerical jute farming in Assam is, however, not very long. The partition of India in 1947 affected the jute industries of the Hooghly side of West Bengal badly as the major jute-producing areas went to the erstwhile East Pakistan. Therefore, efforts were made in India to extend jute cultivation in some new areas. Extension of its cultivation to Assam was a direct impact of large-scale immigration of peasants from the neighbouring overcrowded districts (Mymensing, Pabna and Rangpur) of East Pakistan (present Bangladesh). Bhuyan (1977) goes as far as to say that Assam would never become prominent as one of the principal jute-growning states had there been no immigration from East Bengal.

Jute cultivation is concentrated in the lower and central Brahmaputra valley districts of Assam. The area under this crop amounts to 0.88 lakh ha in 1995-96.

Sugarcane : Sugarcane has been grown in Assam since long. In the historic past also the people of the state knew the art of producing various types of molasses from sugarcane. In the *Harsha Charit,* it has been mentioned that among the various presents sent by king Bhaskarvarma of Kamrup to

Harshavardhana, *guda (molasses)* was one. The muslim travellers and historians who came to Assam during the Ahom period referred to sugarcane and confirmed its cultivation in the state. Quazim, for instance, wrote that sugarcane of Assam was soft and sweet and was of three colours-red, black and white (Choudhury, 1987).

Sugarcane is cultivated in all the districts of Assam, but with varying degrees. Its cultivation is mainly concentrated in the Nagaon and Golaghat districts of the Brahmaputra valley and Karimganj district of the Barak valley. In 1995-96, the total area under sugarcane was 36,000 ha.

In addition to the above mentioned important crops, certain other crops having lesser hectareage strength are cultivated in the state. They are potato, sweet potato, banana, papaya, maize, tobacco, chilli, ginger and several kinds of vegetables. The crops like chilli, ginger and tobacco are now cultivated only by a small number of farmers involving very little agricultural land. Occupying very small area these crops fetch more money than some of the major crops because of their higher market price. Moreover, the gradual emergence of certain new crops like soyabean and sunflower in the cropping pattern of Assam during the recent years, has been found to enrich the agricultural economy of the state.

Crop Output and Yield

Crop output and yield depend partly on natural circumstances and partly on human manipulation of the land resources. The level of agricultural productivity or yield as a concept means the degree to which the economic, cultural and techno-organisational variables are able to exploit the abiotic resources of an area for agricultural production (Singh, 1979). It is dynamic as any modification in the physical factors and improvement in the non-physical bases definitely affects the output of crops per unit of area (Singh and Dhillon, 1984).

Among the food crops, rice has the highest demand in Assam as it is the staple food of the people of the state. Presently the foremost challenge before the state is the appalling increase in population and a corresponding upsurge in the demand for food.

The state would be able to meet this challenge only through properly planned sustainable agricultural development programmes. The state recorded an increase of 143.88 per cent in its rice output during the period 1951-96 as againts the country's 259.22 per cent. In 1995-96, the state produced 3.39 million tonnes of rice. Of this, 2.62 million tonnes were constituted by *sali,* 0.51 miilion tonnes by *ahu* and the rest 0.26 m tonnes by *boro* (Fig. 12.3). In the last few years, Assam witnessed a remarkable increase in *boro* rice output. From mere 0.02 lakh tonnes, its output reached the present level. Recently *boro* rice has gained unprecedented popularity, especially in the flood-affected areas of the state.

So far per hectare yield of rice is concerned, Assam lags far behind the national status. The average yield of rice in the state increased from 919 kg in 1950-51 to 1, 421 in 1995-96 as against

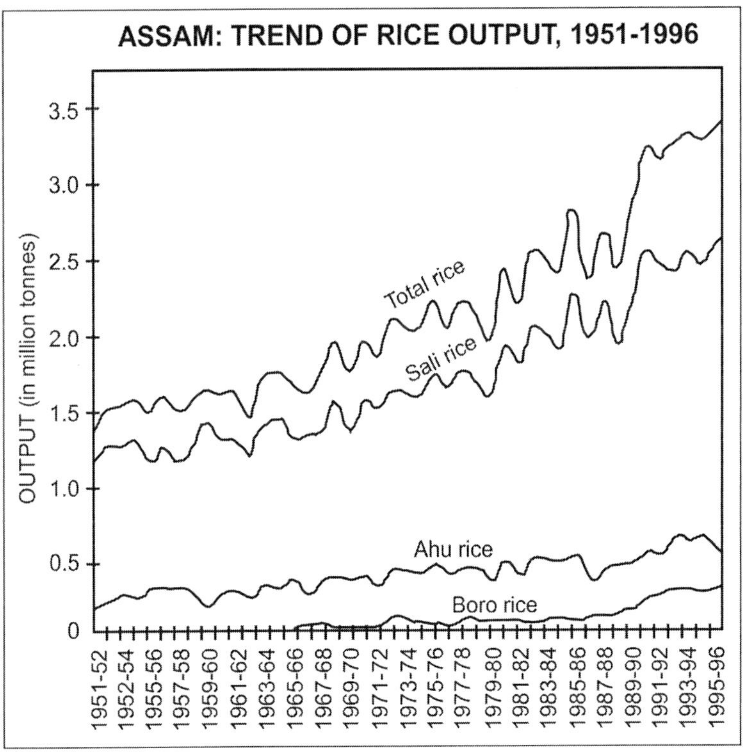

Fig. 12.3

the country's average of 668 kg and 2,617 kg respectively. When compared with the agriculturally advanced states like Punjab (3,500 kg), the position of Assam appears to be quite dismal.

An analysis of the triennium yield data at the district level reveals that the rice yield during 1993-96 on an average was 1,384 kg per ha. The district of Golaghat recorded the highest yield (1,796 kg) followed by Sibsagar (1,787 kg), N.C. Hills (1,786 kg), Dibrugarh (1,667 kg) and Cachar (1,616 kg). Most of the lower Brahmaputra valley districts on the other hand, recorded a relatively low yield (878-1282 kg). Interestingly, the two hill districts- Karbi Anglong and N.C. Hills had average yields much higher than the state's average. In these two districts, wet rice cultivation is practised with all possible care as almost all other major crops are raised through the age-old system of *jhumming* with a very low yield. The yield level in the Barak Valley districts is also found to be relatively high mainly due to extensive *boro* rice practice as well as higher level of adoption of HYV rice. It is important to note that the average yield of *boro* rice has increased from 901 kg in 1951-52 to 1,619 kg in 1995-96, an increase substantially higher than *sali* and *ahu.*

The output of wheat in the state at present stands at 95,000 tonnes and its average yield is 1,107 kg per ha. The production of maize on the other hand is 12,966 tonnes with an average yield of 681 kg. The district of Karbi Anglong shows the highest prodution of maize. In 1995-96, 57.38 per cent of the state's total production of maize came from this district.

The present output of pulses in the state is 6400 tonnes. Nonavailability of best seeds in time and the acidic character of the soil are the main constraints to the growth of their output in the state. The National Pulses Development Programme (NPDP) is being implemented since 1986-87 to increase the production of pulses. However, the desired effect of the scheme is yet to be obtained due to various reasons including shortage of fund. The average yield of puleses in Assam stands at 584 kg. With 970 kg per ha the Golaghat district exhibits the highest yield.

The production of edible oil seeds at present is around 1.8 lakh tonnes against a requirement of 8.0 lakh tonnes. Its production is a little above 1 per cent of the country's total oilseeds output. The central and the state goverments have put stress on making the country self-sufficient in the production of oilseeds. The prospect of *rabi* oilseeds in the state is bright, but the scope for increasing their output during the *kharif* season is very limited due to adverse weather condition. *Kharif* output contributes only 10 per cent of the total oilseeds output of the state. The average yield of oil seeds is 560 kg, the Golaghat district exhibiting the highest yield (880 kg/ha).

The table 12.2 gives the yield rate of some selected crops cultivated in Assam.

Table 12.2
Yield rates of selected crops (in kg/ha) of Assam

Crop	1990-91	1995-96
Autumn rice	873	839
Winter rice	1446	1513
Summer rice	1556	1619
Wheat	1248	1107
Jute	1632	1715
Sugarcane	42510	41553
Potato	7240	6816
Rape and Mustard	535	514

Source : Directorate of Economics and Statistics, Assam

Sugarcane production in the state amounts to 1.5 million tonnes with the yield rate of 41, 533kg/ha. Considering the potentiality of sugarcane cultivation in the state, the government of India introduced a centrally sponsored scheme called "Sustainable Developement of Sugarcane Based Cropping System" during 1995-96 covering 8 districts. This scheme does not intend to increase the area under sugarcane but aims at enhancing its yield so that overall income of the farmers per unit area may be increased.

Levels of Technification

Technification as a determinant of agricultural development implies the use of modern techniques in agriculture. As a process, it includes functions like irrigation, fertilizer input, adoption of HYV seeds, use of insecticides and pesticides and other forms of mechanisation (Bhagabati, 1997). The levels of application of these functions , vary depending upon a variety of natural and socio-economic factors. The section to follow discusses the levels of technification in the agrculture of Assam in respect of certain modern techniques adopted.

Irrigation

In the monsoon lands, where winter cropping is difficult due to lack of rain water, irrigation plays a decisive role in respect of number of crops raised during an agricultural year. Irrigation, in fact, forms the principal component of agricultural modernisation which is basically more a package than a unitary process.

It is disheartening to note that the gross irrigated area in

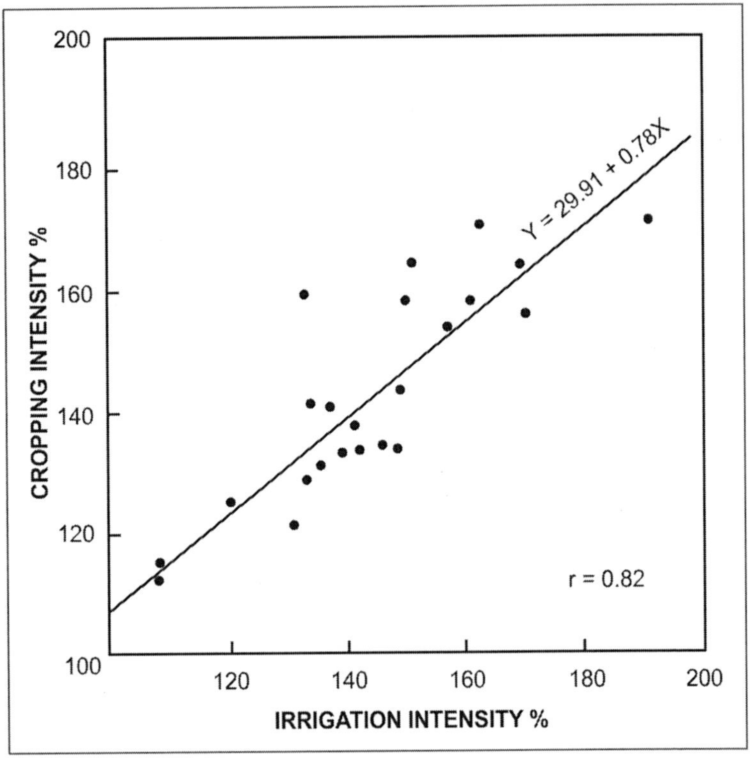

Fig. 12.4

Assam constitutes only 15.06 per cent of the state's gross cropped area (1990-91). It is here worth mentioning that about 3.5-4.0 lakh ha of land in the state are annually affected by flood. The standing *kharif* crops (mainly rice) are usually damaged by floods. Under such circumstances the only alternative left for the farmers of the flood-prone areas is to take up *rabi* cropping in order to compensate the loss caused by flood during the summer. Extensive *rabi* cropping, however, may not be economically viable if irrigation facilities are not adequately provided. It is a fact that most of the flood-prone areas in the state do not have any sort of irrigation facility. As a result, the lands either remain fallow during winter or render very low yield. On the other hand, the areas where there is regular and timely supply of irrigation water, particulary in winter , show substantial progress in productivity. The chronically flood affected Malaybari area of Kamrup district sets a good example in this regard. The cropping intensity is also found to be higher where intensity of irrigation is more. There is thus a strong positive relationship between the two variables (Fig. 12.4).

It is to be noted that the medium and large irrigation projects have not been found to be successful in Assam because of either non-completion of the projects or mismanagement of the systems. However, the agriculture of the state has benefitted mostly from the minor irrigation schemes. Schemes having a culturable command area upto 2000 ha individually are classified as minor irrigation schemes. There are pratically six types of minor irrigation schemes:(1) surface flow, (2) surface lift, (3) deep tubewell, (4) shallow tubewell, (5) dugwell and (6) mobile diesel pump.Of these, the shallow tubewells have proved more successfull as the size of the landholding is generally small and the farmers are unable to invest more capital. There is hope that the massive installation of shallow filters undertaken by the government recently may substantially change the state's agricultural scenario, if implemented properly.

Fertilizer Input

Application of modern technology generally enhances the frequency of cropping and the overall crop output. When more than one crop is raised from the same land, obviously the natural fertility of the soil declines. This necessitates the use of fertilizer. Fertilizer is, in fact, one of the basic ingradients of the agricultural development package without which it fails to give the expected results. Now-a-days, however, application of biofertilizers is more preferred as these are eco-friendly and sustainable.

In Assam, the level of fertilizer input is very low. The per hectare input of fertilizer is only 10.5 kg. The commonly used chemical fertilizers are Nitrogen, Phosphorus and Potassium. In 1995-96, the total fertilizers consumption in the state amounted to 49, 141 metric tonnes. The use of fertilizer is maximum in the Jorhat district, the per hectare input being 20 kg. In the districts like Kokrajhar, Bongaigaon, Lakhimpur, Dhemaji, Hailakandi, Karbi Anglong and N.C. Hills, the use of fertilizer is minimal (only 4 kg/ha).

Today, however, more stress has been laid on the application of biofertilizers like compost, rhizobium culture, azolla, etc. A scheme called 'Biofertilizer and Organic Manures' under the state plan is being implemented since 1991-92. The distribution system of fertilizers in the state is channelised through AAIDC, STATEFED, ASAMB, GPSS and other private agencies or dealers under the control of the Department of Agriculture, Government of Assam.

HYV Seeds

The high yielding seed breeding technology is considered as a revolutionary transition from the age-old tradition to contemporary innovation in the practice of agriculture. The everincreasing population demands biotechnological research to invent high yielding varieties (HYV) so that more food could be produced. Introduction of these varieties in the soils of north-western India during the late sixtees of the 20th century as a part of the green revolution, brought about significant changes in the level of foodgrain output of the country. This revolutionery change in the states like Punjab, Haryana and Western UP generated a spread effect which gradually reached eastern India including Assam.

In Assam, HYV rice was introduced during the late 1960s and presently nearly 50 per cent of the state's total rice area is under these varieties (Table 12.3). Important HYV rice varieties adopted in Assam are IR-8, IR-20, Jaya, Ranjit, Bahadur, Pankaj, Biplav etc. Among the districts of Assam, Golaghat, Morigaon, Hailakandi and Karbi Anglong have more than 60 per cent of the total rice area under HYV rice. On the other hand, districts like Kokrajhar, Tinsukia, Lakhimpur and Dhemaji exhibit a very low percentage of area (less than 15 per cent) under HYV rice (Fig.12.5).

In the case of wheat, the area under HYV wheat in the total wheat area has always been high. As a matter of fact from 1972-73 onward all the varieties of wheat grown in Assam have been of HYV strain. However, it is important to note that with only 3.17 per cent of the state's total cropped area, wheat does not have a significant position in the agriculture of Assam. Moreover, whatever may the area be, the yield of wheat usually remains far below the country's average. Besides ecological constraints, the fact that wheat cultivation in Assam is a recent phenomenon, is adequately reflected in its low area and yield.

Implements and Machineries

It has been realised by all concerned that there is no alternative of mechanisation for agricultural growth at this moment. No field can be properly irrigated unless water is raised by electrically driven pumps or diesel generators. No pesticides can be applied properly to the plants except by means of manually or mechanically operated sprayers. Soil moisture cannot be conserved unless the soil is ploughed deep with iron-made soil scratchers. Thus, it is now commonly agreed that a farmer has to make use of modern implements and machines of some form in order to boost his agricultural output.

Machines are superior and efficient as compared to manual labour and draught power. It is, however, to be remembered that topography and edaphic factors largely control the use of farm implements and machinery in an area. In addition, the quantum of irrigation available, the size of holding, the degree of farming and the farmer's way of life further determine the use of modern farm implements.

In Assam, the strong attachment of the farmers to the wooden plough, an ancient soil scratcher, prevents him from using other means of ploughing. The wooden plough, however, can be

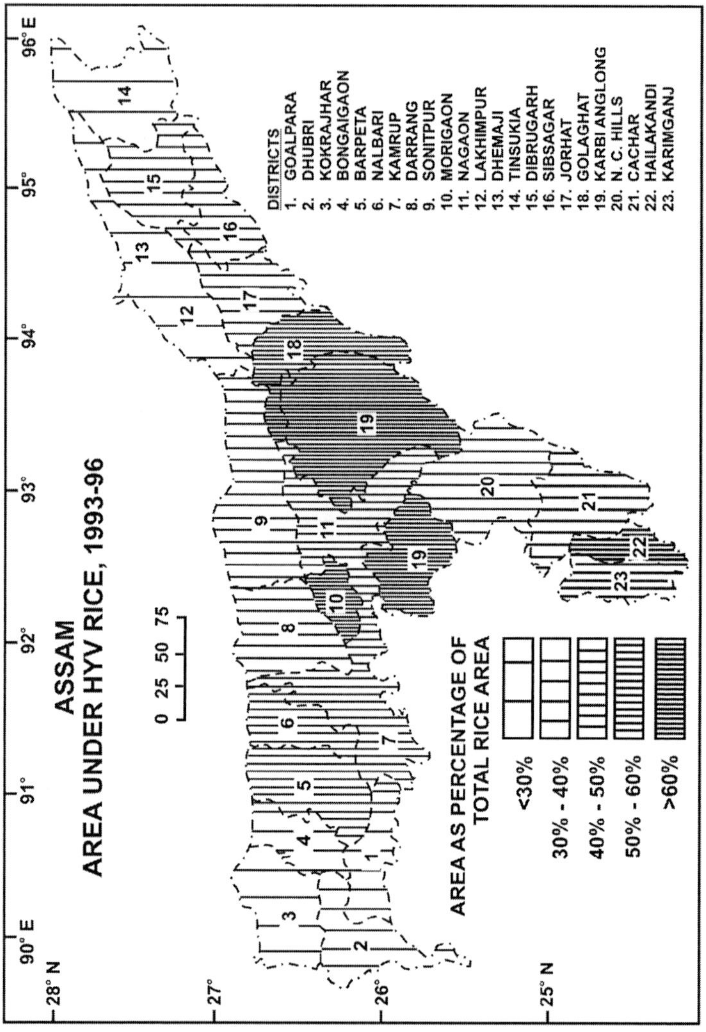
ASSAM
AREA UNDER HYV RICE, 1993-96
0 25 50 75
AREA AS PERCENTAGE OF TOTAL RICE AREA
<30%
30% - 40%
40% - 50%
50% - 60%
>60%
DISTRICTS
1. GOALPARA
2. DHUBRI
3. KOKRAJHAR
4. BONGAIGAON
5. BARPETA
6. NALBARI
7. KAMRUP
8. DARRANG
9. SONITPUR
10. MORIGAON
11. NAGAON
12. LAKHIMPUR
13. DHEMAJI
14. TINSUKIA
15. DIBRUGARH
16. SIBSAGAR
17. JORHAT
18. GOLAGHAT
19. KARBI ANGLONG
20. N. C. HILLS
21. CACHAR
22. HAILAKANDI
23. KARIMGANJ
90° E
91°
92°
93°
94°
95°
96° E
28° N
27°
26°
25° N

Fig. 12.5

improved upon to increase its efficiency without much difficulty and expense. Agriculture, for further improvement, requires appliences which make it possible to reduce harvest losses and enable farmers to complete farm operation within a short span of time. This may help them in intensifying the farming practice.

Table 12.3
Percentage of HYV rice area to total rice area in Assam (1993-96) (area in hectares)

	Districts	HYV rice area	Total rice area	P.C. of HYV rice area to total rice area
1.	Goalpara	24636	68503	35.96
2.	Dhubari	40686	127606	31.88
3.	Kokrajhar	20715	94948	21.81
4.	Bongaigaon	30026	94828	31.66
5.	Barpeta	119530	201483	59.32
6.	Nalbari	77122	144200	53.48
7.	Kamrup	79441	157906	50.30
8.	Darrang	69165	171133	40.41
9.	Sonitpur	60472	146941	41.15
10.	Morigaon	58311	92631	62.94
11.	Nagaon	131038	253475	51.69
12.	Lakhimpur	26214	110377	23.74
13.	Dhemaji	17522	69599	25.17
14.	Tinsukia	12949	64120	20.19
15.	Dibrugarh	37561	87580	42.90
16.	Sibsagar	42576	105188	40.47
17.	Jorhat	41514	97256	42.68
18.	Golaghat	53684	81324	66.01
19.	Karbi Anglong	75373	122351	61.60
20.	N.C. Hills	6238	14909	41.84
21.	Cachar	42438	100466	42.24
22.	Hailakandi	25764	42200	61.05
23.	Karimganj	45722	79843	57.26
	Assam	**1137949**	**2529658**	**44.98**

Source : Basic Agricultural Statistics, Directorate of Agriculture, Government of Assam.

The Department of Agriculture in Assam through its

engineering wing has been trying to develop and popularise certain implements suitable for small and marginal farmers who constitute more than 90 per cent of the state's total farmers. Through its 12 workshops, the department has popularised MB plough, paddy weeder, harrow, sprayer, puddler, etc. among the farmers of the state. The present status of availability of farm machineries and implements in the state along with the population that use them has been presented in Table 12.4.

Table 12.4
Type of farm equipments and the number of people using them in Assam

Type of equipment	Population
Manual	
Paddy weeder	17,926
Dry land weeder	2,032
Seed drill	722
Paddy cultivator	4,723
Sprayer	1,68, 449
Duster	52
Other minor implements	1,67,449
Power operated	
Tractor	1,178
Power tiller	4,785
Sprinkler	58
Pumpsets	17,742
Animal drawn	
Mould board plough	1,10,477
BD Puddler	149
Three tyre cultivator	200
BD harrow	4,723

Source : Department of Agriculture, Assam.

The centrally sponsored scheme of "Promotion of Agricultural Mechanisation" was launched in the year 1992-93. Accordingly an amount of Rs. 37.6 lakh has been provided to the State for supply of low HP tractors to the farmers at a subsidised rate. The total number farmers benefitted from this scheme in the state stands at 151 only. Thus it has been seen that the use of modern implements and machineries by the farmers of Assam is highly limited.

Table 12.5
Total cropped Area and production of Rice Area and HYV Rice Area in Assam at District Level, 2014-15

District	Total Cropped Area (in Hectare)	Rice Area (in Hectare) and Proportion of Rice Area to total Cropped	Area under HYV Rice (in Hectare) Proportion of HYV Rice Area to Total Rice Area (%)
1. Kokrajhar	175290	108382 (61.83%)	74815 (69.02%)
2. Dhubri	163033	86058 (52.78%)	84142 (97.77%)
3. Goalpara	133626	84268 (63.06%)	55123 (65.41%)
4. Barpeta	216243	135966 (62.87%)	103937 (76.44%)
5. Morigaon	128241	89922 (70.11%)	66900 (74.39%)
6. Nagaon	319708	201733 (63.09%)	155034 (76.85%)
7. Sonitpur	272729	173087 (63.46%)	134927 (77.95%)
8. Lakhimpur	182050	124166 (68.20%)	83725 (67.42%)
9. Dhemaji	131859	85539 (64.87%)	58946 (68.91%)
10. Tinsukia	154767	63513 (41.03%)	43297 (68.17%)
11. Dibrugarh	163530	74728 (45.69%)	48118 (64.39%)
12. Sivsagar	149901	97859 (65.28%)	66329 (67.78%)
13. Jorhat	155994	91893 (58.90%)	58803 (63.99%)
14. Golaghat	201281	123624 (61.41%)	70544 (57.06%)
15. Karbi Anglong	207725	127996 (61.61%)	82585 (64.52%)
16. Dima Hasao	54371	14700 (27.03%)	9733 (66.21%)
17. Cacher	165406	107882 (65.22%)	75683 (70.15%)
18. Karimganj	87454	77402 (88.50%)	51174 (66.11%)
19. Hailakndi	72531	45343 (62.51%)	32238 (71.09%)
20. Bongaigaon	87310	52849 (60.53%)	35383 (66.95%)
21. Chirang	88414	45525 (51.49%)	29016 (63.73%)
22. Kamrup	171513	113627 (66.24%)	109797 (96.62%)
23. Kamrup Metro	49472	29210 (59.04%)	
24. Nalbari	115497	76375 (66.12%)	52143 (68.27%)
25. Baksa	144670	99920 (69.06%)	67423 (67.47%)
26. Darrang	119319	69073 (57.88%)	54697 (79.18%)
27. Udalguri	170660	94657 (55.46%)	60020 (63.40%)
Assam	**4082594**	**2495297 (61.12%)**	**1764532 (70.71%)**

Source: Statstical Handbook, Assam, 2016, directorate of Economics and Statistics, Govt. of Assam

Prospects for Future Development

The development of agriculture in Assam still depends mainly on the performance of rice cultivation. Rice is the dominant crop for all the farming families, no matter how low its yield may be. The agricultural land use of the state as has already been mentioned is characterised by an extraordinarily high proportion of rice hectareage (Das. 1984) and as such, in any plan for the development of the agricultural sector, rice is expected to receive topmost priority.

During the last few years, debates on sustainability and long-term effectiveness of development measures renewed the tendency to look to the traditional systems as potential ways of promoting sustained development (Ehlers, 1997). But, for meeting the immediate and expanding demand, the traditional practices are often denied and the question of sustainability of the modern practices is simply neglected.

In Assam, the agricultural practice is, by and large, tradition-bound. The present status of farm implements, fertilizer and irrigation water used, methods of crop protection and harvesting, storage adopted clearly indicate that there is yet a long way to modernise the state's farming system. This is no less true in the case of rice farming too. Rice is cultivated in a framework of highly fragmented landholdings, the average size being 1.28 ha as against the country's average of 1.68 ha. Interestingly, more than 80 per cent of the operational holdings in the state are below 2 ha size. Moreover, more than 25 per cent of the state's net sown area is regularly affected by flood during June-September which coincides with the *sali* rice season. Under such circumatances, it is really hard to prescribe a alternative rice farming system which would be environment friendly, economically viable and socially acceptable. The existing rice farming system in the state has to be reoriented in the line of what is known as 'ecofarming' in which the plant biomass, weeds, animal dung and rain water can be constantly integrated into the farming process following a befitting crop-calender in consonance with the seasonal weather changes and the rhythms of the behaviour of the Brahmaputra river system, in order to make it sustainable. The main thrust of rice farming should be shifted from *sali* to pre- flood early *ahu* and post-flood *boro* rice according to local ecological suitability. There is thus a bright prospect of increasing yield and output of rice in the state, if its cultivation is properly guided from a right perspective. The prospect of other *rabi* crops, especially vegetables, is equally bright in the state.

It is a fact that small-scale agriculture alone cannot help accrue sufficient income for a aminimum standard of living and for investment in agriculture. Therefore, there should be scope for subsidiary works in allied occupations such as poultry farming, piggery, fishing and other small-scale agro-based industries. However, it can be generalised that the small-scale agriculture forms a wide but strained base on which the economy of Assam and the future of the toilling masses largely rest. The necessary reinforcement of such a weak base can be made through proper implementation of the agricultural development plans constructed within the state's geographical reality.

References

Bhagabati, A.K. (1990) : Spatial Analysis of Small-Scale Agriculture in Assam : A Case Study of Nalbari District, unpublished Ph. D. dissertation, Gauhati University, Guwahati.

Bhagabati, A.K and Dutta,D. (1999): 'Innovation in Farming Practice and Dynamics of Crop Productivity in Malaybari Area of Kamrup District, Assam, *Journal of Geography,* Gauhati University, Vol.2, pp.18-32.

Bhagabati, A.K. (1997): Levels of Technification in Agriculture of N.E. India', *Survival 2000 Plus,* pp. 38.45.

Bhuyan, M.C. (1977) : Immigrant Population in Assam, unpublished Ph.D. dissertation, Gauhati University, Guwahati.

Das, M.M. (1984): *Peasant Agriculture in Assam,* , Inter- India Publication, New Delhi.

Dutta, M. and Bhagabati, A.K. (1998): 'Dominance of Rice in the Agriculture of Assam and Changing Pattern of its Productivity,' Proceedings of National Conference on Environment and Sustainable Development, Cotton College, Guwahati.

Ehlers, E. (1997): 'Traditional Environmental Knowledge and Consciousness and the Problem of Sustained Agricultural Development,' *Applied Geography and Development,* Vol. 49, pp. 79-95.

Estal, R. (1976) : *A Modern Geography of United States,* Penguin Books, Middlesex, England.

Singh, J (1979) : 'A New Technique of Delimiting Agricultural Production Typology in Food-Crop Dominated Economy', *Geographica Polonica,* Vol. 40, p.21.

Singh, J. and Dhillion, S.S. (1984) : *Agricultural Geography,* Tata McGraw- Hill Publishing Company, New Delhi.

13 Mineral Resources

Introduction

Assam is a treasure home of mineral resources. Lots of work have been done by the state Directorate of Geology and Mining for assessing and locating these mineral resources since the mid 19th century. Jenkin (1835,1853) was the first man who gave an account of limestone deposits of the state. Hannay (1845, 1856) presented the picture of petroleum-bearing horizon of the state. Later on, Dalton and Hanny gave an account of the gold-yielding deposits of upper Assam in 1853. But, the gold extraction from the Subansiri river dates back to earlier than eighth century. In 1932, AOC (Assam Oil Company) did a comprehensive work regarding the sedimentary rocks of Assam. But the story of oil goes back as early as the first part of the 19th century. In the case of coal, regular mining was started in 1882 by Railway engineers. In 1882, Ledo Collieries was started followed by Tikak (1884) and Namdang (1896). Besides oil and coal, Assam possesses vast quantities of mineral resources, viz. Limestone, sillimanite, base-metals, bery1, building and road metals, asbestos, fire clay, kaolin, fuller's earth, mica, quartz, silicon, dolomite, etc. But the extraction of most of these is not economically viable except coal, petroleum, limestone and sillimanite (Fig 13.1).

Geological Background

The gneissic group is the oldest rock group in Assam whose age has been fixed as Archaean which mainly occupies major parts of Karbi Anglong district, parts of North Cachar Hill district and some parts of Darrang, Nagaon, Kamrup and Goalpara districts. Quaternary deposits are represented in Assam by the older and newer alluvium covering major parts of the plain areas.

The Tertiary sediments comprise the Jaintia, Disang, Barail, Tipam, Surama, Dupi Tilla and Dihang groups. The limestone,

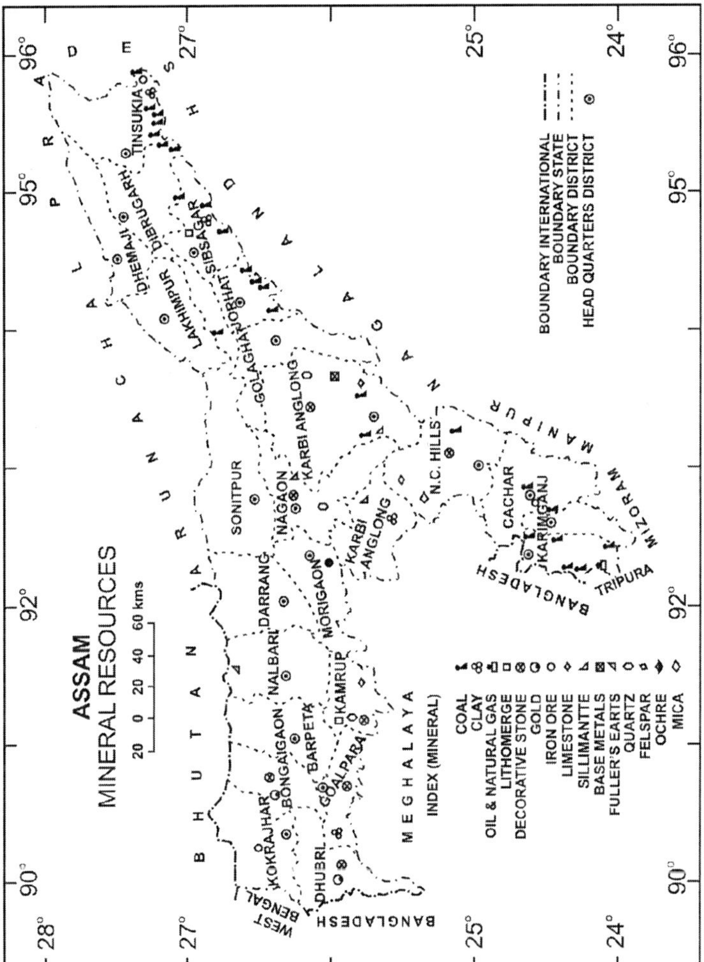
ASSAM
MINERAL RESOURCES
20 0 20 40 60 kms
INDEX (MINERAL)
COAL
CLAY
OIL & NATURAL GAS
LITHOMERGE
DECORATIVE STONE
GOLD
IRON ORE
LIMESTONE
SILLIMANTTE
BASE METALS
FULLER'S EARTS
QUARTZ
FELSPAR
OCHRE
MICA
BOUNDARY INTERNATIONAL
BOUNDARY STATE
BOUNDARY DISTRICT
HEAD QUARTERS DISTRICT
BHUTAN
ARUNACHAL PRADESH
NAGALAND
MANIPUR
MIZORAM
TRIPURA
BANGLADESH
MEGHALAYA
WEST BENGAL
KOKRAJHAR
DHUBRI
BONGAIGAON
GOALPARA
BARPETA
NALBARI
KAMRUP
DARRANG
MORIGAON
NAGAON
SONITPUR
KARBI ANGLONG
N.C. HILLS
CACHAR
KARIMGANJ
GOLAGHAT
JORHAT
SIBSAGAR
LAKHIMPUR
DHEMAJI
DIBRUGARH
TINSUKIA
90°
92°
95°
96°
24°
25°
27°
28°

Fig. 13.1

sandstone and shale assemblages of the Jaintia group are exposed in Garampani of North Cachar Hills and Sheelvetta, Koilajan, Deopani and Nambar area of Karbi Anglong district. In upper Assam, the Surama Group is completely absent and the Barail Group is directly overlain by the Tipam Group.

Assam possesses vast quantities of different mineral resources. The economically viable mineral resources including the energy resources are described below.

Coal

Coal is one of the important mineral resources belonging to the "energy group". Assam coal belongs to two distinct stratigraphical horizons Tertiary formations and Gondwana formations. In upper Assam, the two important Tertiary coal fields are : (i) Makum coal field and (ii) Dilli-Jaipur coal field. The Makum coal field is the oldest coal field and it is about 30 km long and 5 km wide. It has

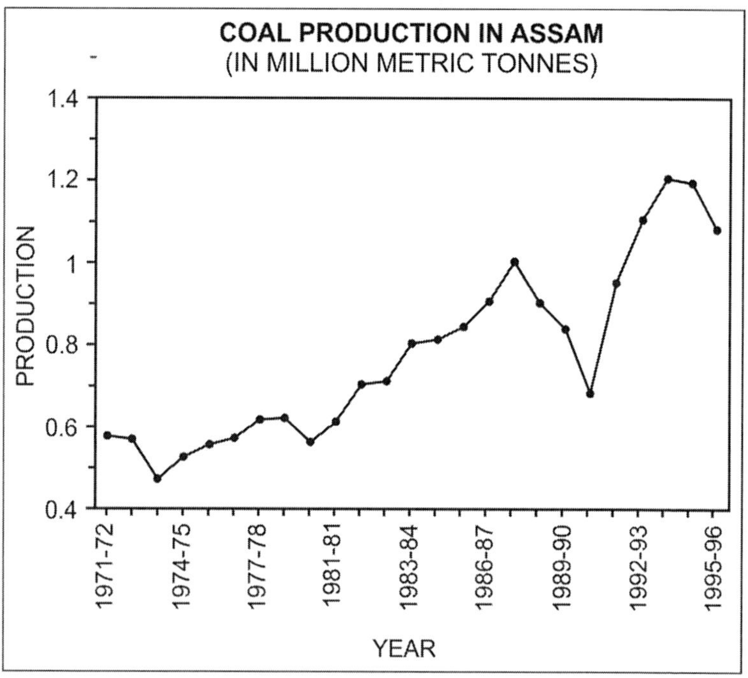

Fig. 13.2

four main mines-Namdang, Bargolai, Ledo and Tipong. The coal division of the Geological Survey of India (GSI) has been conducting regional exploration by drilling work. There are more than 210 million tonnes of high quality coal to be extracted from the Margherita- Ledo coal belt. According to GSI, coal reserve in Makum coal field is 235.66 million tonnes. The Dilli-Jaipur coal fields are located on the east of Dibrugarh and Sibsagar districts. The thickness of the coal seams varies. The total reserve has been estimated by GSI at 54.02 million tonnes.

Table 13.1
Coal reserves in different fields of Assam

Name of coal field	Depth (in meter)	Reserve			Total (in million tonnes)
		Proved	Indicated	Infured	
Tertiary Coal field					
1. Makum	0-600	123.66	50.19	61.81	235.66
2. Dilli-Jaipur	0-300	9.03	14.19	30.80	54.02
3. Karbi Anglong District	0-300	0.69	-	2.02	2.71
Gondwana Coal					
Singrimari	0-300		2.79		2.79
Total		**133.38**	**67.17**	**94.63**	**295.18**

Source : Directorate of Geology and Mining (DGM), Assam.

Table 13.2
Mine-wise production of coal in Assam

Name of Mine	Annual production (Metric Tonnes) 1992-93	1993-94	1994-95
Tipong underground	1,53,012	1,55,007	1,47,428
Tirap open cast	6,03,039	5,90,066	6,68,664
Ledo underground	95,179	1,00,258	1,03,212
Bargolai underground	1,18,426	1,27,007	1,27,991
Tikak open cast	1,08,822	2,00,077	1,17,149
Jaipur underground	21,697	26,235	26,545
Namchic	-	-	1,450
Simsong	-	-	905

Source : Directorate of Geology and Mining, Assam.

Table 13.3
Coal production in Assam, 1971-1996

Year	Production (million metric tonnes)
1971-72	0.577
1972-73	0.571
1973-74	0.471
1974-75	0.526
1975-76	0.556
1976-77	0.572
1977-78	0.620
1978-79	0.622
1979-80	0.563
1980-81	0.612
1981-82	0.703
1982-83	0.710
1983-84	0.801
1984-85	0.812
1985-86	0.841
1986-87	0.905
1987-88	1.001
1988-89	0.900
1989-90	0.836
1990-91	0.680
1991-92	0.951
1992-93	1.100
1993-94	1.201
1994-95	1.190
1995-96	1.078

Source : Directorate of Geology and Mining, Assam.

Besides the existing Tertiary coal fields, a number of small isolated coal fields located in Koilajan, Sheelveeta, Khumba, Longloi and Daigrung river bed of Karbi Anglong, and Khota Arda and Dithor in North Cachar Hills district belong to Gondwana formation. At Koilajan a thin coal seam has a thickness of 1.80m. But the coal is of the good quality. The total coal reserve of Koilajan and Sheelvetta is estimated at 2.71 million tonnes. At Singrimari small amount of coal occurs in the Lower Gondwana rocks.

An estimate made by the GSI states that Assam has coal

reserves to the tune of 292.39 million tonnes, of which 40 million tonnes have already been exploited from the earth's crust by open cast mining. The table 13.1 shows the coal reserves and production in Assam. From table 13.3, it is clear that the production of coal has declined during 1994-96, while it was the highest during the year 1993-94. At the beginning of this century, the production of coal was only 2 lakh tonnes which increased to 10.78 lakh tonnes during 1995-96.

Oil and Natural Gas

Assam is well blessed by nature in respect of oil and natural gas, which is termed as 'liquid gold.' The story of Assam oil goes back as early as to the time of Lieut R. Wilcox, who recorded the occurrence of oil seepage in the Burhi-Dihing at Sapkhong in 1825. Later on, drilling was started in 1889 at Digboi. During the pre-independence period the quest for oil was carried out by the B.C.C. and after independence, intensive search for oil has been carried out by the ONGC. During the second quarter of the present century, a considerable amount of prospecting and drilling had been done by the Assam Oil Company Limited. Oil has been discovered also in Naharkatiya-Hugrijan areas, namely Dihing valley in 1938, Moran in 1956, Lakwa in 1965, Rudrasagar in 1961, Geleki in 1968, Borholla in 1970, Amguri in 1971, Changpang in1973, Charali in1974, Demulgaon in 1977, Changmaigaon in 1984, Kuargaon in 1985, Namti in 1986, and a gas field at Adamtola in 1989. The Naharkatiya-Hugrijan areas in the Dihing valley located 40 km south-west from Digboi have the reserves of 45 million tonnes of oil and 210 billion cubic metres of gas. In the Lakwa and Geleki field, the Tipam sandstones contain thick sands which are expected to become good producers. Besides these fields, oil is also located at Borholla, Charali and Demulgaon.

Presently in Assam, the refineries located at Digboi, Guwahati and Bongaigaon are in operation, while the fourth one at Numaligarh is nearing completion. The Digboi refinery is the oldest refinery in India. The Digboi plant has an annual capacity of 0.50 million tonnes of crude which is being expanded to 0.65 million tonnes. The Guwahati refinery has a capacity of 1 million tonnes while the Bongaigaon refinery with a capacity of 1.35 million tonnes, is being expanded to 2.35 million tonnes. The fourth refinery at Numaligarh will have a capacity of 3 million tonnes per year. Tables

13.4 and 13.5 show the production figures of crude oil and natural gas in the state produced by OIL and ONGC.

During 1994-95, Assam's crude oil production was 4.808 million metric tonnes. Of this, 2.195 million metric tonnes and 2.070 million metric tonnes were produced by OIL and ONGC, and till 1996 February, ONGC produced 1.96 million metric tonnes. The natural gas produced by OIL & ONGC was 175.10 million cubic meter during 1994 -95 (Table 13.5).

The ONGC has so far probable 6 structures, of which 2 are hydrocarbon-bearing. At present ONGC has 5 main trunk pipe lines, viz. Lakwa-Moran, Rudrasagar-Lakwa, Geleki-Dikhow Junction, Geleki- Jorhat and Borholla-Jorhat to take crude oil to the terminal at Moran from where it is transported by OIL to the refineries. The prognosticated reserve of 1,05,000 sq. km covered by both OIL and ONGC is about 3900 MMT oil and OEG (Oil Equivalent of Gas).

Table 13.4

Production of crude oil by OIL and ONGC in Assam, 1980-1995

Year	Net production by OIL (in million metric tonnes)	Net production by ONGC (in million metric tonnes)	Total production (in million metric tonnes)
1980-81	1.289	0.420	1.709
1981-82	3.016	1.750	4.766
1982-83	2.809	2.136	4.945
1983-84	2.841	2.129	4.970
1984-85	2.688	2.165	4.853
1985-86	2.592	2.264	4.856
1986-87	2.562	2.510	5.072
1987-88	2.395	2.605	5.000
1988-89	2.370	2.826	5.196
1989-90	2.636	2.835	5.471
1990-91	2.551	2.245	4.796
1991-92	2.416	2.336	4.752
1992-93	2.359	2.344	4.703
1993-94	2.648	2.029	4.677
1994-95	2.738	2.070	4.808

Source : DGM, Assam.

Limestone

Limestone deposits of Assam are located in Karbi Anglong and North Cachar Hills districts. Many limestone beds are interbedded with coal-bearing sandstones. Good quality limestones belonging to the Eocene age are suitable for manufacturing cement and have been commercially exploited for this purpose. In Umrangshu area of North Cachar Hills district, limestone occurs with a thickness of 80 metres. In Karbi Anglong district, limestone occurs in Koilajan and Dillai Parbat. In Koilajan area, the deposit was investigated by State Directorate of Geology and Mining in the year 1962 and proved a reserve of 30 million tonnes of cement–grade limestone. The Dillai Parbat limestone deposit in Karbi Anglong has total reserve of 29.41 million tonnes. The reserve of cement–grade limestone occurs in Sainilagso and Sheelvetta, but has no economic significance. In the North Cachar Hills district the Umrangshu area has a massive bedded deposits with a thickness of 80 meters. In Panimur and Garampani, limestone was discoverd by the State Directorate in late sixties but had no economic importance (Table 13.6).

Table 13.5
Production of natural gas by OIL and ONGC in Assam, 1982-1995

Year	Net production by OIL (in million cubic meter)	Net production by ONGC (in million cubic meter)	Total production (in million cubic meter)
1982-83	663.87	171.10	834.97
1983-84	641.44	89.99	731.43
1984-85	652.73	143.50	796.23
1985-86	685.27	85.14	770.41
1986-87	777.98	112.25	890.23
1987-88	799.36	232.04	1031.40
1988-89	737.16	238.68	975.83
1989-90	770.95	236.78	1007.73
1990-91	812.46	178.56	991.01
1991-92	788.50	170.62	959.12
1992-93	813.51	185.01	998.52
1993-94	711.78	160.01	871.79
1994-95	748.98	175.10	924.08

Source : DGM, Assam.

The production of limestone grew moderately during the last decades. During the year 1995-96, however, there have been some positive change and its production increased to 0.4080 million metric tonnes (Table 13. 6).

Table 13.6
Limestone production in Assam, 1976-1996

Year	NECM (in million metric tonnes)	Binoy Cement (in million metric tonnes)	CCI at Dillai Parbat (in million metric tonnes)	Total (in million metric tonnes)
1976-77	-	-	0.0883	0.0883
1977-78	-	-	0.0820	0.0820
1978-79	-	-	0.1414	0.1414
1979-80	-	-	0.2917	0.2917
1980-81	-	-	0.2007	0.2007
1981-82	-	-	0.2702	0.2702
1982-83	-	-	0.2415	0.2415
1983-84	-	-	0.2017	0.2017
1984-85	-	-	0.2092	0.2092
1985-86	-	-	0.2110	0.2110
1986-87	-	-	0.2190	0.2190
1987-88	-	-	0.2339	0.2339
1988-89	0.0112	-	0.2113	0.2225
1989-90	0.0021	0.0145	0.2239	0.2404
1990-91	0.0228	0.0655	0.2119	0.3001
1991-92	0.0226	0.0801	0.1618	0.2623
1992-93	0.0229	0.0509	0.1619	0.2355
1993-94	N.A.	N.A.	N.A.	0.2812
1994-95	N.A.	N.A.	N.A.	0.3562
1995-96	N.A.	N.A.	N.A.	0.4080

Source : DGM, Assam
N.A. indicates not available

Sillimanite

Assam contains the best quality massive sillimanite - corundum deposits of India. The deposits occur in the banded gneissic complex of Pre-cambrian age. Here, two types of sillimanite deposit occur. One is sillimanite and other is quartz - sillimanite. The important areas are mostly concentrated in Karbi Anglong district, namely-Chippilangso, Ingtigaon, Samelangso, Upper Chelabor, Chailangso and Amterang. At Bumuni in Nagaon district sillimanite deposit is also found. Sillimanites of Chippilangso area, Samelangso area and Ingtigaon are most promising. The reserve of sillimanite rock will be in several million tonnes.

Base Metals

The State Directorate of Geology and Mining recorded the base metals in the form of Pyrrhotite chalcopyrite, and covellite malachite etc. at a number of places. At Pancharatna, the pyrite disseminations of gneissic groups are available, but have no significance. Shelvetta, Makum and Barjuri in Goalpara district are the other important areas. Shelvetta and Makum have pyrite deposits, but Barjuri has sulphide minerals.

Decorative Stone : Building and Road Metals

In Assam, granite and granite gneisses are available which are suitable for use as decorative stone, building stones and road metals. Decorative stones can be used for manufacture of

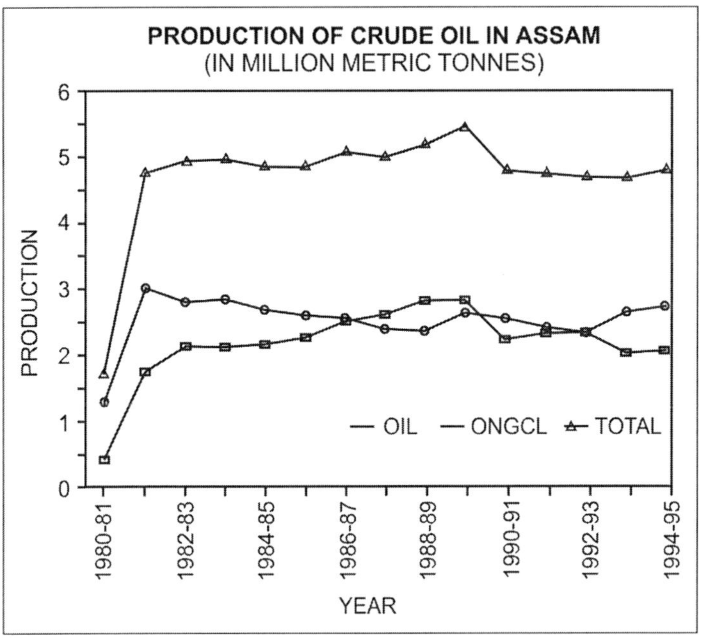

Fig. 13.3

decorative tiles,slabs etc. Deposits of granite occur mainly in Lower Assam, Karbi Anglong and North Cachar Hills districts and, quartzite in Karbi Anglong and Nagaon districts covering about 3,500 sq. km. area. Granites occur mainly in Dhubri, Bongaigaon, Kokrajhar, Goalpara, Kamrup and Nagaon districts of Lower Assam. Massive sandstones occur in Upper Assam below the coal measures. The famous Jagiroad quarry and other quarries around Burapahar area can supply several million cubic meters of metals. But, there is no systematic survey carried out for assessing the metals.

Iron ore

Iron ore deposits occur in Chandardinga Hills, Langapara and Malaigarh of Goalpara district and also in Hahim of Kamrup district. Chandardinga has a reserve of 10 million tonnes in Archaean complex consisting of amphibolite, banded ferruginous quartzites and micaceous anthophyllite. The thickness ranges from 30-74 meters. Iron- ore deposits in Assam are presented in table 13.7.

Table 13.7
Total reserve of iron- ore deposit in Assam

Location	Latitude/ Longitude	District	Reserve	Associated Rocks	Types of Ore
Chandar-dinga Hills	$26^0 12' 30''$N $90^0 20' 30''$E	Goalpara	10 million tonnes	Banded Quartzite & ferruginous phyllite	Haematite Magnetite
Hahim	$25^0 51' 15''$N $91^0 10' 0''$ E	Kamrup	Indicated reserve 0.54 million tonnes	Banded haematite quartzite	Haematite
Langapara	$26^0 3' 55''$ N $90^0 28' 50''$ E	Goalp- -ara	9 million tonnes	Banded haematite, Magnetite, Quartzite	Haematite Magnetite

Clay

Various types of clay deposits occur in Assam but these have not been exploited properly. Clay is used mostly in brick-making and cement manufacture. Fire clay deposits occur at Sheelvetta and Koilajan of Karbi Anglong district. The total reserve is 55,000 tonnes.

Kaolin or china clay which is a soft white residual clay produced by the weathering of granite occurs in Upper Deopani, Sheelvetta and Silonijan area of Karbi Anglong district. The quality of Deopani china clay occurs as lenticular mass within the granite gneiss. It is mostly used in the manufacture of porcelain, files, paper and toothpaste etc. The total reserve of Deopani, Silonijan and Sheelvetta deposits are 0.61 million tonnes, 0.17 million tonnes and 58,000 tonnes respectively.

Fuller's Earth

In Assam, Fuller's earth deposit of clayey rock is located on the left bank of river Pagladia at Subankhata. The total reserve is estimated at 13 million tonnes. The deposit is interbedded with Tipam sandstone and overlain by gravel bed.

Glass Sand and Quartz

Glass sands suitable for optical use have been reported from Jinkyndeng of Karbi Anglong district, Jiajuri and Thanajuri of Nagaon district. These glass sands are associated with clay and silt. The total reserves of Jinkyndeng, Jiajuri and Thanajuri deposits are 13,700 tonnes, 7 million tonnes and 1.78 million tonnes respectively.

Good quality quartz is deposited at Hahim of Kamrup district, Sheelvetta and Rangsali of Karbi Anglong district. Most of the quartz deposits occur in association with felspar. The total reserves of Hahim, Sheelvetta and Rangsali deposits are 20,500 tonnes, 43,000 tonnes and 19,500 tonnes respectively.As regard felspar, the deposits of good quality felspar occur at Hahim and Pancharatna. The felspar contains large crystals.

Placer Gold

Assam was a producer of placer gold since time immemorial. All the river sands of Assam used to bear gold in varying quantity. The river Subansiri in Upper Assam was a best producer. The State Directorate of Geology and Mining carried out geological survey during 1989-90 in Subansiri Gerukamukh, Hallakhbari and

Choudhu area where 18 bore holes were drilled. Out of 18 bore hole samples collected by the geoscientists of DGM, Assam, 13 indicated the presence of gold.

Lithomerge, Ochre and Mica

Lithomerge of white clay occurs at Nambar Nadi valley. The total reserve of Lithomerge is estimated at 35 lakh tonnes. Ochre occurs in the foothills of Kamrup district. Ochre is yellow earth used mainly for colouring matter. Small amount of mica occurs in Barigaon in Kamrup district. Near Tin Tilla village on the bank of Dhir *bil*, mica is located, but it is not extensive and of no economic value.

Sillica Sand

Sllica sand deposits are located at Jiajuri Hills and Thanajuri Hills of Nagaon district. The total reserve at Jiajuri is 10 million tonnes, while at Thanajuri the reserve is 1.78 million tonnes.The glass sand can be used for manufacturing of low grade glasswares.

Environmental Impact of Mineral Exploration

The environmental impact of mineral resources development covering the whole spectrum of operations ranging from exploration through development and mining to beneficiation and utilisation of minerals, needs a comprehensive assessment and serious consideration for proper planning due to the pollution impact on the surrounding environment. Certain negative effects on the surrounding environment through many physico-chemical and biological changes during exploration, production and refining can easily have major pollution impact on the surroundings. Coal is a major source of pollution. When it is burnt in crude state huge quantity of SO_2 is released into the air which damages the atmospheric condition. Coal is mined in two methods: rat hole mining and open cast mining. The former causes least damage to the surrounding ecology, but it has limitations in Assam's soil conditions. Open cast mining causes major scars on nature. Open cast mining in the Margherita area is being opposed by individuals as well as several other organizations due to its degradational impact on environment. The unsystematic and haphazardly oriented coal mining of Makum coal field adversely affects the surrounding ecosystem. The environmental degradation due to the oil based industry is also severe. Release of mine water from Lakowa and Geleki locally known as 'gaspani' has damaged thousands of acres of crop fields rendering unsuitable for cultivation.

Mineral Resources and Economic Development of Assam

The Directorate of Geology and Mining has brought to light the existence of promising mineral wealth besides coal, oil and limestone. In addition to these, there are huge deposits of kaolin, sillimanite, glass, sand, iron and huge quantities of decorative stone. These resources play an important role in the economic development of Assam.

The petroleum resources play an important role in economic growth of Assam. Without petroleum and natural gas modern living is impossible. So, it is termed as 'liquid' gold'. The petroleum consumption in the household sector, transport sector and industrial sector are respectively 20%, 48% and 32%. The national income coming from mining and quarrying sector increased from Rs. 11,690 crores in 1989-90 to 12,098 crores in 1990-91. The value of mineral production in Assam is Rs 1262 crores. In the case of mineral production Assam ranks 5th. During 1990-91 Assam contributed 7% of the total value of mineral production in the country. Petroleum (crude) is the main mineral production of the state contributing 90% value of mineral produced during 1990-91. Oil and Natural Gas Commission (ONGC) is the most profitable public sector organization. The annual profit of ONGC is about 13,000 crores and it has employment capacity of about 43,000. The Government of Assam gets Rs. 250 crores of royalty per year. Besides petroleum and natural gas, coal contributes royalty, which becomes the major source of revenue of the state. It provides enormous employment generation.

Conclusion

During the pre-independence period search for 'economic' as well as 'energy group' of mineral was carried out in Assam. After that the State Directorate searched for mineral deposits in the early sixties. The results are quite revealing and exhaustive informations of metallic and non–metallic deposits of the state are brought to the light.

Geologically, Assam represents a wide stratigraphic spectrum in age, ranging from Archaean to Quaternary. The Tertiary formations are the rich store houses of quite a few economic mineral deposits like oil and natural gas, coal, limestone etc. The state also has the occurrence of kaolin, glass, sand, iron ore and sillimanite besides tracing of base metal occurrence in a few localities.

The minerals play a vital role for the development of economic and comprehensive programme for the region, There are ample scope to develop this region with a number of industries based on the by -products received from the processing of Assam's crude oil. Industries may also be developed for blending of polyester fibres with cotton, endi, muga and wool and spinning of pure and blended fibres for the production of different types of yarns, textile, filter clothes etc. There is also ample scope for industries, such as manufacture of shoe polish, lipsticks and buttons etc. based on the products of Assam crude oil. Some other minerals like asbestos, copper, corundum, gold, gypsum, iron ore, kaolin, salt etc., however, occur in small quantities without much commercial importance at the moment.

The Mines and Minerals Regulation Act, 1975 was amended through an ordinance by the President of India on 25th January, 1994. The main objective of this amendment was to encourage the private entrepreneurs as well as the foreign investments and technology for acquiring foreign trade.

References

Choudhury, J.M. (1993) : Geology and Mineral Resourecs of N.E. India, Souvenir, National Symposium on Remote Sensing Application, ARSAC (ASTEC), Guwahati, Assam, pp. 1-6.

Das, H.P. (1970): *Geography of Assam,* National Book Trust, India, pp. 53-59.

Goswami, P.C. (1988) : *The Economic Development of Assam.* Kalyani Publishers, Delhi, pp.6-8.

Govt. of Assam, DGM (1982) : *Miscellaneous Publications on Geology, Mineral and Ground Water Resources of Assam,* No. 1.

Krishnan, M.S. (1982): *Geology of India and Burma,* CBS Publishers, New Delhi, pp. 40-105.

Trivedi, R.N. (1993): *A Textbook of Environmental Sciences,* Anmol Publications, New Delhi, pp. 10-39.

14 Industries

Introduction

In the sphere of industrialization, Assam is yet to make much headway. Although the base for industrial development in Assam was created more than a century ago with the establishment of tea, coal, oil and plywood industries, it could hardly accelerate the pace of industrial development during the subsequent period. Assam has the distinction of having the first oil refinery, first plywood factory and also the first tea industry in the country. But due to several factors the state continues to remain industrially backward even after more than four decades of economic planning.

At present, Assam has 3 oil refineries, 574 tea factories (848 tea estates), 76 plywood and veneer factories, 3 cement plants, 3 paper mills,out of which one, following a spell of closure, is again in the process of being revived. There are three sugar mills, three jute mills and one spun silk mill. A few Central Government public sector undertakings have been established in the region. Moreover, there are a number of steel rolling mills. Recently a number of engineering industries have also been set up in the state.

Till the publication of Annual Survey of Industries (ASI) 1985-86, Assam had a total number of 1885 registered factories which constituted only 1.9 per cent of the total factories in the country. The number of persons engaged in the state's factory sector totalled 1.17 lakhs (or 1.6 per cent of the country's total) during 1985-86. The value of output and net value added during the year were to the tune of Rs. 1786.7 crores and Rs. 458.7 crores, the percentages to the country's total being 1.5 and 2.0 rspectively[1].

It is, however, encouraging to note that of late efforts are being made from different quarters for gearing up industrial activities and to create a favourable industrial environment in the state through provision of various stimuli steps for removal of existing infrastructural inadequacies. It is expected that these efforts would bear fruits and promote the industrial growth in the state in near future.

Type of Industry

Industry refers, as in common usage, to manufacturing sector and excludes agriculture and mining and most services, but may include building and public utilities (Florence, 1964). In this sense, tea and other plantation industry, coal mining, oil extracting etc. cannot be termed as industries. But in a wide sense, the term 'industry' is quite flexible and includes all those activities whereby (i) desirable goods are extracted from the earth; (ii) they are manufactured into finished products and given more useful form; (iii) they are taken away from one place to another, (iv) they are kept in warehouses in order to provide them time utility and then ultimately, (v) they are placed in the hands of those who pay for them. To be more precise, industrialisation implies–

(a) the use of power, machinery, modern techniques and organisation;
(b) capital investment on a large scale;
(c) division of labour and exchange and distribution of goods with the help of developed monetary system;
(d) the transformation of the whole economy; and
(e) also industrialisation of agriculture, i.e. production of implements for agriculture to facilitate intensive agriculture, manufacture of fertilizers, modern irrigation and artificial rains, etc. (Sharma & Chauhan, 1986).

In short, an industry may basically be defined as an enterprise involved in activities for manufacturing goods, processing of materials and rendering some specified services to others[2]. The value addition is another important aspect of an industry.

Industrial units can be classified on the basis of their investment limit on plant and machinery, types of organisation and basic raw materials processed:

a. Small-scale industries, ancillary units and export oriented units with investment in plant and machinery upto Rs. five lakhs (irrespective of their location) are termed as Tiny Enterprises.

b. The fixed investment in plant and machinery (original value) whether held on ownership terms or on lease or by hire purchase that does not exceed Rs. sixty lakhs is termed as Small Scale Industry.

In case of an industrial undertaking reffered to in (b), the limit of investment in fixed assests in plant and machinery shall be Rs. seventy five lakhs provided the unit undertakes to export at least 30 per cent of the annual production by the end of third year from the date of its commencing production[3].

c. The limit of investment in fixed assests in plant and machinery will be less than Rs. seventy five lakhs for ancillary units. These units manufacture parts, components and should render services to some other small, medium and large-scale industries at least 50 per cent of their production for making the final products[4].

d. Industrial units having its investment on plant and machinery more than Rs. seventy five lakhs but less than Rs. five crores are termed as Medium Scale Sector Industries.

e. Industrial units whose investment on plant and machinery is more than Rs. five crores are termed as Large Scale Sector Industries[5].

On the basis of type of organisation the industries of Assam can be classified into four categories–

1. Central Public Sector Undertaking (Industries)
2. State Sector (Joint Sector Project)
3. Industrial Unit under Co-operative Sector
4. Private Sector.

The Techno-Economic Survey of Assam (1962) conducted by National Council of Applied Economic Research (NCAER) has classified the industries into the following groups according to the basic raw materials processed[6].

1. Agro-based industry :
   - (a) Tea industry
   - (b) Food processing industry
   - (c) Sugar industry
   - (d) Textile

2. Mineral-based industry :
   - (a) Metallic industry
   - (b) Non-metallic industry
   - (c) Chemical industry

3. Forest - based industry :
   (a) Saw milling
   (b) Plywood industry
   (c) Pulp industry

4. Miscellaneous industries

In recent years various types of industries have been set up in the state. Encompassing all the industries, a seperate classifcation can be made. It is based on the basic raw materials processed and their nature. Assam's industries can be classified as follows–

I Agro-based industry : (a) Tea (b) Sugar (c) Refine and Edible Oil (d) Jute (e) Rice mills (f) Flour mills (g) Fruit and Food Processing (h) Packaging of Food Products (i) Manufacture of Beverage, Soft drink (j) Tobacco and Tobacco Products (k) Rice - bran Oil (l) Caffaine (m) Grain - mill, Spice grinding mills etc.

II Mineral - based industry : (a) Coal (b) Petroleum (Crude) and Natural Gas (utilised) (c) Lime-stone

III Forest-based industry : (a) Plywood (b) Paper and Pulp (c) Saw mills (d) Match Factory (e) Hard Board (f) Wood products, furniture and fixture.

IV Chemical-based industry : (a) Fertilizer (b) Synthetic Fibre (c) Industrial Gases (d) Plastic (e) Rubber (f) Pharmaceuticals.

V Textile-based industry : (a) Cotton and Synthetic Yarn (b) Spun Silk (c) Power loom (d) Weaving and Manufacturing (e) Processing, Dying, etc.

VI Animal-based industry : (a) Tannery (b) Leather and Fur goods (c) Woolen mills.

VII Engineering and allied : (a) Tea processing machineries (b) Sewing machines.

VIII Building materials: (a) Cement (b) Asbestos (c) Basic Metal and Alloys (d) Dolomite (e) C.I. Sheet (f) Pipes and Wires (g) Brick and Tile (h) Metal Products and parts.

IX Electrical-based industry: (a) Electrical machinery (b) Apparatus and appliances (c) Electrical consumer durable (d) Electronic goods.

X Miscellaneous industry: (a) Paper products, Printing, Publishing and allied (b) Railway bulk fitting (c) Non-metallic mineral products (d) Agricultural implements (e) Manufactures of transport equipments (f) Education, Scientific and Research Services, (g) Repair Services (h) Welding, Servicing (i) Confectionery and Bakery etc.

XI Cottage industries: (a) Cotton and Silk, handloom (b) Bamboo and cane products (c) Handicrafts (d) Bell-metal (e) Black-smithy (f) Pottery etc.

Registered Factories

Assam had a total of 2438 number of registered factories (i.e. factories registered under the Factories Act, 1948) during 1993 which was, however, lower than the figure of 2670 in 1991. The decline in number of registered factories during 1991-93 was mainly evident in respect of industrial groups like manufacture of food products, wood and wood products, non-metallic mineral products, basic metal and alloys industries, metal products and parts, electrical machineries, apparatus, appliances etc. along with repair of capital goods and repair services. Table 14.1 shows the growth of registered factories and their employment pattern in Assam over the past few years.

Finance and Employment

According to the Annual Survey of Industries (ASI) 1987-88,

Assam had a total number of 1598 registered factories during that year and the fixed capital was Rs. 629.76 crores. The total amount of working and investor capital was Rs. 344.74 crores and Rs. 1018.19 crores respectively. The number of workers employed in these industries was 84,916.

Table 14.1
Number of registered factories and their employment in Assam, 1980-1992

Year	Number of factories	Average number of workes employed daily
1980	1897	87,695
1988	2523	90,126
1990	2677	N.A.
1991	2670	N.A.
1992	2402	N.A.

Source : Chief Inspector of Factories, Assam
Note : N.A. means not available

A detailed picture of the capital employed, persons engaged and net income in the factory sector obtained from the ASI during the period from 2008-09 to 2015-16 is given in Table 14.2.

It appears from Table 14.2 that capital investment in industries of Assam is gradually increasing and it stood at Rs. 3960000 lakh rupees in 2015-16. The same is the case for employment too.

Growth of Large and Medium Scale Industry

Assam had four large and medium scale industries (LMSI) at the beginning of planning viz. - an oil refinery at Digboi, a match factory at Dhubri, a plywood factory at Margherita and a tea processing machinery manufacturing unit at Tinsukia. Till 1980 there were only 62 LMSI units. But during the last four decades the number of LMSI units has increased upto 124 (1995). The sectoral distribution of the units is given in table 14.3.

Table 14.2
Annual Survey of Industries (summary results) in Assam

Item	Unit	2008-09	2009-10	2010-11	2011-12	2012-13	2013-14	2014-15	2015-16
Factories	Number	2211	2247	2795	3019	3303	3518	3717	3890
Fixed Capital	In Lakh	940427	980140	1074713	1515746	1310951	1444632	1616940	3204850
Working Capital	Do	335954	519604	716304	474331	588926	713478	1113789	722520
Invested Capital	Do	1361928	1466773	1739924	2185871	1927084	2135656	2582876	3960000
Productive Capital	Do	12764	14997	17910	19901	18999	21581	27307	Nil
Total Workers	Numbers	126338	125759	141274	150065	140896	157687	163384	173328
Total Employees	Numbers	149256	148731	166657	180531	167835	186483	195567	208520
Wages to Workers	In Lakh	54440	62040	88639	92298	95366	133191	123895	143573
Total Emoluments	Do	124530	156016	209009	212530	232203	257679	277732	272177
Total Input	Do	3297786	3090111	3470513	4522147	4243426	4840084	5188761	4807434
Value of Products	Do	3244850	3329994	3886280	4958072	4526780	5437627	5753185	5708322
Value of Output	Do	3676790	3667156	4236700	5292766	4985210	5761714	6176083	5977564

Source : Statistical Hand-book of Assam, 2014, 2016 and 2018, Govt. of Assam, Guwahati.

Table 14.3
Number of large and medium scale industries in Assam, 1980-1995 (sectorwise, excluding plantation industries)

Sector	Number of LMSI Units (as on 31st March)		
	1980	1991	1995
Central Public Sector	7	14	12 (9.7%)
State Joint Sector	6	12	14 (11.3%)
Co-operative Sector	2	6	7 (5.6%)
Private Sector	47	84	91(73.4%)
Total	**62**	**116**	**124 (100%)**

Source : The Directorate of Industries, Assam
Note : Figures within bracket indicate the percentage of total LMSI units in 1995.

Table 14.3 shows that there are 33 LMSI units under the central public, state joint and co-operative sectors (1995) which constitute 26.6 per cent of the total number of LMSI units in the State. The remaining 91 (73.4 per cent) are in the private sector. The number of units under the central public sector is only 12 (9.7 per cent). The two industries viz. The Cachar Sugar Mill and Assam Spun Silk Mill have been closed down due to sickness. Only two namely Assam Petrochemicals Ltd. and Assam Gas Company Ltd. are running in profit; the rest are loss making[7].

It is evident from table 14.4 that the chemical-based industries are the largest in number (24.1 per cent) followed by forest-based units (21.9 per cent) and agro-based units (20.7 per cent). There are altogether 17 engineering and allied industries. The number of textile-based industries is 13. The percentage of other units is less. The spatial growth of LMSI units is shown in table 14.5.

Table 14.4
Categorywise distribution of LMSI units in Assam (as on 31st March 1991)

Category	Number	Percentage
Agro-Based	24	20.7
Forest- Based	25	21.9
Textile-Based	13	11.2
Engineering and Allied	17	14.4
Building Materials	2	1.7
Chemical-Based	28	24.1

Animal Husbandry-Based	1	0.9
Others	6	5.1
Total	**116**	**100.0**

Source: Mali, D.D., Economic problems and Planning in Assam, 1993, p. App. IX-4.

Table 14.5
Districtwise growth of LMSI units in Assam
(as on 31st March 1991)

District	Number of unit	Percentage
Kamrup	48	41.4
Dibrugarh	16	13.8
Tinsukia	15	12.9
Sibsagar	6	5.2
Sonitpur	5	4.3
Nagaon	4	3.4
Bongaigaon	4	3.4
Morigaon	3	2.6
Cachar	3	2.6
Jorhat	2	1.7
Goalpara	2	1.7
Karbi Anglong	2	1.7
N.C. Hills	2	1.7
Barpeta	1	0.9
Dhubri	1	0.9
Golaghat	1	0.9
Kokrajhar	1	0.9
Total	**116**	**100.0**

Source : Ibid.

As can be seen from table 14.5, Kamrup District has the highest concentration of LMSI units (41.4) followed by Dibrugarh (13.8) and Tinsukia (12.9). All together these three districts constitute 68.1 per cent. Darrang, Nalbari, Hailakandi and Karimganj districts have one LMSI unit each. Two districts of the state (Lakhimpur and Dhemaji) do not have any LMSI unit.

Over the last couple of years a number of LMSI units have been established in the state and a few more are in the offing. Some such projects under various stages of implementation in the state are - Industrial Papers (Assam) Ltd. at Dhing, galvanised plain and corrugated sheet project at Changsari, the fourth oil refinery at Numaligarh and the massive gas-cum-naptha cracker project at Kathalguri (Fig. 14.1).

Growth of Small-Scale Industry

The growth of the small scale industries (SSI) has been one of the most significant features of the industrial development in the state. The small scale industries are playing a key role in the process of industrial growth.

Though there has been a rapid growth of small scale industries in the country, yet the growth in Assam is comparatively slow. In 1961, Assam had only 197 registered SSI against 35,783 in India. The number went up to 3,703 in 1981 against 1,39,577 in the country. The number of small scale industries in Assm is 17,048 (1995). The growth of SSI in the state is shown in Table 14.6.

Table 14.6
Growth of small-scale industries in Assam, 1961-1994

Year	Assam	Unit registered during the year	India	Percentage
1961	197	-	35,783	0.55
1971	1,648	-	1,39,577	1.18
1981	3,703	-	5,23,185	0.71
1989	15,392	-	11,58,765	1.30
1992	21,280	2855	-	-
1994	17,048	1,751	-	-

Source : Mali, D.D., Economic Problems and Planning in Assam, 1993, p. 115 and Statistical Handbook, Assam, 1995.

It appears from Table 11.6 that Assam accounts for less than 2 per cent of the total number of SSI in the country. The number of registered units was 2,885 in 1992 and 1,751 in 1994. According to the last SSI census, generally 45 per cent of the units are either closed or untraceable[8].

The type of products manufactured in the small sector in Assam has also undergone a change over the years. Industrial units for the production of electronic goods, pharmaceuticals, plastics, electrical consumer durables, sewing machines etc. have come up in the small sector in the state (Mali, 1989, p. 116).

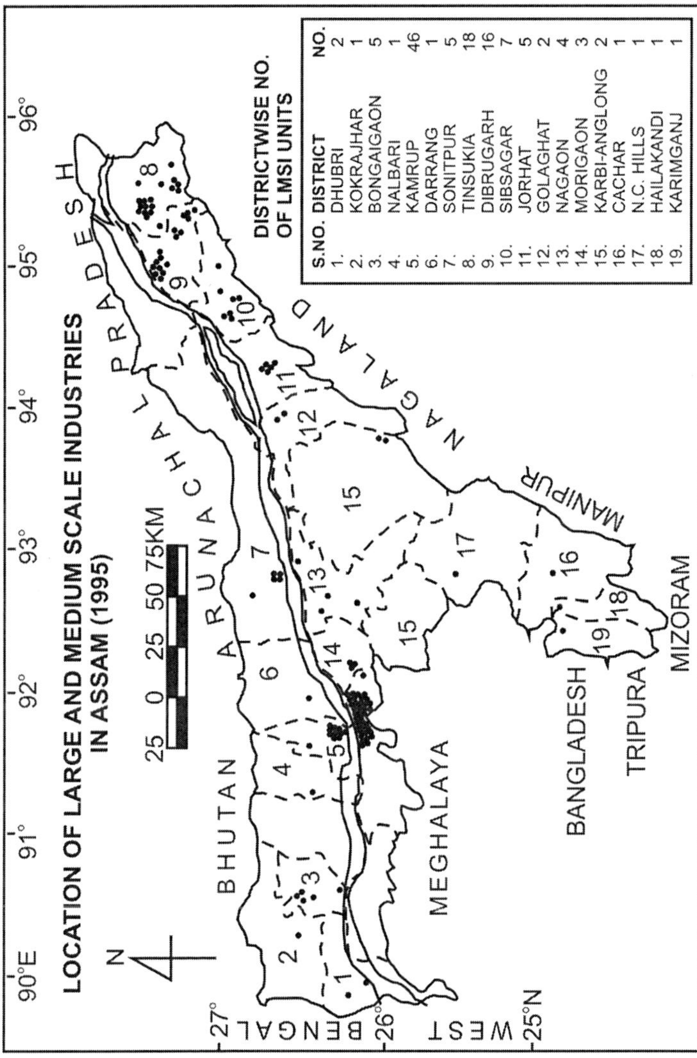
LOCATION OF LARGE AND MEDIUM SCALE INDUSTRIES
IN ASSAM (1995)
25 0 25 50 75KM
N
90°E
91°
92°
93°
94°
95°
96°
27°
26°
25°N
WEST BENGAL
BHUTAN
ARUNACHAL PRADESH
NAGALAND
MANIPUR
MEGHALAYA
BANGLADESH
TRIPURA
MIZORAM
DISTRICTWISE NO.
OF LMSI UNITS
S.NO. DISTRICT NO.
1. DHUBRI 2
2. KOKRAJHAR 1
3. BONGAIGAON 5
4. NALBARI 1
5. KAMRUP 46
6. DARRANG 1
7. SONITPUR 5
8. TINSUKIA 18
9. DIBRUGARH 16
10. SIBSAGAR 7
11. JORHAT 5
12. GOLAGHAT 2
13. NAGAON 4
14. MORIGAON 3
15. KARBI-ANGLONG 2
16. CACHAR 1
17. N.C. HILLS 1
18. HAILAKANDI 1
19. KARIMGANJ 1

Fig. 14.1

As SSI products, Assam can export commercial plywood, tea processing machineries, caffeine anhydrous and citrate, cane and wooden products along with decorated items, veneer, black board, laminates (sheets) etc[9].

Growth of Ancillary Units

In Assam, only 34 SSI units have been functioning as ancillary units, as per record with the Small Industries Service Institute (SISI). Of the 34 units, only 3 have been declared officially by Hindustan Fertilizer Corporation as its ancillaries; while the others are supplying their products to some other large scale units. In addition, another 25 units promoted by Assam State Electricity Board (ASEB) are under consideration for declaration as ancillaries. Though there is an ample scope for setting up ancillaries, its progress so far has not been satisfactory (Bari, 1987, p. 57). At least two hundred and twenty types of ancillaries can be grown on the basis of tea undustry alone in the state (Neog, 1997, p. 142.).

Some Important Industries

As mentioned earlier tea, coal and petroleum were the only large scale industries in Assam till 1950, and of them only the tea industry employs a large labour force. In real sense, none of them were a 'manufacturing' industry. The tea is essentially an agro-based industry and coal and petroleum are mining-based industries. Of course, the principal manufacturing industries in Assam were a few plywood factories, a cement and a match factory, whilst all other industries could be identified as small scale and cottage industries (Fig. 14.2) Now, the picture has changed and diversified LMSI and SSI units are growing in the state.

Tea Industry

The tea industry of Assam is the largest export-oriented sector and the biggest foreign exchange earner employing about 5.5 lakhs workers. Assam occupies a prominent position in tea industry. It accounted for 55.3 per cent of the total area and 53.9 per cent of the total production of tea in India (1990). In Assam tea industry was the first capitalist enterprise introduced by the British in the early part of the nineteenth century. Tea plant was first discovered in Assam by Major Robert Bruce, brother of A.A.

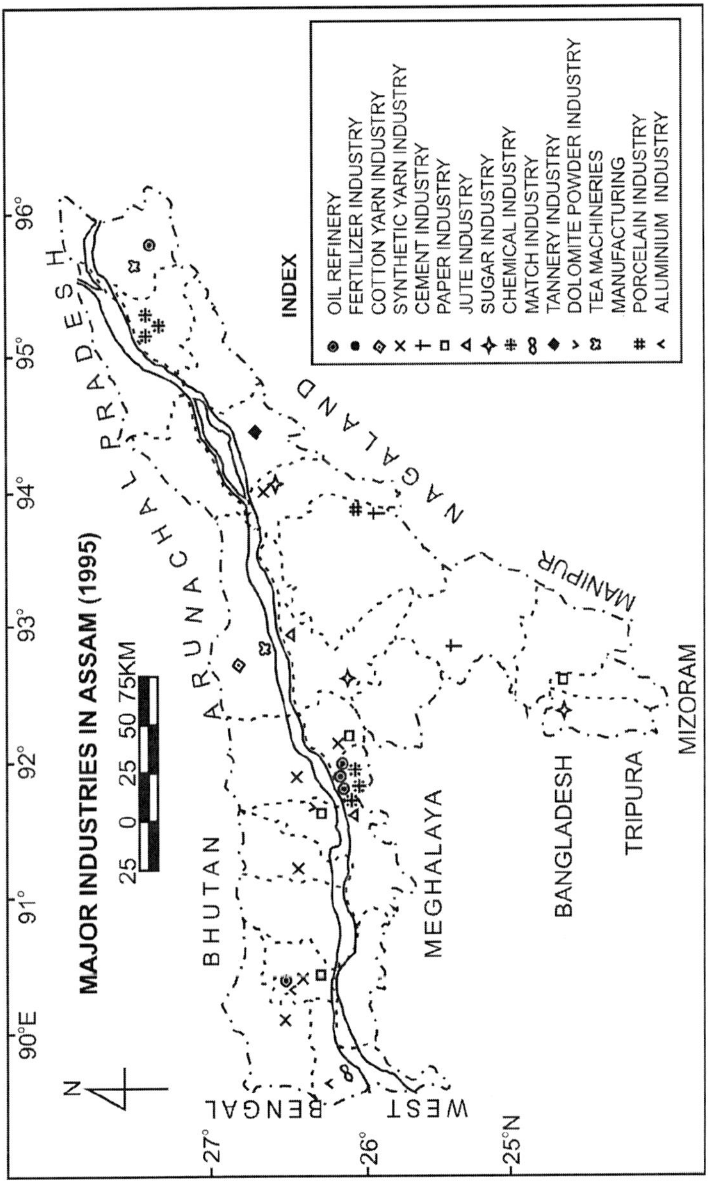
MAJOR INDUSTRIES IN ASSAM (1995)
25 0 25 50 75KM
N
90°E
91°
92°
93°
94°
95°
96°
27°
26°
25°N
BHUTAN
ARUNACHAL PRADESH
NAGALAND
MANIPUR
MIZORAM
TRIPURA
BANGLADESH
MEGHALAYA
WEST BENGAL
INDEX
OIL REFINERY
FERTILIZER INDUSTRY
COTTON YARN INDUSTRY
SYNTHETIC YARN INDUSTRY
CEMENT INDUSTRY
PAPER INDUSTRY
JUTE INDUSTRY
SUGAR INDUSTRY
CHEMICAL INDUSTRY
MATCH INDUSTRY
TANNERY INDUSTRY
DOLOMITE POWDER INDUSTRY
TEA MACHINERIES
MANUFACTURING
PORCELAIN INDUSTRY
ALUMINIUM INDUSTRY

Fig. 14.2

Bruce in 1823 at Balipara near Tezpur. But the cultivation was started in a systematic manner in 1835.

In 1938, the first commercial sample of 8 chests of tea was sent to London and sold at prices ranging from 21 Shillings 8 pences to 34 Shillings per lb. The first tea concern-The Assam Tea Company was formed in 1839 with a capital of $ 2,00,000 and the government tea estate at Jaipur was handed over to it in 1840 (Goswami, 1963, p. 141). Under the Assam Company, the growth of the tea plantation was very slow. When plantations were opened as private enterprises in 1859, it expanded rapidly. There was a tremendous growth of tea industry during the last three decades of the nineteenth century. The total amount of capital invested in the industry increased from $ 1 million (Rs. 15 million) or even less than that in 1872 to an estimated $ 14 million (Rs. 210 million) by 1903.

Table 14.7 shows the progress of tea industry in Assam from the year 1850 to 1990.

Table 14.7
The Progress of tea industry in Assam, 1850-2016

Year	Number of tea estates	Area under cultivation (in hectare)	Production (in '000 kg.)
1850	1	809	99
1900	804	1,36,379	14,012
1950	999	1,56,208	1,52,459
1990	848	2,30,363	3,88,181
2000	39139	2,67,000	451,236
2005	49102	3,01,000	487,487
2011	78856	3,22,210	589,110
2012	69220	3,22,210	590,120
2013	69220	3,22,210	629,050
2014	77710	3,04,400	610,970
2015	83566	3,16,409	526,185
2016	85344	3,04,400	642,180

Source : Economic Survey Assam 2009 - Tea Statistics, Tea Board of India and Statistical Handbook of Assam, 2014, 2016.

It appears in Table 14.7 that in 1850 the number of tea estate was only 1 with 809 hectares of land. But the number went up to 804 at the end of the nineteenth century and the area under tea cultivation rose to 1,36,379 hectares. The number of tea estates in the state decreased to 848 in 1990 as against 999 in 1950, as tea cultivation had been severely affected by drought after 1960. It is also partly the result of industrial sickness and merger of small tea estates. Some large tea companies had purchased small tea estates to increase their hectareages and some other tea estates were declared as sick and unviable. After 1990 the number of tea estates increased to 39,139 in 2000 with increase in area under cultivation and production due to establishment of many small tea estates. Again in 2005 the number increased to 49,102 and accordingly the area under tea cultivation and production went up to 3,01,000 hectares and 487,487 thousand kg. respectively (Table 14.7). It has been observed that since 2011 the number of tea estates and area under tea cultivation alongwith production increased significantly. In 2011 the number increased to 78,856 and consequently area under tea cultivation and tea production increased to 3,22,210 hectares and 589,110 thousand kgs. respectively. Almost an increasing trend can be seen from the table 14.7 upto the year 2016 and this year had the record increase of number of tea estates (85,344) and accordingly the area under tea cultivation and production went up to 3,04,400 hectares and 642,180 thosand kgs. respectively.

Distribution of Tea Industry : Tea is grown almost all over Assam in both the valleys, the Brahmaputra and the Barak. About 75 per cent of the total tea estates and tea hectareages are concentrated in upper Assam comprising the districts of Lakhimpur, Dhemaji, Dibrugarh, Tinsukia, Sibsagar, Jorhat, Golaghat, Darrang and Sonitpur, 19 per cent in the Barak valley, N.C. Hills, Karbi Anglong districts and 6 per cent in lower Assam (Fig. 14.3).

Ownership Pattern : In the teething period, tea industry was dominated by the British planters. Upto 1953, 76 per cent of the total capital invested in tea in Assam were sterling capital, about 13 per cent non-Indian rupee capital, and only 11 per cent Indian rupee capital[10].

The ownership pattern of the tea industry of Assam has undergone basic changes since 1947. A large chunk of the capital invested in the tea estates had passed from British to Indian hands. Before that the entry of the Indian enterpreneurs was few, partly because of lack of capital and partly because of non-

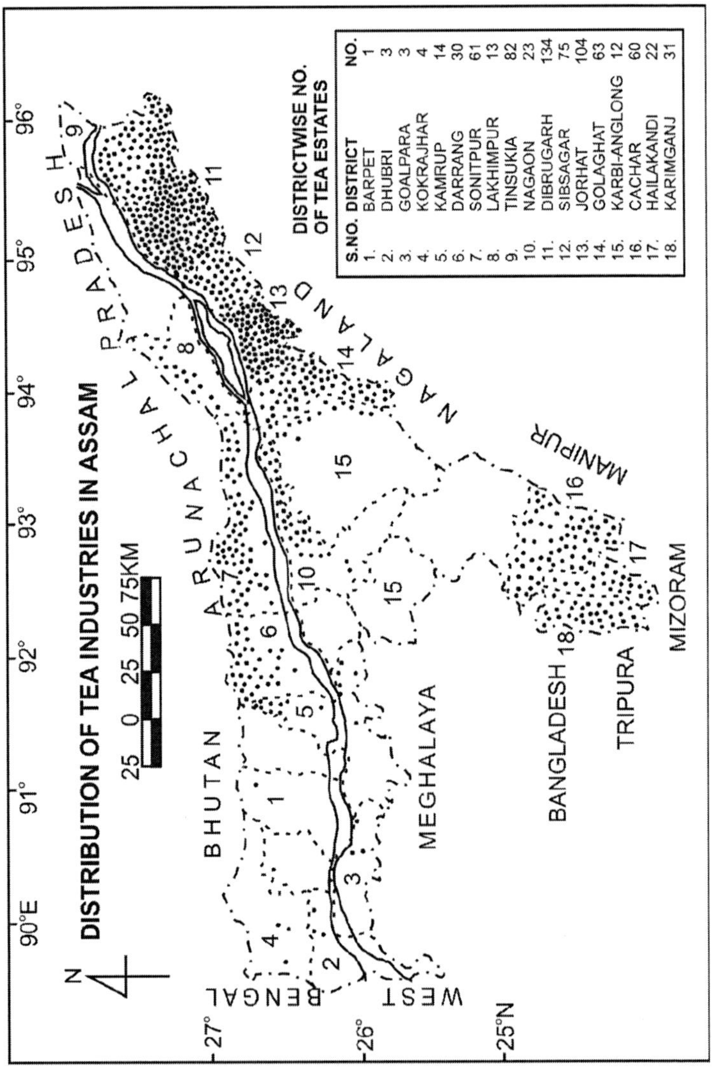
DISTRIBUTION OF TEA INDUSTRIES IN ASSAM
N
25 0 25 50 75KM
90°E
91°
92°
93°
94°
95°
96°
27°
26°
25°N
WEST BENGAL
BHUTAN
ARUNACHAL PRADESH
NAGALAND
MANIPUR
MEGHALAYA
BANGLADESH
TRIPURA
MIZORAM
DISTRICTWISE NO. OF TEA ESTATES
S.NO. DISTRICT NO.
1. BARPET 1
2. DHUBRI 3
3. GOALPARA 3
4. KOKRAJHAR 4
5. KAMRUP 14
6. DARRANG 30
7. SONITPUR 61
8. LAKHIMPUR 13
9. TINSUKIA 82
10. NAGAON 23
11. DIBRUGARH 134
12. SIBSAGAR 75
13. JORHAT 104
14. GOLAGHAT 63
15. KARBI-ANGLONG 12
16. CACHAR 60
17. HAILAKANDI 22
18. KARIMGANJ 31

Fig. 14.3

patronizing character of the colonial government for the Indian. After independence, some Indian enterpreneurs have been purchasing tea estates in Assam. A number of native Assamese enterpreneurs also took up tea cultivation. Many of the reputed industrial houses of India like the Tatas, Birlas, Khaitan, Poddars, Pals, Sethias etc. have become the main producers of tea in Assam. The formation of Assam Tea Corporation Ltd. (ATC. Ltd) in 1972 is also an important landmark in the history of the industry of Assam. The Government of Assam constituted ACT with an authorised capital of Rs. 2 crores to look after the weak tea estate. The organisation has purchased 15 tea estats so far.

Hectareage and Production : Assam has the unique distinction of having the largest area under the plantation producing a little more than half of the total tea production in India. Generally the tea estates in Assam are bigger than those elsewhere in India. The average size of an estate in Assam is 268.5 hectares while it is only 30.5 hectares for India. With an area of 2.33 lakh hectares under cultivation, Assam produced 396 million kg. tea in 1991. The average yield is 1700 kg per hectare in Assam which is the highest among all the estates of northern India. But it occupies the fourth place in India after Tamil Nadu, Karnataka and Kerela. The average yield of Assam was reduced to a great extent due to a very low rate of production in the Barak valley.

Cost of Production : While the yield per hectare and quality mainly depend upon climate, soil, topography, cultivation, plucking practice etc. the cost of production is influenced by some other factors like labour, adoption of modern techniques, etc. The cost of production includes the establishment, cultivation, plucking, manufacturing, packing, freight and transport, duties and taxes, general charges,office and selling expenses.

The cost of production has been going up steadily in almost all industries and tea industry is no exception. According to an estimation of tea centre (Indian Tea Association, Guwahati) the average production per worker is 6336 kg. annually and production cost per kg. is about Rs. 30.00 to Rs. 40.00.

Employment : The tea industry is a major employer of labour and it is highly labour intensive. It provides both direct and indirect employment. In respect of direct labour employment,

Assam employs 55.64 per cent of the total labour employed in tea plantation of India in 1991.

The average daily number of labourers actually employed in the tea industry in Assam was 554536 in 1991. One of the most important features of tea industry is that the female and child labour constitute a significant portion of the total labour force. In 1990, out of the total strength of the labour force, the female, adolescent and child labour were 2,33,486; 15,170 and 41,494 respectively.

As regards indirect employment, a large number of persons find employment in a range of ancillary industries linked with the tea plantation. A number of ancillay industries and services have developed due to tea industries e.g. plywood tea chests, aluminium foilsmetal, fillings, fertilizers, insecticides, transport, warehousing, cane and bamboo craft, trade and business, contract, etc.

Contribution of Tea Industry : The contribution of tea industry to the economy of Assam can not be ignored. The production of tea in Assam during 1990-91 was estimated to be 388 million kg. and with an average auction price of Rs. 44 per kg. contributed to a turn over of Rs. 1700 crores. At an annual average rate of Rs. 4.44 per kg after paying tax, the total profit of the tea industry of Assam may not be less than Rs. 172 crores. Being an earner of foreign exchange through export of its produce nearly Rs. 600 crores are earned by the industry annually (Saikia, 1994, p.113).

The Government of Assam collected Rs. 104 crores approximately by way of agricultural income tax, 10 crores as sales tax from the Gauhati Tea Auction Centre (GTAC) and 70 lakhs as land revenue in the assessment year of 1990-91. But the Assam Government had reduced the rate of agricultural income tax for tea from 75 per cent to 60 per cent in 1994. However, this has eroded the state exchequer.

The GTAC (established in Sept. 1970) is now the world's second largest auction centre in terms of total tea, and in 1989, the centre handled 151.8 million kg of tea valued Rs. 555 crores.

Petroleum and Petrochemical Industry

Oil is known as 'liquid gold'of Assam. Its existence in upper Assam was recorded as far back as 1825. Oil and petroleum fields in Assam are concentrated in several places of Dibrugarh, Tinsukia, Sibsagar and Jorhat districts - viz. Digboi, Naharkatiya, Hugrijan, Doom Dooma, Moran, Rudrasagar, Lakowa, Teok and Gelaki. Till 1889 oil was not commercially extracted, though attempts were made to drill in a few oil wells during 1866-68 near Margherita and Jaipur in Dibrugarh District, but drilling could not continue due to transport difficulties. It was only in 1889, the exploitation of oil was started by the Assam Railway and Trading Company, when oil fields at Digboi were discovered during the construction of railway line from Dibrugarh to Ledo. The first oil refinery in the country was set up at Digboi in 1901 by the same company.

During 1952-54, new oil fields were discovered at Naharkatiya, Hugrijan and Moran near the Digboi fields. To process the newly found crude oil, the second refinery in Assam was set up in 1962 at Nunmati (Guwahati) in the public sector, which was started with the assistance of the Government of Romania. Subsequently several oil companies were formed. In 1956, the Oil and Natural Gas Commission (ONGC) was established to explore and search out new oil fields. The ONGC has drilled (as on January'87) 18,83,306 metres and completed 569 wells of which 371 are oil wells and 9 are gas wells (Mali, 1989, p. 133).

The third refinery and petrochemical complex was set up at Bongaigaon (BRPL) in 1979. The BRPL became fully operational in 1981. The fourth one is also being constructed at Numaligarh of Golaghat District with a capacity of 3 million tonnes annually, the foundation stone of which was laid on 3rd July, 1992.

Assam produces about 5 million tonnes of crude petroleum, which accounts for half of the production of on shore oil in India. Table 14.8 shows the production of crude oil in Assam and India as a whole in selected years since 1974.

Table 14.8
Production of crude oil (in million tonnes) in Assam, 1974-2016

Year	Assam	India
1974	3.74	7.49
1975	4.19	8.22
1976	4.29	8.45

1977	4.47	11.09
1978	4.08	11.27
1979	4.07	12.84
1980	1.06	4.90
1981	4.39	-
1987	5.14	-
1989	5.77	-
1990	5.00	-
1993	4.74	-
2010 - 11	4.72	-
2011 - 12	5.03	-
2012 - 13	4.86	-
2013 - 14	4.71	-
2014 - 15	4.47	-
2015 - 16	4.19	-

Source : Potwari, S., 'Asomar Tel Sampad', Asomor Arthanity 1997, Buniad, Guwahati, 1997, p. 137 and Statistical Handbook, Assam, 1994 and www.dghindia.gov.in

Crude oil in Assam registered a growth of 5.5 per cent in production in 1989 as compared to 1988. Its production stood at 5.77 million tonnes in 1989 as against 5.14 million tonnes in the previous year. In 1993, the production of crude petroleum decreased to 4.74 million tonnes. Thereafter, almost similar trend of crude oil production was observed during the period from 2010-11 to 2015-16 (Table 14.8).

The production performance of the three refineries of the state was very encouraging during 1989. The total volume of output stood at 2.51 million tonnes in 1989 as against 2.49 million tonnes in 1988 which showed an increase of 1.16 per cent. The estimated reserves of crude oil in Assam are 144.95 million tonnes (as on 1990). About 26 oil bearing areas have so far been identified by ONGC in association with Oil India Ltd. (O.I.L.).

Natural Gas : Along with the discovery of oil fields natural gas reserves have also been discovered in Assam. The estimated availability of associated gas from the Oil India Ltd. and ONGC oil fields is 72.8 billion cubic metres. As per the Gas Task Force Report (June, 1984), the estimated free gas reserves in Assam are around 40 million cubic metres (Mali, 1989, p. 134).

The natural gas is a boon for the people of Assam, but the state is not in a position to utilise it fully. An estimated 3 million cubic metres of gas are being loosely flared up daily. The Oil India Ltd. (O.I.L.), ONGC and Assam Gas Company are responsible for production of Liquid Petroleum Gas (L.P.G.) at Nazira and Duliajan whereas BRPL is responsible for production of both LPG and compress gas at Bongaigaon.

On the basis of locally available natural gas LPG bottling plants have been set up - one at Changsari by B.S. (Assam) Industry (P) Ltd. and another at Pathinboragaon by North Eastern Cylinder (P) Ltd.

The production of natural gas (utilised) in 1992 was 1030 million cubic meters as against 838 million cubic meteres in 1993 which showed a decrease of 18.6 per cent. The current production of associated gas in Assam is estimated at about 5.5 million standard cubic meters per day.

Petrochemical Industry : Based on petroleum and natural gas, several petrochemical industries have been established in Assam. The natural gas contains high components of plastic products which provides raw materials for the plastic industry. The natural gas can also be used for the production of nitrogeneous fertilizer. In 1969, the first natural gas-based fertlizer plant in Assam was established in Namrup by the Hindustan Fertilizer Corporation (H.F.C.). It has three units, the second and third units were established in 1976 and 1987 respectively. The production of three units during 1988-89 and 1989-90 is shown in Tables 14.9, 14.10 and 14.11 respectively.

Table 14.9
Production of fertlizers (No. 1 unit) in Assam, 1988-1990

	Production (in tonnes)			
Year	**Amonium sulphate**	**Sulphuric acid**	**Amonia**	**Nitrogen**
1988-89	19,455	18,570	20,370	4,086
1989-90	11,175	11,975	14,690	2,347

Table 14.10
Production of fertilizers (No. 2 unit)

	Production (in tonnes)		
Year	**Urea**	**Nitrogen**	**Amonia**
1988-89	1,14,440	52,642	64,405
1989-90	1,41,020	64,869	81,415

Table 14.11
Production of fertilizers (No. 3 unit)

	Production (in tonnes)		
Year	**Urea**	**Nitrogen**	**Amonia**
1988-89	1,96,806	90,531	1,15,107
1989-90	2,55,959	1,17,741	1,56,340

Source : Dutta, Dr. B., "Abibhokta Dibrugarh Jilar Taila Udyog - Ek Samu Khatian", ACTA Journal, 1994-95, p. 25.

It is evident from the tables that though the production of fertilizers in the first unit had declined in the year 1989-90 the production in other two units had increased. But interestingly enough, the Board of Industrial and Financial Reconstruction (BIFR) has registered this as a sick industry.

However, besides fertlizer, several thermal power projects have also been established on the basis of natural gas of Assam. The Thermal Power Project of Namrup had started its production in 1965-66. This power station receives about 8,87,000 standard cubic metre natural gas daily from OIL. Its productive capacity is 133.5 Megawatt and this project supplies electricity to nearby areas of Upper Assam.

For the exploitation of natural gas, the Assam Petrochemical Ltd. was set up in Namrup and it started its production in June 1976. The Assam Gas Company supplied about 0.14 million standard cubic metres of gas to this industry. Its products are Mithanol, U.F. Powder (Urea Formaldehyde Molding Powder) etc. Almost all the plywood industries of Assam used Formalin and U.F. as adhesives which are being produced in this industry. Based on natural gas, a massive Gas -cum- Naptha Cracker Project has been set up at Tengakhat.

Moreover, a number of petrochemical-based industries have

been growing in the state. Using the produsts of BRPL several synthetic yarn producing industries have been developed. Besides H.F.C., a joint sector fertilizer industry (Assam State Fertilizer Corporation, has been established in Chandrapur where sulphuric acid etc. are produced. Another fertilizer industry been has established in Bonda, Guwahati (Fertichem Ltd.), which is now declared as sick. Several other small and medium scale industries viz. - plastic, peraffin, pesticides, soap factories etc. are growing in the state.

Coal Industry

The coal industry is a mineral - based industry. Coal is one of the important minerals found in Assam. Though the existence of coal in Assam was known as early as 1825, lack of its demand and transport facilities stood in the way of its commercial prospects till 1882.

Assam has vast reserves of coal,but its sulphur content is higher (about 2.0 to 7.9 per cent) in contrast to 2 per cent in the Jharia and Raniganj coal fields. So, it is not suitable for metallurgical purposes except for smelting sulphuric ores. On the other hand, the very low ash content, 2 to 4 per cent as compared to 15 per cent in Jharia is a distinctive quality of Assam's coal. The most important and potentially rich coal deposits are found in the Sibsagar and Dibrugarh districts adjoining Nagaland and in the hills of Karbi Anglong. The notable coalfields of Assam are Jaipur, Dilli (Dibrugarh District), Margherita, Mukum, Ledo, Tipam, Borgolai (Tinsukia District); Silbheta, Koilajan (Karbi Anglong District); Garampani, Adra (North Cachar District); and Singimari (Goalpara District). These coalfields were formed in different geological periods. The Ledo, Dilli, Jaipur coalfields of North Cachar and Karbi Anglong districts are much older (Eocene Epoch). But the oldest (Corboniferous Epoch, 200 million years) coal fields of the region lie in Goalpara District.

The Jaipur - Lekhapani region along with hill the region of Karbi Anglong together produced on an average 6 lakh tonnes coal annually. The estimated reserves of coal in Assam is 864 million tonnes, out of which 266 million tonnes are proved reserved, 58 m. tonnes are indicated reserves and 540 million tonnes are inferred reserves.

Assam accounts less than 1 per cent of the all India coal production. During 1987, Assam produced 0.92 million tonnes valued at Rs. 31 crores. The production went up from 1.07 million tonnes in 1992 to 1.25 million tonnes in 1993 showing thereby an increase of 0.17 per cent.

Plywood Industry

The term 'plywood' denotes fabricated wood in which thin layers of wood called 'veneers' or piles are cemented together with the grains of each layer more or less at right angle to the grains of the layer above and below it. The veneers are thin sheets of wood varying in thickness from 1/100" to 1/4" (0.25 mm to 6.35 mm) and at varying angles to the axis of tree growth (Requoted, Das, 1970, pp. 108-109.)

For the manufacturing of plywood light but solid wood are required. The plywood industry is a forest - based industry and advantageous for the state in terms of both raw material and market. Timber is the most important raw material for this industry and suitable timbers are available in the jungles of Assam. The main demand for plywood in Assam is for making the chests.

Till 1960, plywood was known for packaging purposes. But diversification of plywood products started since sixtees. With the innovation of processing techniques, the plywood industry started production of all commercial grades of plywood and other allied products like black boards, flush doors etc.

The first plywood factory was set up in the state in 1882. The number of factories went up from merely 2 in 1938 to 47 in 1986. Assam accounts for 64.75 per cent of the toal production of India, with an annual turn over of Rs. 150 crores. In 1994, Assam produced 42.72 million sq. metres of plywood as against 38.59 million sq. metres in 1983. The industry is now facing the problem of raw materials. It has to depend on raw materials of other states.

Paper and Pulp Industry

The paper and pulp is another forest - based industry which has considerable natural advantages in Assam. From the point of view of raw materials there is a vast scope for the development of this industry. At present, Assam has four paper mills, out of which two were set up in public sector, one in state joint sector and one in private sector.

The first parer industry, the Ashok Paper Mills was established in Jogighopa in the Bongaigaon District as a state joint sector enterprise. But the mill became sick during 1981-82. The production declined from 14,000 million tonnes in 1981 to 5,000 million tonnes in 1982. The mill was closed down for more than a decade. In 1995, the Government of Assam negotiated with a private company for reopening of the mill.

The other two big paper mills- one at Jagiroad in the Morigaon District and one at Panchgram in the Hailakandi District were established in public sector by the Hindusthan Paper Corporation.

The Kamrup Paper Mills (P) Ltd., a private sector factory is located in Amingaon.

Jute Industry

Jute milling is an agro-based industry. Assam provides suitable agro-climatic condition for the cultivation of jute. In jute production, Assam occupied second position. Its production in 1992-93 was 10.3 lakh bales (one bales means 180 kg.). Jute cultivation in the State is mainly concentrated in Nagaon, Morigaon, Darrang, Barpeta, Kamrup and Nalbari districts which covers an area of about 1 lakh hectares.

The first jute mill in the state was established under the co-operative sector in 1970 at Silghat in Nagaon District with 100 looms for the production of sacks and 50 looms for hessians. The mill could run successfully till 1981. During this year it produced 6000 tonnes of jute textile. But later on by 1983-84 the mill became sick and had to close down its operation in 1984. It is only in 1986 the mill was reopened and started prodution. Its production was 5348 tonnes in 1991. Two more jute mills (Jute Mill at Bijaynagar, Palashbari and Brahmaputra Jute Mill at Kalapahar, Guwahati) in the private sector were established in the state, the production of which is, however, not significant.

Sugar Industry

Like jute, sugar is also an agro-based industry. The climatic condition of Assam is suitable for the cultivation of sugarcane, and it is widely grown in almost all areas of Assam. But sugarcane fields are not concentrated in certain localities which is a major drawback for the development of this industry. Sugarcane cultivation occupies about 40 thousand hectares of land. It produced 1.5 million tonnes of sugarcane during 1992-93.

The first sugar mill was established under the co-operative sector in 1955 at Dergaon, but its operation started in December, 1958, with a crushing capacity of 800 tonnes per day. There are other two sugar industries– one in state joint sector at Chargola in Karimganj District and another in co-operative sector at Kampur (Nagaon). The Cachar Sugar Mill (Chargola) was closed down and the Nagaon Co-operative Sugar Mill is also facing trouble from its inception. The total production of sugar in the state was 1.6 thousand tonnes in 1994.

Cement Industry

Assam possesses favourable condition for the development of cement industry in terms of raw materials and market. In several places of North-East India good quality limestone deposits are available. Locally available limestone and coal can also be used in this industry. The state has 100 million tonnes of limestone. The production of this mineral went up from 2.12 lakh tonnes in 1988 to 2.33 lakh tonnes in 1989 showing an increase of 9.9 per cent over the year.

There are two cement factories in Assam. The Bokajan Cement Factory of Karbi Anglong was set up in central public sector with a capacity of 600 tonnes per day and its production started in 1976. The production of this factory in 1982 was 1.75 lakh metric tonnes. Another mini cement plant is being set up in North Cachar Hill District. The total production of cement in Assam is about 2.84 lakh tonnes annually (1993).

Handloom and Silk Industry

Handloom weaving is a traditional and the largest cottage industry in Assam which occupies a place next to agriculture. It has both cultural and economic impact on the people of Assam. About 8 lakh persons both male and female are engaged in this industry. This is practised both as a part time and a full time activity.

There are more than 7 lakhs looms of different kinds in Assam. Table 14.12 shows the type-wise number of looms.

Table 14.12
Number and types of looms

Type of loom	Number	Percentage
Fly Shuttle	1,57,204	22.45
Throw Shuttle	5,20,492	74.35
Chittranjan	1,058	0.15
Loin and other looms	21,346	3.05
Total	**7,00,100**	**100.00**

Source : Mali, D.D., Economic Problems and Planning in Assam, Omsons Publications, New Delhi, 1993, p. 106.

Though handloom weaving industries are distributed in almost all parts of the state, undivided Kamrup District occupies the top position in number of weavers as well as production of fabrics. This is because of the fact that besides part time weavers, there are a large number of full time professional weavers. The important centres are Sualkuhi, Nalbari, Hajo, Rampur, etc.

The total production of handloom clothes was 65 million metres in 1983, out of which 8.42 lakhs metres are Endi, 5.51 lakh metres Muga, 8.97 lakh metres patsilk and the remaining portion is cotton fabrics.

Khadi and Village Industries

The khadi and village industries (KVI) play an important role in the economy of Assam. The KVI are very much diversified in nature and include the production of cotton, silk and woolen khadi, bee-keeping, village oil, non-edible oil, soap, pottery, processing of cereals and pulses gur and khandsari (molasses), carpentry and blacksmithy, cane and bamboo work, lime, fibre, handmade paper, cottage match, village leather, medicinal plant, fruit preservation, etc.

The contribution of KVI to production and employment is significant. According to information available from the state Directorate of Sericulture, as many as 8127 villages of the state practised sericulture during 1993-94. The number of families engaged in sericultural activities totalled nearly 1.95 lakhs during the year comprising engagement of 1.28 lakhs families under eri-culture, 28.49 thousand families under muga-culture and about 38.76 thousand families under pat-culture. So far as the yield of cocoons is concerned, the achievements during 1993-94 were as follows : eri-cocoons 5.82 lakh kg., muga-cocoons - 4316.8 lakh numbers and mulberry reeling cocoons - 2.82 lakh kg. As regards the production of silk yarn during the year 1988-89, it has been recorded as 3.32 lakh kg. in respect of eri raw silk, 9.45 lakh kg. in respect of muga raw silk and 0.12 lakh kg. in respect of pat raw silk (Economic Survey, Assam, 1989-90). Table 14.13 shows the KVI commission's achievement at a glance.

Table 14.13
Khadi and Village Industries Commission Achievement at a Glance (1992-93, 2013-14, 2015-16, 2017-18)

Year	Item	Unit	Khadi	Village Industry
1992-93	Production	Rs. in Lakh	265.36	1986.02
	Sales	Rs. in Lakh	285.00	2132.90
	Employment	Number	23,465	93,778
2013-14	Production	Rs. in Lakh	Nil	2161.74
	Sales	Rs. in Lakh	67.88	2304.81
	Employment	Number	Nil	73,819
2015-16	Production	Rs. in Lakh	1034.86	141397.5
	Sales	Rs. in Lakh	1374.00	256890.9
	Employment	Number	4,831	3,98,117
2017-18	Production	Rs. in Lakh	1126.74	144733.8
	Sales	Rs. in Lakh	1031.12	263460.9
	Employment	Number	6,865	4,10,017

Source : *Statistical Handbook*, Assam, 1993, 2014, 2016 & 2018.

During 1992-93, the value of production of khadi industries was Rs. 265.36 lakhs and it generated employment for 23,465 persons. The value of production of village industries was Rs. 1986.02 lakhs and it generated employment for 93,778 persons. But after 1992-93 increase in production, sales and employment was recorded in respect of both Khadi and village industry. Especially during 2015-16, the value of production of Khadi industries was Rs. 1034.86 lakhs and it could generate employment for 4,831 persons. The value of production of village industries was Rs. 1,41,397.5 lakhs and it generated employment for 3,98,117 persons. Similar increasing trend in production and employment generation can be seen in respect of Khadi and village industries (Table 14.3).

Though the KVI are distributed in almost all parts of the state yet there are places of concentration. Generally cotton khadi is concentrated in Kamrup, Darrang and Nagaon districts. Village oil industry and village pottery are concentrated in Goalpara, Kamrup and Nagaon districts. In Sibsagar district also there is a concentration of village oil industry. Processing of cereals and pulses is confined to Kamrup, Darrang, Nagaon and Sibsagar districts. Likewise, carpentry and blacksmithy are found in Kamrup, Nagaon and Sibsagar districts. Cane and bamboo work is concentrated in Kamrup, Nagaon and Cachar. Gur and khandsari are concentrated in Darrang and Cachar districts (Mali, 1993, p. 109).

Districtwise Distribution of Industries

The industrial establishments in Assam are unevenly distributed. The industrial establishments are heavily concentrated in Kamrup and Kamrup Metro districts. Table 11.14 shows the distribution of registered industries in different districts of Assam in 2015.

Among all the districts of Assam, Kamrup and Kamrup (M) districts appear as the industrialised zone of Assam with largest number of registered industries (1505). The principal industries are manufacturing, engineering, chemical, food processing, electrical, building-materials producing, etc. The industries are largely concentrated in and around Guwahati because of its suitable location in respect of transport and communication. Besides this, Guwahati is a capital city and headquarters of almost all offices and establishments.

The Kamrup (Metro) district has one oil refinery (Noonmati), one vanaspati plant (Amingaon), one fertilizer and petro-chemical unit, several iron and steel rolling plants, one corrugated asbestos producing industry, a chemical and a spinning unit at Chandrapur, two mini jute mills, one paper mill, one dolomite powder producing unit, a methanol producing unit, two woolen mills, one polyester mill, a wearing and manufacturing unit, one LPG cylinder bottling plant, one railway bulk fitting factory, a multipurpose plastic film and bags producing industry, a dye house, a miniature battery unit, one cattle feed unit and a number of floor and saw mills.

The only tobacco industry of the State is also located in Kamrup district. It has also 14 tea estates. The district is very well developed in agricultural processing and small scale metal-based industries. Hajo is the traditional centre of the bell-metal industry.

Table 14.14
Districtwise distribution of Industries registered under the Directorate of Industries, Assam (2015)

District	Number of registered industries	Number of industrial workers
Dhubri	95	2762
Kokrajhar	42	1812
Bongaigaon	67	5559
Goalpara	23	598
Barpeta	71	2069
Nalbari	33	1086
Kamrup & Kamrup (M)	1505	50,673
Nagaon	373	16,920
Marigaon	64	2352
Darrang	92	2773
Sonitpur	475	19567
Lakhimpur	70	2148
Dhemaji	8	124
Dibrugarh	695	28,495
Tinsukia	641	25,192
Sibsagar	388	13,639
Jorhat	348	9483
Golaghat	317	8185
Cachar	343	11,808
Hailakandi	50	3450
Karimganj	121	3950
Karbi Anglong	40	1254
Dima Hasso	27	591
Chirang	21	1139

Baksa	16	1047
Udalguri	74	3227
Assam	**5799**	**2,19,903**

Source : Office of the Chief Inspector of Industries, Assam.

Dibrugarh District is well known for tea industry. The highest number of tea estates (134) of the state is concentrated in this district. Moran is an important centre of tea industry. The district is also rich in mineral. The oil fields of the district are located at Moran and Naharkatiya and coal is found at Dilli and Jaipur area. On the other hand, Duliajan is known for natural gas. In and around Dibrugarh, Jaipur, Naharkatiya plywood factories are growing. The important industrial centres of the district are Dibrugarh, Naharkatiya, Duliajan, Namrup, etc.

The third important area for factory cencentration is Tinsukia District. This district specializes in plywood (Makum, Margherita, Tinsukia), tea, and enginnering and allied industries. Mining is also an important industry of this district. The coal mines are located at Ledu, Tipam, Lekhapani, Borgolai and Margherita. The oldest oil refinery of the state (Digboi) is located in the district. The important oil-fields of the district are Doom-Dooma and Hugrijan. Doom-Dooma is also an important centre of tea industry. The district has a tea processing and manufacturing unit. Digboi, Tinsukia, Makum are important industrial centres the district.

Sibsagar District is very rich in tea industry. Besides tea, this district is endowed with mineral resources– coal, petroleum and natural gas. Galeki, Sonari, Namti, Lakowa are notable oilfields. Coal fields are found in Namdang. Natural gas is available in Amguri, Rudrasagar, Lakowa, and Sonari.

Jorhat is also a leading industrial town of upper Assam. The oldest industry of Jorhat dsitrict is tea. The district has oil and natural gas reserves in Borhola. Plywood industries are concentrated at Marioni. Titabar is known for hand-made brass, copper and bell-metal industries.

Golaghat District is rich in tea industry. It has 63 tea estates. The oldest sugar mill of the state is located in this district. Recently two oil fields– Kharaghat and Uriamghat have been discovered in this region. A polyester viscose blended yarn mill has been set up in the district.

Sonitpur District is also famous for tea industry. There are 475 registered industries in the district. The notable industries are cotton yarn mill, tea processing factory, rice bran oil mill, saw mills, flour mills, rice mills and brick industries.

Agriculture-based industries are predominant in Nagaon District. There are 373 registered industries. The oldest jute mill of the state is located in this district. One sugar mill has been set up at Kampur. Other industries of the district are engineering workshops, power generation, rice mills and handloom.

Morigaon District has 64 registered industries. These are Hindusthan Paper Mill, Spun Silk Mill and a Polyester Yarn Unit (Assam Polytex Ltd.).

Bongaigoan is an important industrial district of lower Assam. The district has 67 registered industries. The notable industries of the district are Petro-Chemical Complex (BRPL); Ashok Paper Mill, Naopara Spinning Mill, caustic soda, chlorine producing unit, railway repairing centre and aluminium casting factory.

In industrial sector Nalbari District is not very developed. It has one Polyester Viscose Blended Yarn (P.V.B. yarn) producing unit at Nathkuchi (Tihu) and one processing unit at Ghagrapar. The district is rich in cottage industries (KVI) and handloom weaving. This district has 33 registered industries.

Dhubri, Kokrajhar and Goalpara districts are industrially backaward. The oldest match factory of the state is located in Dhubri District. This district has also an aluminium industry at Salakati. A cotton yarn mill (Baitamari) and a few saw mills, rice mills, oil mills are operating in Goalpara District.

The two hill districts of the state-Karbi Anglong and Dima Hasso have one cement factory each. There is also a concentration of tea industry in Karbi Anglong district. The only porcelain industry of the state is located in Karbi Anglong district.

Barpeta district is also industrially backward. There are 71 registered industries in the district. But the district is famous for ivory produsts and bell-metal industry.

The three districts of Barak Valley, i.e. Cachar, Hailakandi and Karimganj are rich in agro-based industries. Cachar District has 60 tea estates whereas other two districts - Karimganj and Hailakandi have 31 and 22 tea estates respectively. There is a paper mill at Panchgram in Hailakandi District. One sugar mill is located at Chargola in Karimganj District. Crude oil and natural gas have been discovered recently in the district at Badarpur and Adamtila. Cachar has 343 and Hailakandi has 50 and Karimganj district has 121 registered industries.

Darrang district is predominantly an agricultural region. It has 30 tea estates. The Pragbosimi Synthetic Ltd. has been set up at Sipajhar. This district has 92 registered industries as of 2015.

The other two districts of upper Assam viz. Lakhimpur and Dhemaji are industrially more backward. There are 70 and only 8 registered industries in these two districts respectively. Some tea estates (13) are located in Lakhmipur District.

As per announcement of the State Industrial Policy, 2008, the government of Assam has recently come forward to establish certain industrial growth centres, infrastructure development centres and industrial parks and centres at different places of the state (Table 14.15).

Table 14.15
Initiatives of Assam government for industrial development under the State Industrial Policy, 2008

Initiatives	Location
(A) Industrial Growth Centres	Balipara (Sonitpur), Matia (Goalpara), Chhaygaon Patgaon (Kamrup)
(B) Integrated Infrastructure Development Centres	Silapathar (Dhemaji), Bandardewa (Lakhimpur), Dalgaon (Darrang), Rangia (Kamrup), Malinibil (Cachar), Demow (Sibsagar), Naltoli-Bhomoraguri (Nagaon), Parobtipur (Tinsukia), Titabor (Jorhat), Tihu (Nalbari), Cherafanggur (Kokrajhar)
(C) Industrial Parks and Centres	(a) Software Techno-Park at Borjhar and Azara (Kamrup) (b) Bio-Technology Park at IIT, North Guwahati

(c) Export Promotion Industrial Park at Amingaon
(d) Mega Food Park at Nathkuchi, Tihu
(e) Food Processing Park at Chhaygaon
(f) Bamboo Park at Chhaygaon
(g) Plastic Park at Gelapukhuri (Tinsukia)
(h) Tea Park at Chhaygaon
(i) Jute Park at Dhing (Nagaon)
(j) Air Cargo Complex at Borjhar
(k) Marketing Infrastructure for Agro-Processing at Silapathar, Dalgaon, Raniganj (Dhubri, Samaguri, Pawoi (Tinsukia)
(l) Border Trade Centre at Sutarkandi (Karimganj), Monacachar (Dhubri), Golokganj (Dhubri) Darrang (Baska), Jagoon (Tinsukia)

Problems of Industrial Development

Though nature has been bounteous to Assam in respect of natural resources, yet the industrial growth in this region is very much sluggish. According to Das (1970, p. 102), the industrial development of the state is mainly handicapped by her isolation from the rest of India, a bad transport system, a very small local market and lack of sufficient capital and labour. It is true that the geographical isolation and poor transport system are still a major constraint but the condition of other three hurdles has been ameliorated to some extent. With a population of 31.6 million (1991) the North Eat India comprising seven states provides a vast local market. Besides growing a meagre local capital, the avenue for tapping money from different financial institutions and outside agencies has been widening. There are more than 1200 bank branches in the state in addition to the regional offices of all the leading nationalised banks,although the credit - deposit ratio is still low in Assam (51.6 per cent) as against 61.0 per cent in the country in 1987. And so far as labour is concerned, there is no dearth of unskilled and semi-skilled labour in this region, yet the problem of skilled labour is still an unsolved question.

In this regard Mali and Bandopadhyay (1993) noticed that the industrial backwardness of this region is reflected in the fact that the contribution of the manufacturing sector to the state domestic product of the region is much smaller than the contribution of this sector to the national product in the country. The industrial backwardness is still prevailing in spite of the fact that the region has rich potential for industrial development and there is also good market, both internal and external.

However, Mali (1993, p. 124) observed that the problems are mainly geographical, infrastructural, financial, marketing, entrepreneural etc. particularly as regards small scale industries.

Geographically Assam is located in the north - east corner of India and connected with the rest of India through a bottleneck of about 16 km. It is exposed to foreign countries only through Bhutan and Bangladesh.In true sense, Assam is a landlocked state. The only outlet is the mighty river Brahmaputra which is passing through a foreign country(Bangladesh). Moreover, about two-thirds of the total boundary of the state are mountainous. The great physical distances from the main consuming centres of India have been multiplied by the bad transport system to and from, and within the state.

Infrastructural facilities refer to power, government and administrative assistance, technical educational institution etc. The state has experienced often unexpected power failures. The unscheduled load sheding has caused great problem to industrial units. The total power available in the state at present is around 315 MW which is less than the peak demand of 375-400 MW.

From time to time the Government of Assam has been formulating industrial policies for the growth of industries in the state, but sincere effort is lacking. It is the failure on the part of the Governement to create a conducive industrial climate. Moreover, red tapism, corruption etc. are major hindrances in the path of development. Unfortunately uptil now the state is not in a position to cater to the needs of the skilled labour with excellent quality due to inadequate technical educational facility.

Another chronic problem for industrialisation in Assam is the lack of local capital. Assam is mainly an agricultural region.

Agriculture is the main-stay of the local people. As mentioned earlier, the ownerships of the principal industry - the tea industry, were under the hands of the British colonialists and later on in the hands of the big industrial houses of India, who were not interested to invest money for other industries of the state. So, the local capital formation is very much meagre. Without the assistance and generosity of the Central Government it is difficult to establish heavy and large scale industry. But at present a number of financial institutions have been set up in this region to help and forward financial assistance to interested entrepreneurs at least for the development of medium and SSI.

For centuries the state continues to remain a trader's paradise. Assam is a good importer of indutrial finished products and consumer goods. The local products can hardly compete with the goods coming from other parts of the country. For a prolonged period of time, the nearest industrial giant 'Calcutta' has been capturing the market of this region. With meagre capital and age-old technology, it is difficult to capture market. So long the government dose not take effective measures to protect the interest of the enterpreneurs, healthy industrial growth will not be possible in the state.

Another most hindering factor for the growth of industries in the state is insufficient supply of skilled manpower. Modern industries require skilled labour and excellent technicians. The age-old ITI 's of Assam could not satisfy the demand of the day. So, entrepreneurs very often take recourse to supply of skilled manpower from outside Assam.

Last but not least is the problem of intrepreneurship. Local entrepreneurship is developing very slowly. The numbers of entrepreneurs is few who are coming willingly to this sector of economy. Rather it will not be incorrect to say that local people come to this sector not by choice but by compulsion.

Prospects of Industrial Development

In the midst of so many odds, there are some glimpses of hope. Firstly, Assam has unique location for which it may be termed as 'gateway' of North East India. All sorts of transport of goods and services to other states of North East India are made through

Assam. Moreover, Assam is situated at a position which has tremendous potential from the international point of view so far as world trade is concerned. It is only one hour distance by airways from Assam to South-East Asian countries which together comprise over 50 per cent of the worlds' population and include some of the fastest developing countries. The access to the large market of these countries is possible from Assam as the state is close to these rapidly growing economies. So, Assam can take advantage of this situation.

Secondly, Assam is endowed with various resources, both renewable and non-renewable like agricultural, forest, mineral, livestock, etc. Abundant natural resources along with cheap manpower makes the region most lucrative for industrial investors.

The possibilities of industrial growth and development are immernse. These can be considered under two heads: establishment of (i) resource-based and (ii) non-resource based industries. Again resource-based industries can be grouped into four categories on the basis of type of raw materials processed, such as (a) agro-resource based (ancillary to tea) (b) forest resource-based (c) animal resource-based and (d) mineral resource-based. Non-resource-based industries are chiefly : (a) metal manufacturing and (b) engineering goods, etc.

Agro-resource-based industries may be developed on the basis of agricultural products such as rice, sugarcane, jute, rubber, cotton and various fruits. Based on rice, rice mills, rice bran oil mills, cattle feed unit etc., based on sugarcane a number of producing units such as sugar, polish, industrial liquer, preparation of medicines, ethyl actitate, cloroform, ether etc. can be prodced. In jute production Asssam has second position in the country. On the basis of jute, one or two jute mills can be developed. There is a vast demand of jute products such as sacks, hessians etc. A number of items can be produced locally viz. - gloves, foam rubber. Likewise one or two cotton mills based on cotton, a number of fruit-processing industries e.g. orange, pineapple etc. based on fruits can be developed.

Though a number of forest-based industries have already been established in the state, yet there is a possibility to grow industries for the production of newsprint, manufacturing of sports materials, electrical appliances, weaving materials, photo-frame etc.

There is a vast demand for leather products in the state. Animal resource-based industries such as tanning, prepared leather, glue, industrial leather, production of footwears, sports materials etc. can be developed. Along with these industries several SSI units such as confectionery, dairying, bone mill, caffeine manufacturing can also be grown.

Assam is endowed with rich mineral resources such as coal, petroleum and natural gas, limestone, dolomite, bauxite, kaolin, glass sand, china clay, fuller's earth etc. Surprisingly enough, till 1960, the development of minerals was confined mainly to coal and petroleum. After the establishment of the Directorate of Geology and Mining, Assam in 1960, important mineral deposits of the state have been delineated. These are coal deposits of Saraipung, Garampani and Koilajan, iron ore deposits of Chandardinga and Lengapara; glass sand deposits of Jiajuri, Thamajuri, quartz deposits of Halim and Sheelvetta, china clay of Koilajan and Sheelvetta; and fuller's earth of Subankhata.

As regards the minor minerals, the Directorate has delineated few grey, pink and black varieties of granite deposits in the districts of Nagaon, Kamrup, Goalpara, Kokrajhar, Bongaigaon, Dhubri and Karbi Anglong, which have good use in the manufacture of decorative tiles, slabs and polished block. The granites have good demand in domestic as well as the world market (Dutta, 1996).

It is obvious that various mineral-based industries can be developed in the state. Besides production of cement, a number of products such as precipitated calcium carbonate, bleeching powder, lime kilns, dehydrated limes etc. can be produced. Likewise from coal, several products viz. - fuel oil, fenol, amonical liquor, volatile gases, pit's coke, carbon black etc. can be found. Dolomite powder, refractories can be produced from dolomite. Based on bauxite, aluminium sulphate, plaster of paris producing factories can be developed. Assam can support a few ceremic industries based on her clay deposits.

Using the chemical products of BRPL a number of down stream industries can be developed. From the by-products of the

Bongaigaon refinery, polyester clothes, net, carpet etc. can be produced. The other items such as sleeping bag, rain coat, rope, x-ray film, insecticides, etc. can also be produced from D.M.T.C.-9 and calcined coke[12]. Assam has good scope for the development of plastic industries.

Apart from resource-based industries several non-resource-based industries can be established. Though Assam is deficient in metallic minerals, yet various industries may be set up in different parts viz. - rolling mill, wire drawing units, spun - pipe plant, manufacturing of agricultural implements, enamel wires, electrical motors, diesel generating sets, etc.

Conclusion

At present, industrialization is the corner stone of economic growth. Almost all countries of the world are trying to strengthen their economies through industrialization. Assam is no exception. But as the modern industrial culture being alien to the indigenous people of Assam, the know-how of the local people about the recent development in the field of modern industries is developing very slowly. On the other hand, the industrialists of outside are less enthusiastic to invest money for the industrial development of this region. Any way, the situation is changing to some extent due to initiatives and efforts of both government and non-governmental organisations.

It is worth mentioning that the process of industrialization is not confined to the establishment of manufacturing industries alone, but involves a change in the whole economic structure of the country. In an underdeveloped state like Assam, the process of industrialization goes hand in hand with the development of agriculture. A well developed agricultural sector can provide a good market and raw materials for manufacturing. Therefore, to achieve a balanced economic development a comprehensive planning combining both the sectors must be launched.

Notes

1. Economic Survey, Assam 1989-90, Government of Assam, 1991, p.42.
2. The Quest, Directorate of Industries, Government of Assam, 1990, p. 5.

3. Annual Report, 1993-94, Ministry of Industry, Government of India, p. 159.
4. Industrial Policy of Assam, 1997, Government of Assam, p. 17.
5. The Quest, Directorate of Industries, Government of Assam, 1990, p. 6.
6. Small Industries Service Institute, Guwahati, 1996.
7. Report of the Advisory Committee on Industries, Vol. I, Assam, August 1996, p. 18.
8. Industrial Policy of Assam, 1997, p. 13.
9. Directory of Export Worthy Units, SISI, Guwahati, 1996.
10. Report of the Plantation Enquiry Commission (1956), pp. 449-51.
11. Industrial Policy of Assam, 1997, Government of Assam, p. 11.
12. Udyogik Nirdeshana (Assamese), NISIET, Guwahati, 1992.

References

Bari, M.A. : 'Industrial Development in Assam', in *Industrial Development in North East India (Ed.),* Omsons Publication, Guwahati, 1987, p. 57.

Das, H.P. : *Geography of Assam,* National Book Trust, India, New Delhi, 1970, p. 102.

Dutta, A. : 'Trend of Mineral Development in Assam', *The Assam Tribune*, Guwahati, June 8, 1996.

Florence, P.S. : *Economics and Sociology of Industry,* 1964, p. 1.

Goswami, P.C. : *The Economic Development of Assam,* Asia Publishing House, Bombay, 1963, p. 141.

Mali, D.D. and Bandopadhyay, R. : *Small Industry Development in North East India* (Status and Strategy), Omsons Publications, New Delhi, 1993, p. 12.

Neog, C.K. : 'Anusangik Udyogar Sambhabaniata aru Assom', *Asomor Arthaniti 1997* (ed.), Buniyad, Guwahati, 1997, p. 142.

Saikia, B. : 'Tea Plantation Labour of Sonitpur District, Assam : A Geographical Study', unpublished Ph.D. Thesis, G. U., 1994, p. 113.

Sharma, T.R. and Chauhan S.D.S. : *Industrial Economics,* Shiva Lal Agarwala and Company, Agra, 1986, pp. 3-4.

15 Tea Industry

Introduction

Tea (Camellia Sinesis), a tropical cash crop, is the most commonly used beverage in the world. Produced in more than thirty two countries of the hot and humid regions, it is consumed by nearly half of the world population. India occupies the second top position in the world as the largest producer of tea, but it holds the highest consumable market of tea for decades. Contributing more than 15 per cent of the world's total tea, India commands worldwide both in terms of quantity and quality in the international tea market. However, due to local variations in climate, soil, topography, drainage and various social and political attributes, only few states of India successfully grow tea. The major tea growing areas of India are Assam, West Bengal, Tamil Nadu, Kerala and Tripura. Among all these states, Assam alone contributes slightly more than 53 per cent (2020) of the country's total area under plantation and accounts for nearly 40 per cent of all India produce.

Assam tea constitutes the oldest surviving and dominating industry of the state, employing more than one million population, the majority being women workers. As a single largest sector, tea industry adds significantly to the nation's economy, both in terms of GDP and foreign exchange earnings. Although the production of tea in the state has recorded significant increase depending on variation in physical and socio-economic factors including area coverage and modern inputs, it witnesses considerable variation in both spatial and temporal terms. The Assamica variety of tea is grown both in the Brahmaputra and Barak valleys of the state. About 75 per cent of the total tea estates and tea areas are concentrated in Upper Assam, 19 per cent in Barak valley and only 6 per cent in Lower Assam (Lahkar, 1996). Among the tea producing districts in Assam, the undivided districts of

Dibrugarh, Sibsagar, Darrang and Cachar (Fig. 15.1) contribute more than 88 per cent of the total area coverage under tea and more than 93 per cent of the state's tea production. Hence, the state of Assam occupies an important place both in the country and world in terms of production and quality of tea.

Trend of Tea Production in Assam

Assam witnessed a rising trend of area coverage and production of tea including the yield rate during the period 1970-2010. It is worth mentioning that area under tea and its production in Assam increased from 180 thousand hectares and 215 thousand tonnes respectively in 1970 to 318 thousand hectares and 484 thousand tonnes in 2005 with corresponding increase of yield rate from 1178 kg/hec. to 1620 kg/hec. However, the proportion of area and production to the country's total has increased marginally from 50.85 per cent and 51.41 per cent in 1970 to 57.54 per cent and 50.07 per cent in 2010 respectively. It means that Assam alone produces little more than half of the country's total tea. The phenomenal increase in area and production of tea in Assam can be directly related to the shift in the agricultural practices from

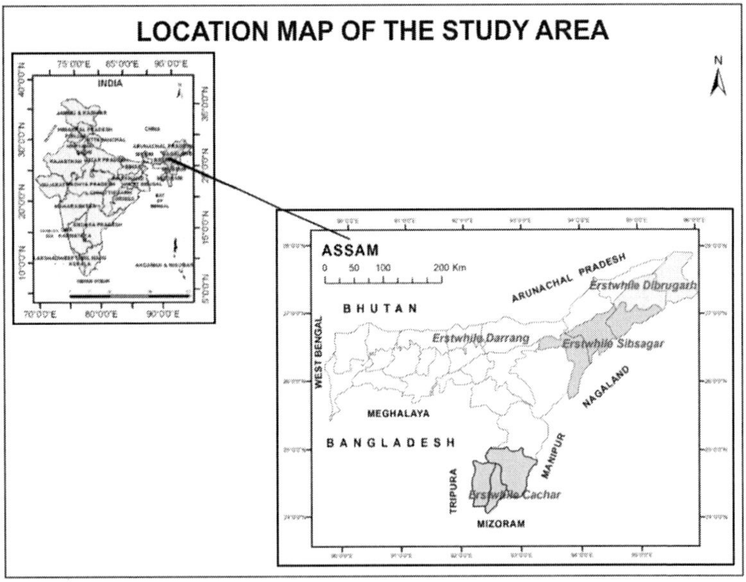

Fig. 15.1

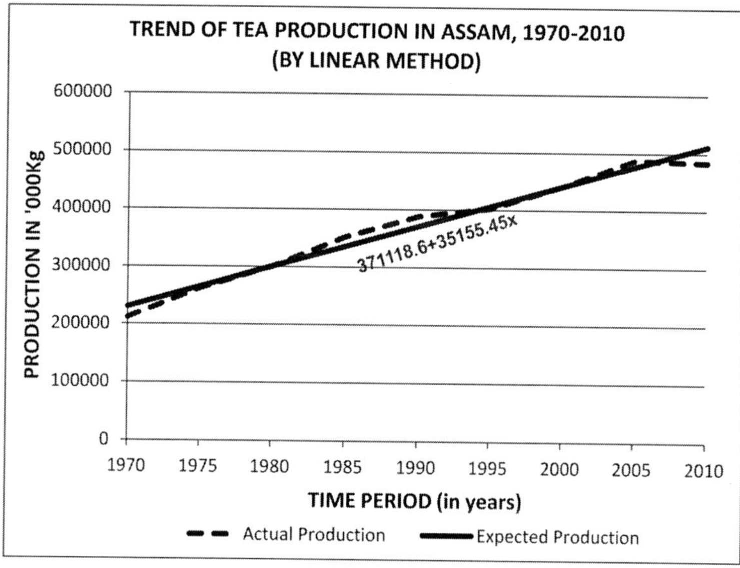

Fig. 15.2

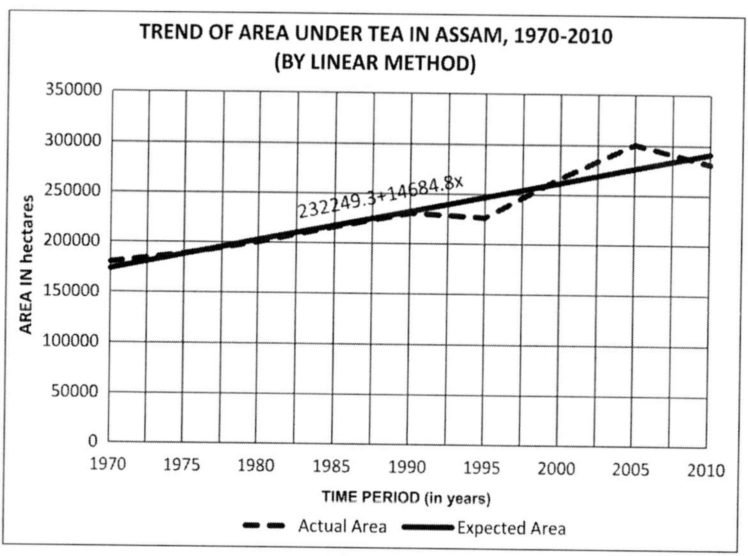

Fig. 15.3

Table 15.1

Area, Production and Yield of Tea in India and Assam, 1970-2020

Year	India			Assam		
	Area (in hec)	Production (in million kg)	Yield (Kg/hec)	Area (in hec)	Production (in million kg)	Yield (Kg/hec)
1970	354,133	418.52	1182	180065(50.85)	215.16(51.41)	1178
1975	363303	487.14	1341	188793(51.96)	263.05(54.00)	1393
1980	381086	569.17	1474	200569(52.63)	300.70(52.83)	1537
1985	399936	656.162	1641	216117(54.03)	352.52(53.72)	1631
1990	416269	720.34	1731	230363(55.34)	388.18(53.89)	1685
1995	427065	756.02	1770	226280(52.98)	402.62(53.26)	1779
2000	504366	846.92	1679	266512(52.84)	449.22(53.04)	1689
2005	521403	945.974	1703	300502(57.63)	487.48(51.53)	1622
2010	553421	967.132	1674	281043(50.78)	484.25 (50.07)	1723
2015	615364	1233.14	2004	304400 (47.17)	526.18 (42.67)	1728
2020	636895	1257.53	1974	343256 (53.89)	514.04 (40.87)	1497

Source: Statistical Handbook, Assam, 1970-2020.

102nd Report on Performance of Plantation Sector-Tea and Coffee Industry,Department Related Parliamentary Standing Committee on Commerce.

rice to tea by many individual households. In fact, the number of small tea growers in Assam increased from 4,203 in 2010 to 5,053 in 2012 covering an area of 12.6 thousand hectare, thereby leading to a subsequent increase in tea production in the state (Table 15.1 & 15.2 and Fig. 15.2 & 15.3). It is worth mentioning that the small tea growers in Assam alone contributes about 29 per cent of the total tea produced by Assam, which is approximately 14 per cent to the total tea produce of India as a whole (AASTGA, 2013).

Table 15.2
Trend of Area under Tea and Production of Tea in Assam, 1970-2020

Year	Actual Area (in hec)	Expected Area (in hec)	Actual production (in' 000kg)	Expected Production (in' 000kg)
1970	180065	173510.1	212027	230496.8
1975	188793	188194.9	263055	265652.2
1980	200569	202879.7	300700	300807.7
1985	216117	217564.5	352528	335963.1
1990	230363	232249.3	388181	371118.6
1995	226280	246934.1	402617	406274.0
2000	266512	261618.9	449219	441429.5
2005	300502	276303.7	487487	476584.9
2010	281043	290988.5	484253	511740.4
2015	304400	-	526180	-
2020	343256	-	514041	-

Source: Statistical Handbook, Assam, 1970-2020.

Again, the yield rate of Assam tea, which was 1195 kg/ha as against all India average yield of 1182 kg/ha in 1970, increased to 1620 kg/ha in 2010 against India's average of 1674 kg/ha (Table 15.3). The increase in the number of small tea growers in the state since the year 1998 has considerably led to increase in rea, production as well as number of tea estates. Consequently with the significant increase in the number of tea estates, small tea growers and area under tea, demand of tea workers has increased. Infact the tea industry is considered to be the second highest employer in the country after the Railways. It provides direct employment to more than a million workers at present, of which a sizeable number are women (Chakraborty, 2013). In addition more than one million people derive their livelihood from ancillary activities associated with the industry.

Table 15.3
Trend of Work Force engaged in Tea industry and Yield of Tea in Assam, 1970-2020

Year	Workers Engaged	Expected Workers	Yield (kg / hec)	Expected Yield (kg/hec)
1970	394,000	383,103	1178	1375.7
1975	402,195	418,619	1393	1424.6
1980	448,949	454,134	1537	1473.4
1985	484,380	489,649	1631	1522.3
1990	541,661	525,164	1685	1571.1
1995	567,398	560,679	1779	1620.0
2000	602,531	596,195	1689	1668.8
2005	618,139	631,710	1723	1717.7
2010	619,450	-	1728	1766.5
2020	676,835	-	1497	-

Source: Statistical Handbook, Assam, 1970-2020.

A close look at the tea producing districts of Assam reveals that Dibrugarh has the largest area under tea (31.91 per cent) with the highest production (39.52 per cent) in 2020, followed by Sibsagar (with an area of 28.65 per cent and production of 30.06 per cent), Darrang (with 19.17 per cent area and 16.80 per cent production). The district with lowest area (only 0.07 per cent) and production (0.001 per cent of total) is Dima Hasao (Table 15.4). The main cause behind such variations is closely related to history and geography of the tea growing areas. Historically, the British discovered and introduced tea first in the Upper Assam region, where majority of the tea gardens are found today. Moreover, the climate and soil including relief of this region is highly conducive for tea plantation. The dense network of transport routes-road, rail and airways when compared to the hilly terrain of Karbi Anglong and Dima Hasao for easy collection of raw materials and distribution of finished products are other associated reasons for such spatial variation in area, production and consequently productivity.

Table 15.4
Percentage share of Area and Production of Tea in Tea growing districts of Assam, 2020

District	Area under Tea (in hec)	Percentage Share	Production of Tea (in' 000kg)	Percentage Share
Darrang	65823	19.18	120351	16.80
Goalpara	5195	1.51	9900	1.38
Kamrup	3380	0.99	3766	0.53
Lakhimpur	7938	2.31	15805	2.20
Dibrugarh	109567	31.92	283152	39.52
Nagaon	10994	3.20	14490	2.02
Sivasagar	98349	28.65	215400	30.06
Cachar	36523	10.64	46399	6.48
Karbi-Anglong	5237	1.53	7210	1.01
Dima Hasao	250	0.07	10	0.001
Assam	343256	100.00	716483	100.0

Source: Statistical Handbook, Assam, 2020.

Productivity Pattern of Tea in Assam

Unlike the area and production of tea which show a continuously increasing trend in Assam, the yield rates have been found to be highly fluctuating over the years (Table 15.3). Although the yield rate increased from 1178 kg/ha in 1970 to as high as 1779 kg/ha in 1995, it again declined to 1620 kg/ha in 2010 (Fig. 15.4). This is mainly

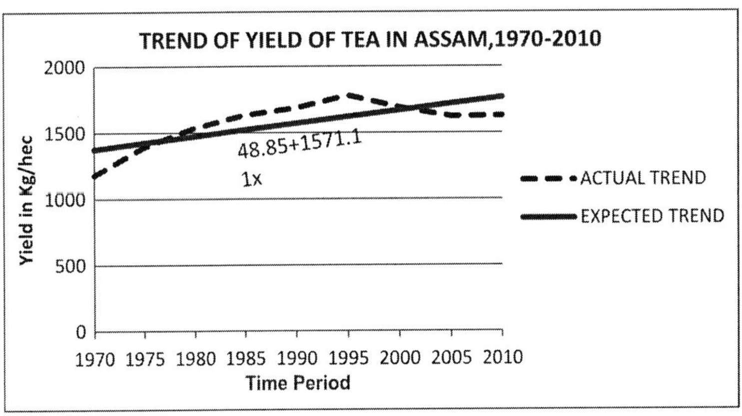

Fig. 15.4

due to the increase in relatively less productive area under tea cultivation in different parts of the state especially with the advent of small tea-growers after 1998. The proportional change in area under tea in Assam has been showing a steady increase over the years, i.e. from mere 9.01 per cent during 1971-80 to 12.91 per cent during 1981-90, 13.88 per cent during 1991-2000 and 15.78 per cent during 2001-06. But, this is not the case with percentage change in tea production. The change rate of tea production in the state which was as high as 31.54 per cent during 1971-80 slowed down to as low as 6.68 per cent during 2000-10 (Table 15.6).

The age of tea bushes is an important factor determining productivity of tea. Normally a tea bush achieves its prime time of output of tea leaves from the age of 5 years till 30 years. But 41 per cent of the tea bushes of Assam had been above 30 years in 1995. This proportion further rose to 46 per cent in the year 2006. It means that fewer tea bushes are being planted to replenish the old. Therefore, enhancement in the efficiency of labourers will have little effect when the tea bushes don't yield the optimum.

Tea Workers in Assam

The tea industry of Assam is highly labour intensive. It provides employment to more than 6 lakhs people as workers, which constitutes around 50 per cent of the total average daily workers

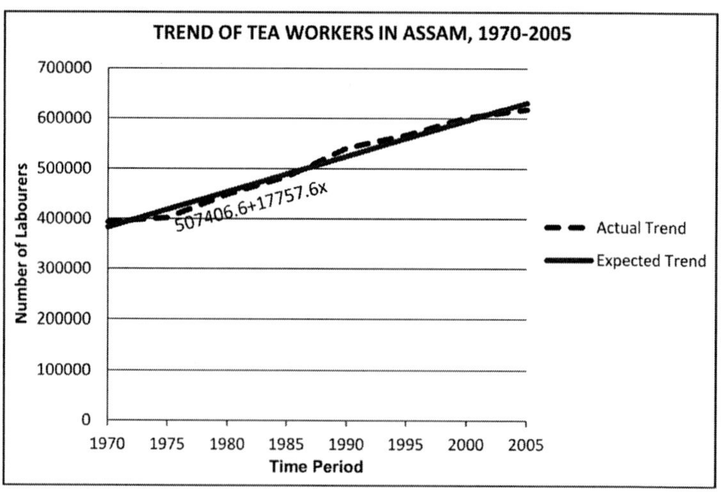

Fig. 15.5

employed in the country (Economic Survey, Assam 2010-11). The employment of total work force in the tea industry of Assam during the period 1970- 2005 shows a rising trend (Table 15.3 and Fig. 15.5).

The average daily number of workers actually employed in the state's tea industry increased from 3.94 lakhs in 1970 to 6.19 lakhs in 2006 through 5.42 lakhs in 1990. One of the most important features of tea industry in Assam is that the female, adolescent and child labour constitute a significant portion of the total work force. In the year 1980, out of the total strength of work force, the female, adolescent and child labour were 1.87 lakhs, 19.5 thousand and 43.9 thousand respectively, which increased to 2.87 lakh female workers (46.3 per cent) and 23.8 thousand adolescent (3.84 per cent), the rest being adult male workers. Although the number of total tea work-force has been showing an increasing trend, the percentage increase in this respect after 1990 had witnessed a decline over the period prior to 1990. This is mainly due to the fact that after 1990, child labour protection enactment was emphasized which led to dropping of child labourers engaged in tea industry.

The labour productivity (i.e. production of tea in kg per labour) in the tea industry of Assam witnessed an increase from 583 kg in 1975 to as high as 789 kg in 2005. The prevalence of high productivity in 2005 has been mainly associated with higher yield of tea leaves due to increased use of modern inputs as compared to 1975. The decrease in the number of labourers can also be examined from the fact that the daily number of workers employed per hectare has also decreased from 1.54 in 1970 to 2.05 in 2005 (Table 15.8).

Spatial Variation in Trend of Tea Production in Relation to Area Coverage and Labour Input

As stated earlier, for showing the spatial variation in the trend of tea production in Assam, four major districts (undivided), i.e. Dibrugarh, Sibsagar, Darrang and Cachar have been considered on the basis of their significant contribution to the total state's production and area coverage (Table 15.5 and 15.6).

Growth of Area under Tea

So far the change in area coverage under tea is concerned, all the selected districts, except Cachar, registered more than 12 per cent increase during the decade 1981-90. As the districts of Lakhimpur and Dhemaji were part of Dibrugarh district prior to 1980, the rate of increase in area coverage in the then district of Dibrugarh especially during 1971-1980 was very low (2.53 per cent).

Table 15.5
Percentage change in Area under Tea in selected districts of Assam, 1970-2020

District	1971-1980	1981-1990	1991-2000	2001-2010	2010-2020
Dibrugarh	02.53	12.73	34.06	1.19	15.09
Sibsagar	11.08	12.37	12.70	1.80	27.34
Darrang	11.75	12.42	02.18	17.15	32.26
Cachar	03.04	08.28	-8.76	-0.42	13.65
Assam	09.01	12.91	13.88	4.48	22.13

Table 15.6
Percentage change in tea Production in selected districts of Assam, 1970-2020

District	1971-1980	1981-1990	1991-2000	2001-2010	2010-2020
Dibrugarh	24.41	21.01	12.40	18.60	43.38
Sibsagar*	30.21	23.01	29.77	-11.13	102.62
Darrang	40.57	42.24	-11.13	6.10	45.01
Cachar	24.15	31.98	19.22	8.48	-14.96
Assam	31.54	27.89	12.00	6.68	6.15

Source : Computed from Statistical Handbook of Assam, 1975 to 2020, Directorate of Economics and Statistics.

* implies the data of undivided Sibsagar till 2000 also included Karbi Anglong and North Cachar districts.

The main reason for expansion of tea acreages in the state is due to the expansion of area under tea coverage in non-traditional areas including in the homesteads of rural peasantry. It may be mentioned that the statistics of small tea-growers have been included in all the figures since 1998. Accordingly, the districts with large number of small tea-growers, such as Dibrugarh and Sibsagar registered high percentage change in area, i.e. 34 per cent and 13 percent respectively during 1991-2000 (Table 15.5). But, the area under tea in Cachar showed a negative change during this period. This negative change was basically due to shut down of some sick tea-estates. This is evident from the fact that the total number of tea-estates in Cachar came down from 121 in 1990 to 115 in 1995, and again from 206 in 2000 to 198 in 2005.

Growth in Tea Production

So far the trend of tea production is concerned, all the selected districts witnessed rapid increase during both the decades, i.e. 1971-80 and 1981-90. But after 1990 there had been significant decrease in production in all the concerned districts (undivided) except Sibsagar, even after inclusion of small tea-growers. However, in the subsequent decade, the sharp decline in production figure of Sibsagar was due to inclusion of the production figures of not only of undivided Sibsagar, but also two other districts of Karbi Anglong and North Cachar till the year 2000. Similarly production of Cachar also showed a declining trend during the time period 2000-10. On the contrary, Dibrugarh district maintained a significantly positive production growth rate throughout the period. The main reason behind such a high rate of increase had been large scale production of tea by the small tea-growers. Moreover, rejuvenation pruning and infilling and uprooting and replacement planting practices have also contributed to the increased production scenario of the district (Singh, 2011).

The decline in production in Cachar district was basically due to low market demand because of increase in its production cost and subsequent increase in its price. According to D.P. Maheshwari (President of Tea Association of India), the poor connectivity of Cachar district, abysmal conditions of roads, extremely slow progress of the railway gauge conversion project between Lumding and Silchar and erratic power supply tend to increase the production cost due to involvement of additional costs in addressing such problems. Similarly, he attributed to the decline in tea production in Upper Assam owing to growing trend of absenteeism among tea labourers in the lure of MNREGA (Mahatma Gandhi National Rural Employment Guarantee Act) works, where they earn more than Rs. 120 per day. He asserted that during the peak plucking season, absenteeism touches above 35 per cent in Upper Assam districts, where the tea gardens are mostly concentrated. However, Mridul Hazarika (Director of Tea Research Association, Toklai) attributed the decline in production to climate change. According to him days with sunshine have become far fewer during the monsoon season, which resulted in a shortfall in production, as damp weather is unfavourable for tea plants. Moreover, dampness aggravates bug attacks on the tea crop. A pest called tea mosquito thrives in such weather and attacks fresh shoots of the tea bushes. Restrictions in the use of chemicals because of environmental and health constraints have added to planters' woes.

Growth in Tea Productivity

So far tea productivity (Kg per hectare) is concerned, it was highest in all the selected districts in the state in the year 1995. During the first two decades (1970-90), the trend of productivity had been positive in all the districts. But, Dibrugarh and Darrang districts witnessed a declining trend in productivity during the decade 1991-2000 (Table 15.7). However, Cachar's percentage change in yield remained high and satisfactory. But, again during 2001-10 the percentage change in yield rate at Dibrugarh and Sibsagar showed a negative trend. During the early part of 2006-09, after sharp fall in prices and over supply situation (due to advent of small tea-growers), the Producers' Association decided to cut down production to the end of the season and the tea-estates tried to produce better quality tea instead of higher quantity by adopting shorter plucking round, using of better inputs, etc to fetch better price per unit. This is evident from the fact that the average yield of tea in Dibrugarh and Sibsagar declined from 2195 kg and 1659 kg in 1995 to 1682 kg and 1552 kg in 2010 respectively.

Table 15.7
Percentage change in Yield of Tea in the major tea growing Districts of Assam, 1970-2020

District	1971-1980	1981-1990	1991-2000	2001-2010	2010-2020
Dibrugarh	20.41	08.32	-14.78	-4.97	53.63
Sibsagar	08.32	16.78	08.02	-1.58	41.11
Darrang	25.52	18.98	-7.70	5.80	-8.92
Cachar	20.52	21.99	31.97	9.68	-25.73
Assam	22.04	13.53	0.06	2.07	-13.12

Source: Computed from Statistical Handbook of Assam, 1975 to 2020, Directorate of Economics and Statistics.

Growth of Tea Workers and Their Productivity

The tea workers associated with the tea plantations of Assam are basically known as Adivasis or simply tea-tribe. They were specifically brought from the Central Plateau region of India by the British to work in tea gardens. The total population of the community is projected to be around 60 lakhs (2011), which constitutes about 19 per cent of the total population of the state. Among the community, Munda and Santhals predominate, although Oraon, Gonds, Kharia, Saora, Bhumji, Kanda, Tati, Kurmi, Karmakar, Teli, Lakra, Tirki, Jojowar, Kujur, Toppo, Minza, Paharia, Tassa, Ghatowar, Majhi, Panika, Guwala, etc. are also found.

So far the growth of tea workers is concerned, it is evident that the total average number of workers engaged in the tea industry had shown a steady increase during 1970-1985. But average number of workers employed in these districts witnessed a significant decline after 1985 due to the reorganization of most of the districts in Assam.

Table 15.8
Labour Productivity of Tea (Kg/Labour) and daily number of workers employed (in '000) in the major Tea Producing Districts of Assam, 1970-05

District	1980		1985		1990		1995		2000		2005	
Dibrugarh	675.4	-	785.2	-	825.8	-	870.0	-	829.0	13	1040	14
Sibsagar	639.0	-	630.6	-	715.5	-	600.0	-	765.0	156	712.0	159
Darrang	749.0	-	842.9	-	965.5	-	806.2	-	672.0	109	698.0	118
Cachar	561.8	-	573.3	-	651.1	-	529.0	-	727.0	58	642.5	70
Assam	682.6	449	721.2	484	755.8	-	700.0	-	745.5	602	788.6	618

It is quite encouraging that labour productivity had remained high during the entire period of 1970-2005 in Dibrugarh and Darrang districts. Although labour productivity in Sibsagar district was comparatively lower than these two districts over the years, it increased to 765 in 2000, which was much above Darrang district (672) (Table 15.8). Cachar, on the other hand, maintained steady increase, but was lower than the state average productivity. Similarly, Sibsagar also witnessed lower productivity throughout the period except in the year 2000.

Tea and the Economy of Assam: A Synergistic Linkage

Tea is not only an important cash crop and a major area of employment, but it is also the important source of revenue collection for Assam. The development of tea industry of Assam led to infrastructural development of other ancillary industries also. The tea industry of Assam contributes to the nation's income by paying cess, duties and taxes as per prescribed regulations/provisions.

The annual sale of Assam tea of 118.16mkg and average auction price of Rs.58.21 per kg (private and other sales excluded) in the year 2003 implied that the estimated annual turnover of the tea industry and the tea trade in Assam was about Rs.687.8 crores. It became Rs. 753.8 crores in 2004 and further increased to Rs. 900.8 crores and 1494.3 crores in 2008 and 2009 respectively.

Table 15.9
Position of Tax Revenue from Tea with respect to total Tax Revenue of Assam, 1996-2003

YEAR	Sales Tax	Total Tax Revenue on Tea	Total Tax Revenue of the State	Share of Tax Revenue of Tea to the Total Tax Revenue of the State (in %)
1996-97	1175	8106.93	76690	10.57
1997-98	1882	15889.94	88194	18.02
1998-99	2418	18770.24	98256	19.10
1999-00	2549	16428.00	122476	13.41
2000-01	3011	12675.01	140969	8.99
2001-02	2521	10163.70	155695	6.53
2002-03	2305	8033.68	193451	4.51

[Figures in Rs. Lakhs]
Source: Office of the Commissioner of Taxes, Assam [Collected by S.D. Lahkar]

It is found that the share of tax revenue of tea to the total tax revenue of the state has increased from 10.57 per cent in 1996-97 to 19.1 per cent in 1998-99 (Table 15.9). However, it dropped down to as low as 4.15 per cent in 2002-03. The total collection has come down from Rs. 18,770.24 lakhs in 1998-99 to Rs. 8,033.68 lakhs in 2002-03. The reason for this fall in collection of revenue has been sharp fall in tea prices since 2000, which continued till 2006.

Table 15.10
Average Price of Tea Sale at Guwahati Auction Centre, 1970-2012

Year	Average Price	Year	Average Price	Year	Average Price	Year	Average Price
1970-71	5.68	1993-94	48.54	2000-01	71.97	2007-08	70.90
1975-76	9.85	1994-95	42.99	2001-02	63.17	2008-09	91.73

1980-81	12.19	1995-96	49.99	2002-03	59.01	2009-10	108.54
1985-86	22.19	1996-97	52.69	2003-04	58.33	2010-11	111.15
1990-91	42.16	1997-98	76.67	2004-05	68.78	2011-12	125.76
1991-92	38.97	1998-99	72.72	2005-06	59.40		
1992-93	43.78	1999-00	79.03	2006-07	68.09		

Source: Statistical Handbook, Assam, 1970-2012.

In addition to the harbinger of taxes, tea industry of Assam also supports other ancillary industries, from which the state's exchequer collects additional revenue. It may be mentioned here that the tea industry of Assam has been pivotal in the development of several ancillary industries, which provide employment to masses and help to develop the economy of the state.

Since its inception the tea industry developed the plywood industry, railways, roadways and waterways in Assam. Although the plywood chests have been replaced by poly-bags, its (plywood industry) contribution to the state's economy was immense for a very long period of time. Moreover, the transport systems were developed by the British to transport the finished products to the market centers, especially foreign countries. In fact, the inland container depot at Amingaon was established to cater to the needs of export of tea to foreign destinations like UK, Ireland, Holland, Germany, Australia, etc.

Today modern tea plantations employ modern cultivation practices in the field and hence consume substantial quantity of fertilizers, organic manures, pesticides and other agro-chemicals. The tea industry has also immensely contributed to the growth of infrastructure in the form of schools, hospitals, etc. in the state. Moreover, it has led to self-employment of local people mostly for supply and other related activities.

Problems and Prospects of Tea Industry in Assam

The major problems faced by tea industry of Assam can be summarized as follows:

Presence of age-old tea bushes (above 40 years) is the most important problem faced by the industry, which directly affects productivity. In fact, the proportion of old bushes in the state has witnessed significant rise from nearly 40 per cent in 1990 to more than 47 per cent during the year 2006.

Pest attack along with mosquito bug, a disease called the bacterial black spot, has also hit scores of plantations. Chemicals to control these pests are not preferred due to fear of pesticide residual content as most foreign countries are very reluctant to tea with pesticide content. In fact, high use of pesticides, fertilizers and other different chemicals to enhance production in lingering season has affected the overall quality of Assam Tea.

The small tea growers have started cultivating tea basically at the expense of the land under rice. So, altogether this has led to the economic development of the state in the context of tea, but somewhere in between these the rice production suffers.

Presence of chemical substances like DDT, Phenol, etc has led to further supervision and examination of the tea, which has ultimately affected export. Thus, decrease in the demand of Assam Tea has adversely affected its price in International market.

Since the fall of Russian market after its disintegration, export of Assam tea has witnessed decline over the years. Moreover, with tough competition from countries like Sri Lanka and Kenya the export of tea from Assam has suffered much. Globally India ranks fourth among the top tea exporting countries of the world.

Sluggish demand in the domestic market and growing preference of the younger generations towards aerated drinks due to attractive promotional measures resulted in fall in prices, in addition to the factors mentioned above.

With the introduction of government welfare policies, especially NREGA, the temporary tea garden workers have started to skip their work and do the manual jobs under Employment Mission. This has resulted in large number of absentees in tea gardens, which has led to low productivity.

Since the small tea growers cultivate in small scale they cannot go for manufacturing and hence they have to sell their output to large estates which leave them with no option, but low price for their produce.

Erratic power supply and inadequate road connectivity in Assam has adversely affected the industry by increasing the cost of output. "In fact, power and fuel account for a significant proportion (21 per cent) in production process. Erratic power supply has forced the industry to rely upon diesel, which is not cost effective" (Poor Infrastructure hits Barak Valley Tea Units, Times of India, April 30, 2012). Similar situation also prevails over the tea producing units of the major Upper Assam tea producing districts. Processing factories of about 45 per cent tea estates in the production centres depends upon natural gas supplies by Assam Gas Company. However, frequent bandhs called by different organizations and extremist groups affect the gas supply and again to resume the gas supply sufficient pressure is to be created in the pipelines which require much time. This ultimately results in stoppage of production process. (Tea Production in Assam affected due to non-supply of Gas, Business Standard, June 2, 2012).

Exploitative nature of the tea industry owners towards the small tea growers is another burning problem. In fact, these owners in order to garner more profit manipulate the market. The year 2011 shows a sharp fall in the price of green tea leaves (about Rs. 3.50 per Kg.), which left the small tea growers with no option but to sit for mass protest.

Despite being the largest producer of tea in the region, it is surprising to note that the per capita consumption of tea in the country is very low compared to other countries. It amounts to only 750 grams, whereas the figure reaches 2.10 kg. in UK, 1.64 Kg. in Russia and around 1 kg. in Pakistan (Yojana, 2011). Therefore, domestic demand for tea is very less.

Shortage of skilled labour is another pressing problem affecting the tea industry. The labourers getting attracted to MGNREGA, the industry is reported facing reverse migration. North-Eastern Tea Association (NETA) Chairman Bidyananda Barkakoti informed that around 300 families have left their jobs for good in tea estates and gone back to Telengana (Reverse Migration is a worrying Trend for Assam Tea Industry, The Hindu, August 7, 2011).

The irrelevant wage system followed in tea gardens of Assam has also led to the deterioration of the quality of tea. Now a days the wage system in the tea gardens is such that if a tea picker cannot pluck tea leaves according to the prescribed quantity (generally 23 Kg./day), then his/her total wage will be deducted by Rs. 2 per kg. This system has led to the plucking of unwanted and old tea leaves which has affected quality directly.

The practice of alcoholism among the tea workers (both men and women folk) is widespread all over the tea plantations, draining a significant share of worker's income. Moreover, this directly affects the health of the workers leading to low labour productivity. Additionally, this is also the root cause of abject poverty and ignorance among the tea workers.

Another burning issue related to the tea sector is the gradual conversion of tea land for real estate and other infrastructural development. The proposed National Institute of Design (NID) has acquired more than 93 bighas of garden land of Tocklai Tea Estate (Tea Workers Protest NID Project on Garden Land, The Times of India, March 3, 2012). Moreover, the Brahmaputra Cracker and Polymer Project at Dibrugarh is another glaring example of this kind.

In addition to the above problems, tea industry of Assam suffers from identity crisis in the world market. Absence of official recognition as a unique variety is still a distant dream. Other problems include old age of gardens, scanty rainfall, lack of infrastructural modernization, underdeveloped transport systems, luke-worm attitude of the tea planters towards the tea garden labourers, labour unrest and inefficient management practices. Moreover, the prevailing general climate of globalization and liberalization, the dismantling of tariff barriers, granting of Most Favoured Nations (MFN) status to other countries have led to stiff competition of tea from outside world.

Prospects of Tea Industry

In spite of so many problems faced by the tea industry of Assam, there exist future prospects to improve the present condition of tea and reduce the magnitude of the problems, if not completely eradicated.

While most of the agricultural products are entitled to the central transport subsidy in the north-eastern states of India, this subsidy is denied for tea. Extending this subsidy to tea would facilitate in bringing down the cost of production and make the product cost competitive in the World market.

Scope of fresh direct and indirect employment in both the organized and unorganized sectors of the tea industry is greatly apparent. In the small tea growers sectors, through Self Help Groups (SHG) and by formation of cooperative societies, thousands of unemployed can self-employ themselves and earn unexpected returns.

Age-old bushes must be removed and replenished with new high yielding breeds of tea plants so as to reverse the trend of low production and productivity.

The establishment of a second auction centre at Upper Assam will ensure sale of more tea in Assam itself instead of diverting to Kolkata auction centre.

E-auction or online bidding recently introduced at Guwahati Tea Auction Centre has unlimited prospects in the better sale of tea and better price realization.

Proper field level training for improvement of quality and production practices can be imparted to the small tea growers so that they produce quality tea with high yield.

Former President APJ Abdul Kalam, while inaugurating the World Tea Science Congress organized by Toklai Tea Experimental Station on 23rd November, 2011, remarked that tea has great potential and therefore its global marketing is highly essential. As it is a cheap and hygienic drink the country will benefit if a proper marketing system is formulated. Pointing out some technical lapses in the sector, he said that better education and training and technical knowledge in tea cultivation, harvesting and marketing are needed. He was of the view that proper and dedicated research work will help in developing this industry, which will also help enhance production and income. For more production on a particular land, Kalam suggested multi-cropping. He said that researchers should find out some tea-friendly crops to increase its production.

Conclusion

The foregoing discussion reveals that the tea industry of Assam, which is the first in the country, has been playing a major role in the economic development of the state. The British government encouraged tea cultivators in the state by granting special concession in the form of grant of land without rent. However, after 1990 the glorious past of Assam tea became unable to sustain because of the many problems discussed above. The relatively lower production and productivity of tea after 1990s coupled with tough competition from outside world have led to decline of tea industry in the state. Nonetheless tea plantation continues to be an important economic activity providing employment to a large section of people including women. Better price realization by tea depends upon increasing market share through better productivity and efficient marketing system. Considering the rich history of origin of tea in Assam and its significance, a serious

Plate 1: A Tea garden with its flush green tea leaves

Plate 2: A tea leaf plucker

Plate 3: Most of the tea garden workforce provide fuel including fodder for their domesticated animals.

Plate 4: Not only employment but tea gardens are women.

effort by the government needs be undertaken to regain the past glory, as this sector still holds major share towards economic growth of the state.

References

Awasthi, C. R. (1975) : *Economics of Tea Industry in India with special reference to Assam,* United Publishers, Guwahati.

Baruah, P. (2008) : *The Tea Industry of Assam: Origin and Development,* EBH Publication, Guwahati.

Chakraborty, S. (2013) : *Tea Garden Workers Are Now Entrepreneurs,* Employment News, Vol. XXXVII, No. 52, 30 March- 5 April, New Delhi.

Choudhury, K. R.(2006) : A Special report on India's Tea Industry and Assam, *North-East News Agency,* Vol 3(15), Nov. 16-30.

Comptroller and Auditor General of India (2012) : *Role of Tea Board in Tea Development in India,* Report No. 10 of 2011-2012, New Delhi.

Das, K. (2012) : *How High is the Road? Structural Changes, Institutions and Well Being in Tea Plantations of South Asia,* Lambert Academic Publishing, Deutschland.

Guha, A. (1988) : *Planters Raj to Swaraj: Freedom Struggle and Electoral Politics of Assam, 1926-1947,* People's Publishing House, New Delhi.

Hussain, M.M. and Hazarika, S.D. (2010) : Assam Tea Industry and its Crisis, *Dialogue, New Delhi,* Vol. 12 (1), pp. 166-174.

Karmakar, K.G. and Banerjee, G.D. (2005) : *The Tea Industry of India: A Survey,* NABARD, Orion Publications.

Katangil, K. (1993) : The Socio Economic Effects of Tea Agriculture in our Country and Evolution, *Tea Enterprises General Directorate,* Caykur Publication, No. 15, pp 128.

Lahkar, S.D. (1996) : Tea Industry in Assam, *The Assam Review and Tea News,* pp 6-11.

Sarma, G. (2001) : A Study of Socio-economic Conditions of Labourers in the Tea Gardens of Jorhat District, Assam", Ph.D. Synopsis, Dibrugarh University, Dibrugarh.

Sen, B. B. (2008) : *Origin and Development of Tea,* EBH Publishers, Guwahati.

Singh, B. (2011) : Reviving the Tea Industry, inR.S. Kouli (ed.), *Yojana-North East Focus:Assam,* Ministry of Information and Broadcasting, New Delhi.

Singh, S.H. (2004) : Tea Plantation in Barak Valley: A strategy for Future Development, unpublished Ph.D. thesis, Manipur University, Imphal.

Directorate of Economics and Statistics (2002) : Assam at a Glance, Guwahati, Assam

Directorate of Economics and Statistics (1970-2020) : Statistical Handbook of Assam, Govt. of Assam.

Tea Board of India, (2007) : Tea Digest 2005-06, Kolkata.

Tea Board of India, (1970-71 to 2001-02 and 2011) : Tea Statistics, Kolkata.

102nd Report on Performance of Plantation Sector-Tea and Coffee Industry, Department Related Parliamentary Standing Committee on Commerce, presented to Rajya Sabha on 9th August 2012 (Accessed online on 27th August, 2014 from www.performance.gov.in)

Website of All Assam Small Tea Growers Association (AASTGA), www.aastga.org (Accessed on 14/09/2014)

www.teaboard.gov.in

www.assamteaexchange.com

www.assamnet.org

www.teaindustry.com

www.teatourindia.com

www.performance.gov.in

www.aastga.org

16 Transport and Communication

Introduction

Assam is a state in the extreme north-eastern part of India with diversities not only in its physical setting, but also in social and economic characteristics. The state represents 2.38 per cent and 2.64 per cent of the nation's total geographical area and population respectively. The present geographical size of Assam is almost one-third of the size of what it was at the time of Independence, because after 1947, Nagaland, Meghalaya, Arunachal Pradesh and Mizoram have been separated. Manipur and Tripura, which were princely states merged into the Indian Union in 1949. Thus, there are seven states in North East India popularly known as the 'Seven Sisters', of which Assam is the biggest in terms of population. Since the state is endowed with a variety of natural resources, an efficient transport system is essential for their proper utilization and speedy development.

Since transport is a basic pre-requisite, no specific elaboratiion is required here to highlight the role played by it in any region's socio-economic change. The close connection between the development of various sectors of economy and a good transport network is, in fact, well established. The movement of goods, people and services, and also the messages from one part to another would have not been possible had there been no evolution of transport and communication system. Lavarishchev (1969) remarks that, 'it effects multifarious connections between production and consumption, between industry and agriculture, between the extractive and manufacturing industries and ensures production connections between economic areas and various branches of the national economy'. In Europe for example, the freight transport which increased rapidly between 1950 and 1980, encouraged significantly the establishment of new industries in a number of formerly backward areas. Similarly, the North

Eastern Industrial Region of the United States (Pittsburgh-Wheeling-Cleveland district, New York-Philadelphia-Baltimore district, Southern New England and Detroit districts etc.) would not have come into existence had there been no excellent transport facilities provided by the Great Lake system, rivers, man-made canals, roads and railways. In India, the dispersal of cotton textile industry, for instance, from its original centres of Bombay and Ahmedabad to many other states like West Bengal, Uttar Pradesh, Bihar, Orissa, etc. can also be attributed mainly to the development of transport, specifically the railways. The growth of iron and steel industry in Durgapur and Asansol in the Damodar Valley which is expected to be the future 'Ruhr' of India, is not only determined by the locally available cooking coal and iron-ores but also by its favourable location on the Calcutta-Delhi rail route plus a navigable canal connecting Durgapur with the river Hooghly providing facilities for the movement of heavy steel goods to Calcutta market. Considering these few instances, it can safely be stated that the transport and communication is as important for the smooth functioning of a region's economy as the 'circulation of blood through human body'. Against this background, the existing transport and communication system of Assam covering its quantitative and qualitative dimensions has been attempted here.

Existing System

Although Assam is still much lagging behind in transport and communication compared to other parts of the country, it is now served by all the modern means of transportation, viz. roads, railways, waterways and airways. They are contributing, more or less, towards internal growth by consolidating the places located in various corners. They help in bringing villages and towns and the remote and more developed regions closer to one another and enhancing productivity, widening the market and introducing new stimuli to economic activity (Medhi, 1998). Of course, the achievement which can be expected from this sector is yet to reach an optimum level. Under these circumstances, it seems relevant to make an investigation of the present transport system along with the factors operating against its expansion. The quantitative change and qualitative (i.e. efficiency of operation) improvement need an elaboration in relation to all the modern means of transportation in the state. Since Assam is a gateway to other states of the North Eastern Region, viz. Arunachal Pradesh, Manipur, Meghalaya, Mizoram, Nagaland and Tripura, an efficient transport system of the state is directly related not only to the development of Assam, but also to the development of the entire region. Considering this specific importance of transport and communication in Assam various means of transportation have been discussed below.

Water Transport

Cheapness is a special quality of water transport. It provides opportunities to move bulky goods over the seas and big rivers, and since it is natural as well as ready-made, heavy dependence on it either at the national or international level remains unavoidable from time immemorial. Before the invention of automobiles or steam engines, waterways were the only means of transportation and the discovery of different parts of the world by the explorers during the medieval period was possible only through it. Coming quickly to the regional level, it has been observed that whatever little communication existed in Assam in the early period was through the rivers and the area is endowed with navigable waterways provided by the Brahmaputra and the Barak and their numerous tributaries. Thus, the inland waterways of the state enjoyed and continues to enjoy significant position in the field of transport, communication and trade.

Historical records reveal that the waterways provided by the main rivers of Assam were utilized by the Ahom kings not only for the purpose of trade and movement but also for warfare. Surprisingly, for the convenience of transport and communication, the kings also dug several artificial canals by diverting the original course of the river. For instance, King Kamaleswar Singha dug the Bhogdoi river by diverting the course of Disoi at the upper course of the river. Similarly, the Rupahi, Sonai, Khanajan and many other in Sibsagar division were dug by different Ahom kings so as to make short cuts to the Brahmputra river and to some other places (District Gazetteer, Sibsagar, 1967). Besides this, it is also found that "Assam's trade with the neighbouring provinces was mainly carried by river transport, the main route to Bengal, Bihar and Orissa being the Brahmaputra and the Ganga "(Medhi, 1978). During the early British period, it was the only means of transport but with the passage of time, the traditional boats used by the kings for communication had been replaced and steamers were used for the first time to ply between Calcutta and Guwahati in 1848. Subsequently, the steam navigation began to expand and by 1884 the steamer started plying between Dibrugarh and Dhubri, besides connecting Guwahati . What is more important that the direct steamer service from Dibrugarh to Calcutta via Sundarban was also opened in the following years and all these together influenced a great deal in minimising transport difficulties between Assam and the outside world (District Gazetteer, Sibsagar, 1967).

The introduction of steam navigation was a major turning point in the history of water transport and communication in Assam. It was estimated that when Calcutta ranked first in the origin and destination of Assam traffic in 1889-90, the total volume of trade of Assam with Calcutta was 71 per cent of the whole province. It was further estimated that steamer "practically monopolised the whole trade with Calcutta block, 94 per cent of the inward and 95 per cent of the outward traffic being carried by them" (Medhi, 1978). This status continued till Independence and the role of water transport remained unique for the state.

Quite a number of perceptible changes had taken place in the state's water transport system after 1947 and in this connection, three notable events can be considered responsible. These are : firstly, the earthquake of 1950 caused considerable changes in the course of the Brahmaputra resulting in deterioration of the navigable channel, particularly in the upper reaches resulting in closures of river services by the joint steamer companies from 1954 onwards. Secondly, the construction of the rail-cum-road bridge known as Saraighat Bridge across the Brahmaputra at Pandu in 1962 greatly improved the overland road and railway transport system, and thus it dealt a great blow to river services. Thirdly, due to Indo-Pak hostilities in 1965, the inland steamer service connecting Assam with the rest of the country was closed down (District Gazetteer, Darrang, 1978). All these events together badly affected the intra and inter-state navigation in Assam. It is found that between 1953 and 1956, the steamer services under the joint steamer companies between Amingaon-Guwahati, Tezpur-Neemati, Dhubri-Goalpara,

Disangmukh-Dibrugarh, Badati-feeder service, Badarpur-Loobe feeder service (in Cachar) were closed one by one (Medhi, 1978). However, in spite of having a host of shortcomings, the Government of Assam did not remain silent, but attempted to rejuvenate the inland water transport system and set up a Directorate of Inland Water Transport (IWT) under the transport department in 1959 and entrusted it to look after the development of waterways along with the training of IWT personnels. Similarly, the Central Inland Water Transport Corporation Limited (CIWTC), a Government of India undertaking was also established in 1967 to resume water services in various parts of the country including Assam. Currently, both these public sector organisations are operating in the state on commercial basis. The operation area of the former organization is limited to the state itself while the latter is operating between Assam and Calcutta via Bangladesh (Economic Survey, Assam, 1989).

At present, there are a large number of ferryghats on the either banks of the Brahmaputra and the Barak and a huge number of cargos and passengers are handled by them. The composition and capacity of waterways fleet and ferries under IWT in Assam is given in Table 16.1.

Table 16.1
Composition and capacity of waterways fleet and ferries under IWT, 1990-93

Composition	Number		Capacity (metric tonnes)	
	1990-91	1992-93	1990-91	1992-93
Ramp powered lighter vessels	-	10	-	-
Modern steel vessels	24	34	1110	315
Marboat	29	17	522	185
Single boat	57	58	586	318
Pantoon	7	9	70	-
Relief boat	4	14	20	70
Total	**131**	**142**	**2358**	**888**

Source : Statistical Handbook, Directorate of Economics and Statistics, Govt. of Assam, 1992 and 1994.

Cargos and passengers handled by ferries on either side of the Brahmaputra river are given in Table 16.2.

Table 16.2
Cargos and passengers handled by ferries on the Brahmaputra, 1990-93

	Item	1990-91	1992-93
1.	Passenger ('000)	5,306	5,471
2.	Goods (tonnes)	13,814	10,464
3.	Vehicles (scooter, motor cycles & cycles)	650,258	641,679
4.	Animals (nos.)	526,775	50,871
5.	Total Receipts (Rs. '000)	6,680	5,701

Source : Statistical Handbook, Directorate of Economics and Statistics, Govt. of Assam, 1992 and 1994.

Both the tables reflect two important features regarding the water transport of the state : firstly, the composition (in terms of number) of waterways fleets and ferries has been increasing, but the total capacity of all these fleets has been reduced from 2358 metric tonnes to 888 metric tonnes; secondly, the number of passengers travelled in ferries has slightly increased from 5,306 thousand to 5,471 thousand between 1990-91 and 1992-93, but the quantity of goods involved in water transport has dropped down (Table 16.2). Now, two probable explanations can be given for this. Firstly, the crossing of the major rivers of the state through various ghats by ferries is still unavoidable because adequate number of bridges connecting either banks of the major rivers are not there. Secondly, although the water transport is cheaper than others in monetary terms, it is time-consuming and since the rivers of the state became unsuitable for navigation after the earthquake of 1950, the role of the waterways fleets has been declining over the years. Besides these, the development of roads and railways also provided quick means of transportation and reduced the role of water transport during the last few decades. One of the recent events, however, in the sphere of water transport is the resumption of cargo service (December, 1996) between Assam and Bangladesh after a gap of six years. The state IWT has already entered into an agreement with Panden Cement Authority Limited (PCAL) of Bhutan to transport the boulders to

Bangladesh. The IWT vessels are already engaged in transhipment of Bhutanese boulders from Dhubri to Jogighopa under contract with PCAL. The most striking feature is that the state IWT, at present, has a 10,000 MT capacity fleet of barges along with high powered tugs to carry out cargo operations on the Brahmaputra. It has also been reported that four 500 MT capacity charges and a 750 MT self-propelled barge are currently under construction, which when completed will strengthen the capacity of water transport to handle more volume of cargo within and outside the state (The Assam Tribune, Dec., 3, 1996).

Road Transport

The present layout of the road transport network of the state, more or less, has developed over the framework laid down by the colonial rulers during their regime. Ahom rulers contributed a lot towards the development of road transport but except for a few of them these roads became useless to the British who were only concerned with the extraction of natural resources from this territory. Consequently, with the annexation of Assam to the British rule through the Treaty of Yandaboo in 1826, the erstwhile road transport structure of the state began to change slowly. The creation of the Public Works Department in 1868 and the Local Boards in Assam in 1880 was a significant step taken by the British to expand and develop the road network of Assam with a view to collecting resources of the state and then to despatching the same to their own land through Calcutta. Yet, the real development in respect of the construction of new roads did not take place until the establishment of the Road Board in 1926-27. A scheme of classification of roads into metalled, gravelled and unsurfaced was prepared by the Board and at the same time recommended for the adoption of mechanical means of construction of roads in the state. In 1928, the Road Development Committee called for a structural change of the road policy of the government and suggested taking over a bigger share of the road building activities. On the basis of recommendations of the committee, the government of India set up a Central Road Fund (1929) which subsequently provided substantial amount towards the development of roads in the state. Towards the end of 1942, Assam Transport Organization was set up and subsequently the organization attempted with full vigour to expand the road transport system of the state by materializing the suggestions worked out at the conference of Nagpur in 1943. Thus, prior to independence an extensive network of road development plan had been prepared, but the main lacuna was that roads were planned to connect only those places which were rich in resources or useful to the colonial rulers. No systematic attempt was made towards the development of the interior areas and as a result indigenous people had to suffer a lot from lack of transport and communication facilities. During this period, roads were constructed as linkages between the centres of commercial importance and the resource producing regions, e.g. between the tea gardens and the railway stations or river ports.

After independence, the Government of Independent India has been emphasizing more on the development of transport and communication facilities within the state in order to expand the British laid network so that the passengers and goods can be transported even from the interior areas. A significant proportion of investment alloted for the development of transport system during the first and the second Five Year Plans (26.4 per cent and 27.0 per cent respectively) indicates the government's seriousness towards the improvement of this sector. The Government of Assam also paid similar attention to change the inefficient means of transportation of the state and consequently, the outlay in transport sector which was just 1.82 per cent of the total expenditure during the first Five Year Plan increased to about 13 per cent in the next plan. During the last four decades, the Government of Assam has also been stressing more on the creation of new links between the interior and main centres, besides improving the existing network of transport in the state. It is seen that during the first Five Year Plan period (1951-56), altogether 9,120 km of motorable roads were completed under the road development programme of the state (this included Mizoram and Meghalaya also). On the other hand, 7,696 km of new roads were constructed in the hill areas and the total nationalized road length increased to about 960 km during the said plan period. Slowly but steadily, the transport and communication scenario of the state has been changing covering almost all parts of the state irrespective of the resource consideration. As a result, the road transport has been emerging as a significant mode of communication in recent years with growing involvement of both inter and intra-state flow of goods and passengers traffic. Statistical records reveal that the total length of the road under Public Works Department (PWD) in Assam increased to above 30,086 km in 1976-77, the annual rate of increase being 4 per cent. The surfaced and unsurfaced roads which were just 4,276 km and 16,046 km respectively in 1976-77 increased significantly to 6,839 km and 23,247 km in 1989-90. Of course, the proportions of surfaced and unsurfaced roads to the total road length have also been remaining constant during the said period. It is clear that the share of surfaced road to the total road length in 1976-77 was nearly 21 per cent which increased to 22.73 per cent in 1989-90 (less than two per cent increase over a period of thirteen years). On the other hand, the extension of National Highway in the state has been rapid; it has gone up from 1,383 km in1976-77to 2,026 km in 1988-89 and further to 2,033 km in 1989-90. Contrary to this, the length of the state highway in Assam has been declining from 2,131 km in 1976-77 to 2,106 km in 1989-90. The nationalization of a few kms of state highway might be the cause of reduction of this category of road in Assam. However, the district and other PWD roads of the state witnessed a large scale expansion over the last few years. This, for instance, has increased from 11,417 km in 1976-77 to 24,953 km in 1988-89 and further to 25,947 km in 1989-90, the annual rate of expansion being 127 per cent between the said periods. This indicates that to improve the transport and communication system in rural areas of the state, more than thousand kms of roads have been constructed by the Government of Assam each year. The types of road, the categories of roads and road length per lakh of population and per hundred sq.km of geographical area of Assam have been shown in Tables 16.3, 16.4 and 16.5 respectively.

Table 16.3
Types of road under PWD in Assam, 1976-1990

Year	Surfaced (in km.)	Unsurfaced (in km) Gravelled	Earthen	Total	Total road length (km)
1976-77	4276	14261	1784	16045	20321
1988-89	6601	21432	1024	23247	29035
1989-90	6839	22223	1024	23247	30086

Source : Statistical Hand Book, Assam, 1980 and 1992

Table 16.4
Categories of roads under PWD in Assam, 1976-1990

Year	National highway (km)	State highway (km)	Other PWD roads (km)	Total (km)
1976-77	1383	2131	16807	20321
1988-89	2026	2056	24953	29035
1989-90	2033	2106	25947	30086

Sources : Statistical Hand Book, Assam, 1980 and 1992.

Table 16.5
Road length (in km) per lakh of population and per 100 sq. km of area in Assam, 1976-1990

Year	Per lakh population	Per 100 sq. km.of area
1976-77	118	25
1988-89	136	37
1989-90	137	38

Source : Statistical Hand Book, Assam 1980 and 1992

So far as the types of road are concerned (as shown in Table 16.3), it is clear that the proportion of surfaced roads to the total road length was only 22.7 per cent in 1989-90 against the national average of about 48 per cent. On the other hand, the proportion of unsurfaced roads is still more in the state (nearly 88 per cent) than that of the nation as a whole (nearly 51 per cent). Whatever may be the quantitative dimension of the total road length, the most staggering point is that the condition of unsurfaced roads is not satisfactory in many parts of the state. Since most of them are confined mainly to the rural parts and acting as feeder lines either to the national or state highways, they need considerable qualitative improvement.

Categorywise, the proportions of National Highway, State Highway and other PWD roads to total road length are 6.75, 6.99 and 84.24 per cent respectively. The share of National Highway in the state is higher than the national average of only 2 per cent. Although the National Highway constitutes only 2 per cent of the total road length, they carry nearly 45 per cent of the road traffic of the country. Similarly, in Assam the contribution of National Highway to the total road traffic is most significant. From the view point of industrial development, the role of National Highway is more pronounced because the link between Assam and the rest of the country is provided by the National Highway No. 31. Realizing the special role played by the National Highways the Government of India set up a National Highway Authority in June, 1989 with a view to expanding roads to almost all areas of the country. In Assam, at present, national highways have been extended to cover all the districts (Table 16.6 and Fig. 16.1).

Table 16.6
Districts covered by National Highways

National Highway No.	Area/places connected	District covered
31	Guwahati to Barahi (Bihar) via Siliguri and Koch Bihar	Kamrup,Nalbari, Barpeta, Kokrajhar, Bongaigaon and Dhubri
31.B	North Salmara to Jogighopa	Bongaigaon
31.C	Chapaguri via Srirampur to West Bengal	Kokrajhar
36	Nagaon via Dabaka to Dimapur	Nagaon, Morigaon and Karbi Anglong
37	Pancharatna via Guwahati Nagaon, Jorhat, Sibsagar to Saikhowaghat (Dibrugarh)	Dhubri, Goalpara, Kamrup, Nagaon, Jorhat, Sibsagar, Dibrugarh, Tinsukia
38	Makum, Jaipur to Lekhapani	Tinsukia
39	Numaligarh via Golaghat Dimapur to Kohima	Golaghat and Kabri Anglong
40	Guwahati to Shillong	Kamrup
44	Shillong viz Badarpur, Karimganj to Agartala	Cachar and Karimganj
51	Paikan (Goalpara) to Tura	Goalpara
52	Baihata Chariali via Mangaldoi, Tezpur, Lakhimpur to Jonai	Kamrup, Darrang, Sonitpur, Lakhimpur and Dhemaji
53	Silchar via Lakhimpur to Imphal	Cachar
54	Silchar to Aizawl	Cachar

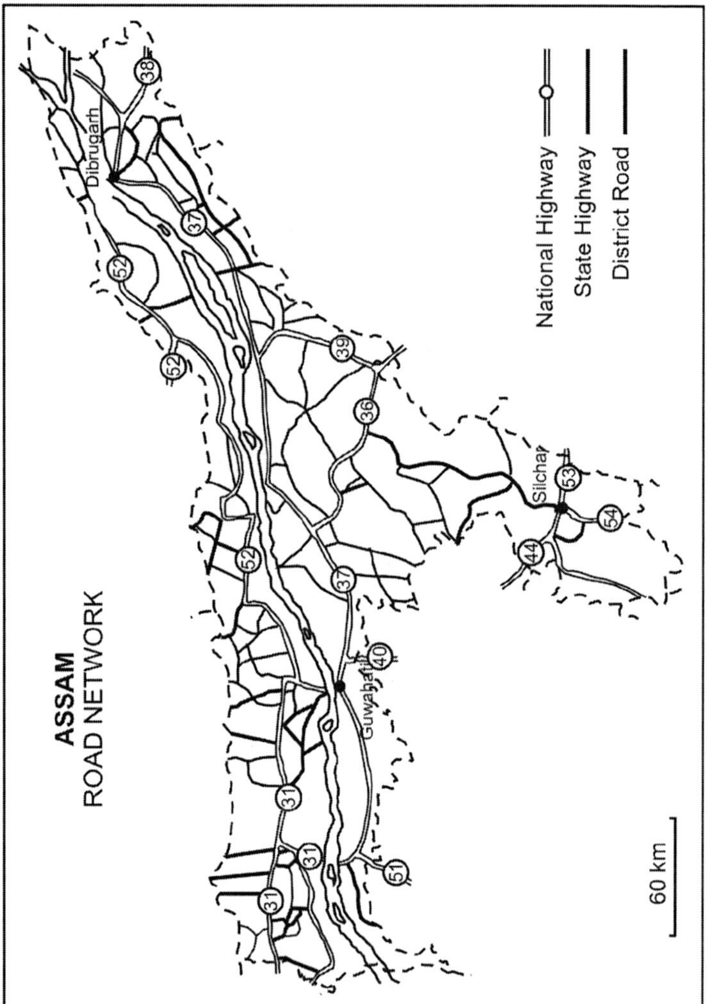
ASSAM
ROAD NETWORK
Dibrugarh
Guwahati
Silchar
38
37
52
52
39
36
53
54
44
52
37
40
31
31
31
51
National Highway
State Highway
District Road
60 km

Fig. 16.1

The existing state highways and other PWD and rural roads are under the state government and they altogether cover nearly 93 per cent of the total road length. In the rural areas, roads are being developed under the Minimum Needs Programme (MNP), the objective of it being to link all villages with a population of 1,500 and above and 50 per cent of the villages with a population of 1000–1, 500 by all weather roads by 1990 (India, 1993). The contribution of this programme to the development of rural roads in Assam is no doubt substantial. Even then, the need of quantitative expansion of roads to serve fully the entire rural areas is still profound. On the other hand, the more construction of new roads in the rural areas cannot serve the needs of the people if it is not all-weathered road. Therefore, required efficiency in terms of construction and management of all-weathered roads in the interior areas is indispensable. Another important aspect is that the road length per lakh of population and per 100 sq. km. of geographical area in the state is 137 km and 38 km respectively. This is lower than the national average of 221 km and 53 km respectively. Thus, it is clear that the state's road transport system is yet to reach even the minimum national standard.

Railway Transport

Rail transport is one of the biggest nationalized undertakings in India with a total route length of more than 62,000 km, of which nearly 6.0 per cent (3,858.14 km) is represented by the North East Frontier Railway serving basically the north-eastern states of India. It is the largest railway system in Asia and fourth largest in the world. The total railway route length of N.F. Railway in 1991-92 was 3,854.14 km, of which about 64 per cent (2,466.83 km) lies within the state of Assam. The total broadgauge route length in Assam in 1991-92 was 266.41 km and the length of meter gauge was 2,200.42 km representing nearly 0.76 per cent and 9 per cent of the total broadgauge and metergauge lines of the nation respectively. The proportion of broadgauge route length to the total route length of railway in Assam is worked out to be around 11 per cent against the national average of 55 per cent. However, the state was favourably placed in respect of the railway route length per 100 sq. km. It was about 30 km against national average of only 19 km by the end of 1989-90.

The age of the rail transport system of the state is now 100 years. Its history started with the construction of a 80 km long Dibru-Sadiya metergauge railway line constructed by the Assam Railway and Trading Company in 1882. The state also witnessed the construction of another 30 km long and 2 feet wide state railway running from Kokilamukh on the river Brahmaputra to Mariani and Titabor of Jorhat district in 1885. Both these lines were originally built for the convenience of tea gardens located in the districts of Sibsagar and Lakhimpur as the unmetalled roads to the river became impassable to wheeled traffic in rains (District Gazetteer, Sibsagar). These were the feeder lines to the river transport because they had no link or contact with any railway system outside the region (Medhi, 1998). Another feeder line was constructed between Pandu and Jamunamukh in 1897. Similarly, the completion of Tezpur-Balipara feeder line in 1895 and Chaparmukh-Silghat in 1930 had finally added to the growth of rail transport in Assam.

Since the beginning of the present century rail transport has been getting changed in its structure. It is observed that "Since 1902 railway development in Assam was meaningful in the sense that Assam was connected with the two great ports by two lines, one the Eastern Bengal Railway (EB Railway) extended from Calcutta to Dhubri in 1902 and the other, the Assam Bengal Railway (A B Railway) extended from Chittagong to Dibrugarh in 1904. Five years later the E.B.Railway came as far as Amingaon and the total railway length increased to about 1,000 km" (Medhi, 1978). Considering the importance of railway in the province as well as to improve the earnings of the A B Railway, the Government of India subsequently paid greater attention to the construction of few branch lines in the state. Consequently, the Simaluguri-Naginimara line (16 km) of 1917, the Chaparmukh-Silghat line, the Katakhal-Lalbazar line of 1924, the Furkating-Badulipara-Jorhat line, the Karimganj-Longai Valley line, and Senshoa-Moirabari line, etc. began to appear in the state. Another significant development was taking place in the north bank of the Brahmaputra river. The E B Railway extended its metergauge tract from Rangia to Tangla in 1912-13 and subsequently to Rangapara in 1932-33. The railway network in the state took a new turn after 1942 when the A.B. Railway, with a route length of about 1133 miles, was amalgamated with the E

B Railway and named it as the Bengal and Assam Railway (Medhi, 1978). But after independence, the organizational set up of the B.A. Railway needed a change. Due to the partition of the country, Assam had to loose direct links with Chittagong and Calcutta ports. As a result, the creation of a link route with the rest of the country for Assam became indispensable. At this critical juncture, the Assam Rail link was started in 1948 and completed in 1949 which was then handed over to the Assam Railway. It is seen that before as well as immediately after independence there prevailed a complicated system of ownership and control of railway in India. As this system was found unconducive from the economic and efficiency points of view, the entire administrative and organizational set up of the railway system in the country was reorganised in 1950 and attempted to materialize the new scheme in 1952. The impact of reorganization was evident in the Assam Railway which, with some minor adjustments was brought under the jurisdiction of the North East Frontier Railway in 15th January, 1958 with Maligaon as its headquarters. The total route length of the N F Railway in 1991-92 was 3,858.14 km of which 772.47 km was broadgauge (BG), 2,998.19 km was metergauge (MG) and 87.48 km, narrowgauge. The major route length of this zone falls within the state of Assam (nearly 64 per cent). The length of broadgauge and metergauge in the state is 266.41 km and 2,200.42 km respectively. After the creation of this zone in Assam, various developmental activities have been undertaken in the railway sector among which the extension of railway tracts to remote areas is more significant. During 1960s a few of these projects had been completed, viz. the extension of broadgauge line from New Jalpaiguri (North Bengal) to Jagighopa via New Bongaigaon in 1965, Metergauge line from Rangapara North to North Lakhimpur in 1962 and from North Lakhimpur to murkongselek in 1966 along the foothills of Arunachal Pradesh, and also the extension of a metergauge line from Kalkalighat in Assam to Dharmanagar in Tripura in 1966.

Currently, the gauge unificatiion process in the country has been in progress. Following this unification process, the N F Railway has been directing its emphasis on the conversion of existing metergauge line to broadgauge line. The Guwahati-Lumding broadgauge line (177 km) has already been completed and opened for traffic. Besides this, the conversion work of

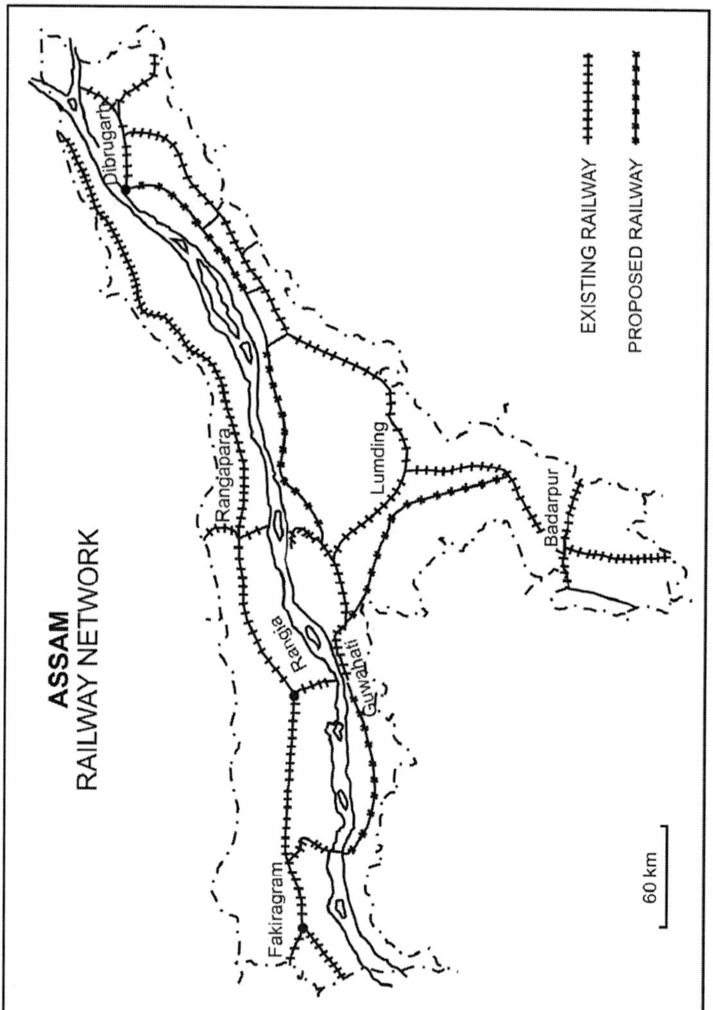
ASSAM
RAILWAY NETWORK
Dibrugarh
Rangapara
Lumding
Badarpur
Rangia
Guwahati
Fakiragram
EXISTING RAILWAY
PROPOSED RAILWAY
60 km

Fig. 16.2

Guwahati-Tinsukia-Dibrugarh (560 km) line has been in progress. The surveys for the construction of an alternative broadgauge connection between Jagiroad and Badarpur (296 km) and Chaparmukh and Dibrugarh (350 km) have been completed by the railway department. The most significant development in the railway transport is the construction of the third rail-cum-road bridge across the river Brahmaputra connecting Jogighopa and Pancharatna and very recently the foundation stone for a fourth bridge has also been laid at Bogibeel (Dibrugarh). On the other hand, plans for the construction of a few feeder lines (MG) like Itanagar-Tipling (34 km), Lekhapani-Kharchang (26 km), Murkongselek-Pasighat (30 km) etc. have been undertaken by the railway department. During the 80s, Balipara-Bhalukpung (34 km), Arunachal-Jiribam (56 km), Lalbazar-Bhairabi (80 km) and Amguri-Tuli (18 km) MG lines have been completed and opened for both passenger and traffic (Fig. 16.2).

It is now obvious that the rail transport system of the state has been expanding steadily and if the proposed plans and projects mentioned above are completed in stipulated time, it would definitely be able to meet a long-felt necessity of the people of the state. In table 11.6, a statistical profile of the NF Railway (major part falls within Assam) for the period of 1987-88, 1988-89 and 1989-90 has been presented to highlight the role of rail transport in the fields of passengers and goods movement. The following characteristics can be observed from the data shown in table 16.7.

1. The gross earnings of the NF Railway have been increasing at a rate of nearly 17 per cent per year. Unfortunately, the net earnings have been remaining negative over the years. It is due to the higher working expenses resulting from the construction of new feeder lines and conversion of MG to BG in the state. However, if the entire projects completed, the railway transport would reach a level of stability after which the net earnings could be expected to be positive.
2. Passengers originating and passengers carried by the NF Railway had slightly declined during the periods as shown in the table 16.6. But the average earning per passenger km and average journey per passenger has been increasing. Generally, people prefer long-distance journey by train

avoiding the short distance journey. It indicates that the movement of people within the state by train has been declining day by day. Although the intra-state movement marginally declined, the inter-state movement increased. Thus, the rail transport has been playing a positive role in bringing the people of this region to the national mainstream.

3. Since the flow of commodities to and from the state has been going up, both the freight tonnes km and the earnings from it have been increasing steadily. Therefore, the dependence on the rail transport for the development of the entire north-east is increasing day by day.

Table 16.7
Statistical profile of NF Railway, 1987-1990

Particulars	Units	1987-88	1988-89	1989-90
A. Financial Position				
a) Gross earnings	(Rs. crores)	201.69	227.09	302.83
b) Working expenses	-do-	396.45	443.00	536.83
c) Net earnings	-do-	(-)197.76	(-)215.90	(-)234.00
B. Passengers				
a) Passengers originating	(million)	30.22	27.26	29.77
b) Passenger carried	-do-	32.03	29.82	31.89
c) Passenger km	-do-	5124.4	5198.6	5597.3
d) Earnings from passenger	(Rs. crores)	49.52	55.04	70.21
e) Average earning per passenger km.	Paisa	9.13	10.59	12.54
f) Average length of passenger journey	km	169.4	194.3	175.5
C. Freight				
a) Freight originating	(m. tonnes)	5.41	5.11	5.84
b) Freight carried	-do-	11.10	11.13	11.16
c) Freight tonnes km	(million km)	5859.6	5915.3	6237.1
d) Earning from goods carried	(in crore Rs.)	134.12	152.51	207.39

Source : Statistical Hand Book, 1992 and Economic Survey of Assam, 1989-90.

Air Transport

Air transport has a special role to play for the movement of passengers and goods in an area like Assam with peculiar topography and geographic location. Of course, its potentiality came to light only after 1949 when Pakistan prohibited transhipment of cargo to Assam by waterways. Consequently, the government of independent India became more attentive towards the expansion of the air transport in the state. At present, Assam is well-connected with the other parts of the country through air link. The Indian Airlines are instrumental in establishing air connection between Assam and the rest of the nation. They are operating regular Air Bus and Boeing services to and from Guwahati, Calcutta and Delhi. Besides this, there are regular air transport services connecting six civil airports located in Guwahati (Borjhar), Tezpur (Salonibari), Jorhat (Rowriah), Dibrugarh (Mohanbari), Lakhimpur (Lilabari) and Silchar (Kumbhirgram). Guwahati is served by three levels of air services. viz. Air Bus, Boeing and Bayudoot. All the capital towns of the states. of North Eastern India have been served by Vayudoot services. Since the air transport of the state is comparatively recent, it has still a long way to cater to the needs of the people of the state.

Conclusion

Considering the present position of all the modern means of transportation in the state, the following conclusions can be drawn:

1. The inland water transport system of the state is providing facility basically for the intra-state flow of goods and people. The glorious tradition of the water transport linking the state with the rest of the country has gradually been declining. The constant upliftment of the beds of the major navigable rivers of the state due to silting every year has badly affected the waterways.

Besides this, the expansion of the roads and the railways in the state has also reduced the importance of the water transport. The declining trends of the goods traffic through water transport during the last decades, as indicated in the preceeding table, can be used as indicator to reflect the declined role of this transport system. However, the steps recently taken by the Government of Assam in order to improve the water transport system are quite impressive. Since the communication between the people of the either banks of the major rivers as indispensable both from the social and economic points of view, the regularity and modernization of ferry services connecting the banks are more essential, i.e.. the qualitative improvement is a must to regain its glorious tradition.

2. So far as the road transport system of the state is concerned, there is abnormal gap between the existing and expected length and efficiency of the roads. In view of the fact that the density of roads per hundred sq km of area and the road length per lakh of population in the state are lower than the national average, the expansion of road network should be more rapid than what it is currently. Greater emphasis is needed to expand the surfaced roads in the rural areas. It should be noted that mere expansion of road network will not fulfil the demands of the people unless it is upgraded through some qualitative improvement. Quite a large number of villages are yet to be connected by all-weathered roads. Whatever road extension programme is going on in the state, it is seen that most of them are guided by political considerations rather than the needs of the villages. As a result, the systematic mobilisation of resources including agriculture in the state is yet to make much headway. Thus, the need of the hour is to make a structural improvement of the existing road transport system keeping in view the state's prospects for future growth and development.

3. It is the rail transport system of the state which is instrumental in channelizing the flow of commodities and people at inter-state level. Unfortunately, this system of transport has been operating with negative net earnings since a long period of time. However, the negative trend is expected to get improved after the completion of various projects currently under progress. But still there is a problem. The major railway tracks of the state were constructed before independence and were oriented basically towards the growth of oil and tea industries. The creation of link between other production and consumption centres was not properly looked into. However, the recent addition of feeder lines in the state is expected to fulfil these shortcomings.

4. The air transport system of the state is yet to reach the expected level. As this system of transport is costly and require constant monitoring, it has covered only a smaller section of the society. Its expansion as well as improvement, especially in the hilly parts, needs due attention.

However, the existing means of transport in the state need better coordination and management in order to optimise the flow of goods and people for integrated growth and development of the state.

References

Lavarishchev, A. (1969) : *Economic Geography of the USSR,* Progress Publishers, Moscow, p. 318.

Guha, J.L. and Chattaraj, P.R. (1978) : *A New Approach to Economic Geography,* World Press, Calcutta, pp. 467-468.

Medhi, S.B. (1978), : *Transport System and Economic Development in Assam,* Publication Board, Guwahati, p.1.

Gazetteer of India, Assam State, Sibsagar District, 1967, Govt. of Assam, pp. 228-229

Medhi, S.B. (1978) : op. cit., p.19.

Sibsagar District Gazetteer (1967), op.cit, pp. 230-231.

Medhi, S.B. (1978) op. cit., pp. 22-23.

Gazetteer of India,(1978) Assam State, Darrang District, Govt. of Assam, p. 281.

Medhi, S.B. (1978) op. cit., p. 28.

Gazetteer, Darrang District (1978) op. cit, p. 281.

Economic Survey, Assam (1989) : Directorate of Economics and Statistics, Guwahati, p.34.

The Assam Tribune, (1996), Dec 3rd, Guwahati, Assam

Gazetteer, Darrang District, op cit. pp. 268-269.

India (1993) : *A Reference Annual,* Ministry of Information and Broadcasting, p. 599.

Gazetteer of India, Sibsagar District (1978), op. cit., p. 236.

Medhi, S.B. (1978) : op.cit., p.56.

17 Tourism

Introduction

Although tourism means travel, in the present day context it stands for the totality of activities and phenomena arising from 'travel' and 'stay' of strangers. However, it excludes remunerated activity and establishment of permanent residence by the strangers. The term includes all interconnected processes, especially economic, that came into play through the temporary movement of strangers into, within and from a certain district, state or country. India has adopted the definition of tourist as a person who temporarily travels to an area other than that in which he has his usual residence and the main purpose of whose visit is other than the exercise of an activity remunerated in some material way.

The potential economic benefit continues to be the driving force behind tourism industry and its development. In 1950, 25 million tourists (international arrivals) were recorded world-wide, and it has grown to 500 million in 1990. Thus tourism has become the second largest economic activity in international trade with an annual growth rate of 6%. Considering the importance of tourism, developing countries like India, have made attempts to attract international tourists to their countries. This is because the amount spent by tourist in a country goes to the exchequer of that country, besides job opportunities. India with its well known cultural heritage and natural beauty, can certainly attract foreign tourists. But tourism is not developed much in India in comparison to some neighbouring countries like Thailand. Thailand received about 2% of world's total income from tourism (in 1993-94 the total receipt from tourism in the world was 3,05,816), whereas the share of India continues to be only 0.48%.

Tourism in Assam : Assam is a pioneer state in north-eastern part of India to introduce a state tourism department in 1958 to promote tourism activities in the state. But the progress is far below expectation. In 1979, Assam was visited by 2320 foreign tourists accounting for 0.3% of foreign tourism to India. Unfortunately, the number declined to 476 in 1995 due to decade-long socio-political instability, starting with the Assam agitation in 1978-79. As a result of this, the R.A.P. regulation (Restricted Area Permit-for foreign visitors introduced since 1962, having been lifted only in the later part of 1995) was made stringent for foreign visitors.

Table 17.1
Tourist arrivals in Assam (1979-1997)

Year	Foreign tourist	Domestic tourist
1979	2320	N.A
1980	1000	5190
1981	50	8686
1982	132	10177
1983	32	7028
1984	23	8481
1985	40	10,383
1986	149	14,777
1987	430	18,722
1988	595	18,470
1989	319	17,092
1990	475	15,382
1991	332	15,016
1992	431	19,262
1993	440	20,154
1994	350	17,648
1995	479	16,056
1996	1000	14,730
1997	723	14,238

Source : Directorate of Tourism, Assam.
N.B. Figures represent records of Government Tourist Lodge, Assam.

In the domestic sphere also, the trend is not so impressive as Assam witnessed only 0.01% of the domestic tourists (17,648) of India (Total domestic tourist in India -100,036,261) in 1994. Table 17.1 represents the trend of tourist arrivals in Assam, which reveals a poor state of affairs in tourism in Assam. Kaziranga is

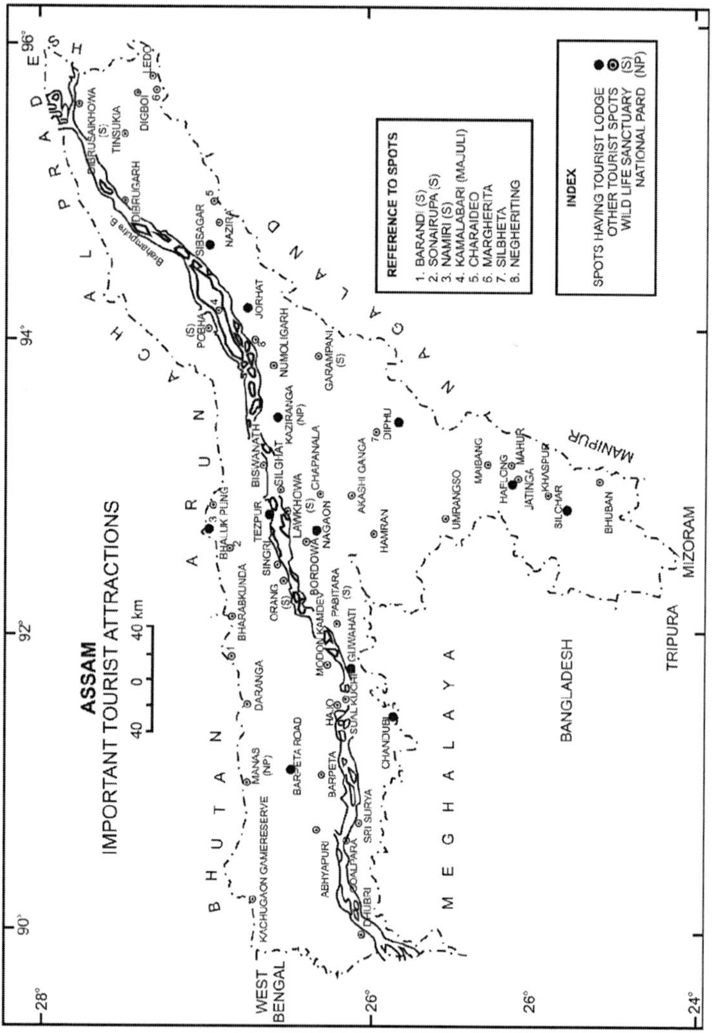
ASSAM
IMPORTANT TOURIST ATTRACTIONS
40 0 40 km
REFERENCE TO SPOTS
1. BARANDI (S)
2. SONAIRUPA (S)
3. NAMIRI (S)
4. KAMALABARI (MAJULI)
5. CHARAIDEO
6. MARGHERITA
7. SILBHETA
8. NEGHERITING
INDEX
SPOTS HAVING TOURIST LODGE
OTHER TOURIST SPOTS
WILD LIFE SANCTUARY (S)
NATIONAL PARD (NP)
BHUTAN
ARUNACHAL PRADESH
NAGALAND
MANIPUR
MIZORAM
TRIPURA
BANGLADESH
MEGHALAYA
WEST BENGAL

Fig. 17.1

the prime destination in Assam as well as North-Eastern India which has been hosting foreign and domestic tour- ists consistently. During the period from 1951-52 to 1994-95, it has recorded an eleven fold increase of foreign tourists, from 71 to 781 in number and 480 fold increase in domestic tourists from 101 to 48, 501 in number (including excursionist) with an average of 274 tourists per day (Source: Department of Forest, Govt. of Assam). The premier tourism agency of Assam, the Department of Tourism has so far developed 12 destinations (Fig. 17.1) along with accommodation facilities having altogether 360 bed capacity (Table 17.2).

Table 17.2
Tourist centres and government accommodation facilities in Assam, 1995

Sl. No.	Destination	District	Significance	Accomm-odation unit	Bed-capa-city
1.	Guwahati	Kamrup	Gateway to N.E. Region, capital of the state, organisations, shopping of handicrafts, sight see-ing facilities, accommodation and entertainment facility, historic and pilgrimage sites, air link etc.	1	56
2.	Chandubi	Kamrup	Around 64 km from Guwahati, ideal picnic site, potentiality for trekking and jungle exploration, tribal life of Rabha people, boating facility etc.	1	16
3.	Barpeta Road	Barpeta	Business centre of the district, gateway to Manas, only tiger reserve in Assam (National Park and World Heritage site)	1	10
4.	Tezpur	Sonitpur	Headquarters of the district, historical monuments, pilgrimage and archaeolgical site, one of the beautiful township of Assam with recreational	1	35

			parks, boating facility, air link etc.		
5.	Bhalukpung-Sonitpur (Lower)		Around 60 km north of Tezpur, one of the important gateways to Arunachal Pradesh, natural beauty, ideal for rafting and angling in Jia Bharali river, Adj-acent to Nameri National Park, hot spring (Pung) etc.	1	12
6.	Kaziranga	Golaghat	Important National Park especi--ally of one horned rhino, eleph-ant safari, teagarden, tribal life of Karbi people, potential site for trekking in Karbi Hills, Deo-pahar Archaeolgical site, etc.	4	108
7.	Nagaon	Nagaon	District headquaters, celebrated pilgrimage site Bardowa at a distance of 12 km, Akashigonga Archaeologial park and pilgrim-age site of Dabaka is located at a distance of 52 km.	1	17
8.	Sibsagar	Sibsagar	District headquarters, socio-cul-tural organisations, Ahom age tanks (Sagar), temples and pal-aces, Ajan pir 'Dargah', A celeb-rated pilgrimage site both for Hindus and Muslims.	2	14
9.	Jorhat	Jorhat	District headquarters, important township of upper Assam, Tea gardens, tea research centre, Regional Research Laboratory, transit point for greatest river island 'Majuli' and its 'Satra' Institutions along with beautiful riverscape, air link.	1	24
10.	Diphu	Karbi Anglong	District headquarters, Silbheta rock dam on Jamuna river-40 km from Diphu, tribal life of Karbi people, potential site for trekking and Jungle exploration.	1	20
11.	Haflong	N.C.Hills	District headquarters, socio-cult-ural organisations, only hill station of Assam, potential trekking site on the Borail		

		hills, nearby important sites, viz. Jatinga (6km) for bird watching and natural beauty, Mahur (26 km)- picnic and nature spot, Maibang (50km) archaeological site.	1	26
12. Silchar	Cachar	District headquarters, socio-cultural organisations, air link, important township in Barak valley, Khaspur archaeological site (24 km.), Bhuban hill pilgrimage site (43 km), tea gardens etc.	1	26

N.B. Accommodation units represent available accommodation facilities under tourism department.

Sources of Tourist Attraction

The state has suitable natural infrastructure for the promotion of tourism. It has wide ranging physical, biotic and socio-cultural potentiality in this regard. Situated beneath and encircled by the Himalayan ranges to the north and north-east, the state is surrounded even to the south by hills. The plains of the river Brahmaputra and Barak account for 81% of the total geographical area. The plains with numerous criss crossing tributaries have provided beautiful riverscape. There are also two hill districts (North-Cachar and Karbi Anglong) ranging from 150m to about 1800m in altitude with unique landscape. Hill destinations like Haflong, Jatinga, Diphu and Hamren can offer a lot in the field of adventure tourism, viz. walking, trekking, jungle exploration, rock climbing, etc. Hot water spring like Garampani (Golaghat district), water bodies in Umrangso (N.C. Hills), water falls at Chapanala (Nagaon dist), rapids along the courses of the river Jamuna at Silbheta (Karbi Anglong dist) and along Kapili river at Panimur have tremendous potentiality to attract tourists. As the whole range of natural infrastructure for tourism is meant for out door recreation, the rainfree and cold season, viz. Autumn (Oct-Nov.), Winter (Dec.-Jan-Feb) and early spring (Mar-April) are regarded as suitable period for tourists. It is found that around 78% of the tourists visit the state within these seasons.

Assam is well known for its exotic flora and fauna. In fact, it forms a part of global bio-diversity 'hot spot' characterised by variety of habitats ranging from swamps and savannah grasslands in the plain to sub-tropical broad leaved forest in the high Barail range. About 180 species of mammals and more than 800 species of birds are found here which is the highest among individual states in India. The greenery of Assam consists of evergreen, semi evergreen, mixed deciduous, riperian, moist savannah and miscellaneaus types of vegetation. It also has ornamental plants like orchids, both common and rare varieties. Though Assam has small coverage of protected area (about 3 per cent of total geographical area of the state), it still possesses 13 important wildlife areas (Table 17.3), which provide room for 31 types of endangered and rare species of mammals and 14 types of amphibians and reptiles . Hence the state can be said to have all the natural infrastructure to develop tourism.

In cultural front, the state is rich enough to attract the attention of tourist. It is a meeting ground of Austro-Asiatic, Tibeto-Burman and Indo-Aryan groups of people. Hence human variety is reflected through each and every sphere of life of its residents. The colourful life of people, their traditional customs, festivals and dances are some of such important sources of tourist attraction. Assam has the tradition of weaving which is rich also among the tribal people. Assam also practices 'muga' 'endi', and 'pat' silk, worm rearing and weaving. It has also the tradition of folk and classical dance forms, mask making, puppetry, wood and ivory curving, bamboo and cane craft, etc. Besides, various forms of performing art of 'Satra' institutions like 'bhawna', 'Jatra', etc. are worth mentioning in this regard.

Assam is also rich in antiquity assets like ruins and monuments. The existing archaeological ruins of Assam go as far back as Gupta period (5th-6th century AD), though their numbers are very few. Ruins of the Post-Gupta period are also to be found in different parts of the state. Stylistically all the remains belong to different north Indian art tradition with local influences. There are as many as 12 archaeological sites (Table 17. 4) and about 25 major monuments dating back from 7th to 17th century AD (Table 17.5).

Despite the variety of attractions the focal point of tourism in Assam continues to be only the wild life covering internationally famous Kaziranga and Manas National Parks along with ruins and monuments, especially the pilgrimage trail to the age -old

'Sakti' shrine Kamakhya. Fairs and festivals, adventure activities, ethnic culture, cuisine are other relatively new segments with potentiality for development of overall tourism scenario of the state.

Table 17.3
Protected areas in Assam

Sl. No.	Name and category	Area in sq. km	Rare and endangered animals and birds
1.	Kaziranga National Park	430	Rhinoceros, Asiatic water buffalo, swamp deer, elephant, tiger, florican, hoolock gibboon, capped langur.
2.	Manas National Park	500	Tiger, goldenlangur, pigmy hog, hispid hare, elephant, gaur florican.
3.	Barnadi Wild Life Sanctuary	26	Pigmy hog, hispid hare, Gaur, elephant, florican hornbil.
4.	Laokhowa Wild Life Sanctuary	70	Rhinoceros, aquatic and migratory birds.
5.	Orang Wild Life Sanctuary	72	Rhinoceros, Spotbilled pelican, greater adjutant stork.
6.	Pobitara Wild Life Sancturay	18	Rhinoceros
7.	Garampani Wild Life Sancturay	6	Elephant, gaur, hollock gibbon and varities of birds.
8.	Dibru- Saikhowa Sanctuary	640	Elephant, white winged, wood duck, aquatic avifauna.
9.	Nameri Wild Life Sanctuary	130	Elephant, gaur, capped langur, golden mahsheer.
10.	Sonairupa Wild Life Sanctuary	175	Tiger, elephant, hornbill.
11.	Pobha Buffalo Sanctuary	49	Asiatic water buffalo.
12.	Deepar Beel Wild Life Sanctuary	4	Aquatic and migratory birds.
13.	Kachu gaon Game Reserve	214	Elephant, gaur, golden langur, spotted deer.

Source: Forest of Assam, 1992, Dept. of Forest, Govt. of Assam
N.B. Nameri and Dibru Saikhowa have been declared as National Park.

Table 17.4
Major archaeological ruins of Assam

Sl. no.	Name of the ruins	District	Nearest town	Period	
1.	Da-Parbatia ruins	Sonitpur	Tezpur	5th-6th	century AD
2.	Suryapahar ruins	Goalpara	Goalpara	8th-9th	"
3.	Madan Kamdev ruins	Kamrup	Guwahati	11th-12th	"
4.	Pandunath ruins	Kamrup	Guwahati	11th-12th	"
5.	Cole park ruins	Sonitpur	Tezpur	7th-12th	"
6.	Bamuni Pahar ruins	Sonitpur	Tezpur	11th-12th	"
7.	Singari ruins	Sonitpur	Dhekiajuli	10th-12th	"
8.	Na-Nath ruins	Nagaon	Hojai	7th-8th	"
9.	Sankha Devi ruins	Nagaon	Hojai	10th-11th	"
10.	Abhayapuri ruins	Bongaig-aon	Abhayapuri	11th	"
11.	Borpukhuri ruins	Nagaon	Hojai	10th-11th	"
12.	Numaligarh ruins	Golaghat	Bokakhat	12th	"

Source : Directorate of Archaeology, Assam.

Despite the decades long developmental efforts by government agencies, tourism activities of the state are concentrated in some selected pockets, viz. Guwahati and Kaziranga. The travel demands of both the domestic and foreign tourist are high in Guwahati, because the city acts as the gateway to the entire north-eastern region and a tourist usually prefers to have an overnight halt here before going to Kaziranga, Shillong and other parts of the region. Apart from this, various government and private agencies of the city provide package tour facilities to destinations like Kaziranga, Manas, Shillong (Meghalaya), etc.

Table 17.5
Major historical monuments of Assam

Sl. No.	Name of monument	District	Nearest town	Period
1.	Raidongia Dol	Dibrugarh	Dibrugarh	18th cent. AD
2.	Bezor Dol	Dibrugarh	"	18th cent. AD
3.	Thaora Dol	Sibsagar	Sibsagar	1663-1685 AD
4.	Siva Dol	"	"	1734 AD
5.	Rang Ghar	"	"	18th century AD
6.	Jai-Dol	"	"	1698 AD

7.	Kareng Ghar	"	Nazira	17th-18th Century AD
8.	Piyaliphukanar Dol	"	Nazira	19th Century AD
9.	Rudrasagar Dol	"	Sibsagar	1780 AD
10.	Gaurisagar Dol	Golaghat	"	1724 AD
11.	Negheriting Dol	Golaghat	Dergaon	18th Century AD
12.	Maghnowa Dol	Lakhimpur	Narayanpur	17th Century AD
13.	Basudev Dol	Sonitpur	Gohpur	18th Century AD
14.	Bar Dol	Sonitpur	Biswanath Chariali	1750 cent AD
15.	Kamakhya Mandir	Nagaon	Silghat	1745-46 AD
16.	Gupteswar Mandir	Sonitpur	Dhekiajuli	12th century AD
17.	Kamakhya Mandir	Kamrup	Guwahati	1565 AD
18.	Aswaklanta Mandir	"	"	1725 AD
19.	Hayagriva Madhava	"	Hajo	1583 AD
20.	Billeswar Mandir	Nalbari	Nalbari	1521 AD
21.	Vishnu math Mandir	Barpeta	Barpeta	18th C. AD
22.	Panbari Masjid	Dhubri	Gauripur	18th-16th C. AD
23.	Maibang Mandir	N.C. Hills	Maibang	1761 AD
24.	Khaspur Palace	Cachar	Silchar	18th C. AD
25.	Sonai Mandir	"	"	1885-87 AD

After Guwahati, Kaziranga continues to witness the bulk of the share of domestic as well as foreign tourists (around 62% of domestic and 81% of foreign tourist), due to gradually growing demand for nature-based environment and associated tourist resources (Table 17.6) . As the domestic tourists prefer less physical effort during the tour period, their presence in the areas of wilderness is found to be lesser than that of the foreign tourist. Moreover, the interest of domestic tourists is diverted to Shillong (Meghalaya) for their general fascination toward hill stations, and a couple of important pilgrimage sites of Assam, viz. Kamakhya, Umananda (both in Guwahati), Bhairabi and Mahabhairab (both in Tezpur). Rest of the destinations have very negligible share of tourist arrival. However, lack of publicity and prevailing law and order situation continue to be the main reasons of such polarized nature of tourism development, though the state possesses fairly developed road-transport service, especially in private sector. For efficient, quick, and comfortable travel, as many as six air ports located at Guwahati, Dibrugarh, Jorhat, Tezpur, Silchar and Lakhimpur (now unused), a well connected railway network (B.G) with major tourists generating centres of India along with interconnecting (M.G.) tract that passes through a hilly district (having 37 tunnels) with length varying from 39.6 m to 586 m. Such infrastructures are not properly oriented to promote tourism. In the same way around 848 tea gardens, with their distinctive tourism appeal, especially in central and upper Assam along with oil town Digboi and coal mines at Ledo are yet to attract attention in tourism front.

Table 17.6
Travel pattern of domestic and foreign tourist in Assam, 1995 (in some selected destinations)

Sl. No.	Destination	% of domestic tourist	% of foreign tourist
1.	Guwahati	100	84
2.	Kaziranga	62	81
3.	Manas	4	11
4.	Sibsagar	5	9
5.	Tezpur (Lower)	8	8
6.	Bhalukpung	2	5
7.	Hajo	2	4
8.	Sualkuchi	2	4
9.	Jorhat	1	6
10.	Majuli	-	3
11.	Nagaon	-	2
12.	Haflong	4	-
13.	Orang	1	-
14.	Madankamdev	3	3

Source : Based on Survey conducted by the author

Problems and Prospects

Though the state has a lot of tourist resources, it has not been able to give due publicity in national and international arena. As a result of this, tourists' mobility is restricted to only a couple of destinations, viz. Guwahati and Kaziranga. Other destinations lack standard government tourist accommodations as well as way side amenities to fulfil the demand of foreign and domestic tourists. Even the well publicized destination, Kaziranga, has only 108 beds under Tourism Department to accommodate tourists. In the private sector, Assam has only 14 approved hotels, with 572 rooms accounting for 1.03% of total room capacity (in approved hotels) of the country (as on 1995, Deptt. of Tourism, Govt of India). These approved hotels of the state are located in four destinations, viz. Guwahati, Jorhat, Kaziranga and Tinsukia. However, besides these centrally approved hotels, there are other 40 hotels recognised by the State Tourism Department to meet the demand of tourists belonging to different economic back- ground. However, out of the total accommodation capacity (expressed in terms of bed capacity), in all categories of hotels around 51 per cent is concentrated in Guwahati itself (based on field survey by the author). Other important places like Dibrugarh, Tinsukia, Tezpur, Silchar and Jorhat belong to the group of destinations having around 33 per cent share of hotel accommodation. As most of the accommodations of the state are 'off site' accommodation from the tourism point of view, it is one of the causes of slow pace of expansion of tourism in the state. In the same way, package tours organised by government and private agencies are running to limited number of destinations, that also without strict regularity. So, bulk of the tourists have to make their own arrangement, which is a difficult job.

It is found that the government initiative in tourism sector is far below expectation in the context of modern day tourism. For example, in the nature tourism segment, only two spots, viz. Kaziranga and Manas have elephant safari facility. Though one horned rhinoceros continues to be the major source of attraction in Kaziranga, there are other four protected areas (Table 17.3) having rhinoceros along with a variety of other rare species of animals and birds which are found to attract relatively less number of tourists. Regarding cultural tourism segment, only a couple of archaeological spots and monuments have been renovated so far. Practically nothing has been done to sell the rich folk and cultural heritage as well as crafts of the state through tourism. Even the 'Bihu' and ' Tea- tourism' festivals (recognised by Government of India in festival calander of tourism) organised by State Tourism Department annually since 1995, in a ritualistic way, have been unable to attract both domestic and foreign tourists. Hence, well planned publicity both in domestic and international spheres is an urgent need to attract more tourists in near future. Consolidation of development activities along with inter departmental understanding as well as co-ordination between government and private agencies will certainly open a new phase of tourism development in the state. Tourism is a seasonal activity of the state. Therefore, in order to overcome the shortage of required accommodation, government and private agencies should pay their attention to alternative accommodations like rented accommodation, paying guest system, etc. at the initiative of local people, tea garden managements and others. Apart from it, government and private agencies can build wood cabins, tree houses, especially in wilderness areas to meet the seasonal demand to avoid huge capital expenditure for building.

By well-planned publicity, trained manpower, improved service quality, informative publicity, literature, tourism can be geared up in the state. The process of diffusion of positive information regarding the state is an urgent need which is basically entrusted to government agencies. Renovation of major monuments, while presenting the unique and natural atmosphere, promotion of crafts and projection of folk and classical dance forms that exist in the rural areas will not only help in promotion of tourism venture, but will also provide economic benefit for local people and craftman as well as to re-evaluate their own cultural root. In that context 'small scale' and 'Local Level' tourism should be encouraged. Considering all these aspects one can say that, though tourism is a budding activity till now, a well planned strategy can make it a major industry like that of the tea and oil industry of the state.

References

Chib, S.N. (1983) : *Essays on tourism* (Edited), Cross Section Publication, New Delhi, pp. 15-52.

Ferranio , F.F. (1981) : *Method Approach for Evaluating Tourist Resource; Case of South Africa*, Paper from Studies in Wildlife

Parks Conservation, Metropolition Book Company, pp 120-122.

Rai, L P. (1993) : *Development of Tourism in India*, Printwell. p.3.

Forest of Assam (1992) : Report of Forest Department, Govt of Assam, p.47.

Bhattacharya, P. (1996) : Status and Potentiality of Tourism in Assam, Paper presented in NEIGS Annual Conference.

18 Geopolitics

Geopolitics of Assam can be understood on the basis of the dialectical relationships between geography, society and politics. The land of Assam with its mighty river, the Brahmaputra and the Barak, possesses enviable fertile plains surrounded by densely forested hills and plateaus. Medieval Assam and its adjoining areas which covered most parts of the present north-east Indian states had vivisections in different phases of history which shaped and reshaped its geography and politics. In fact, the geopolitical forces here are the resultants of the interplay of various relation of people with land and its developmental processes.

Geo-Ethnic Base

Assam was known as Pragjyotisha and later as Kamrupa in the pre-historic and medieval periods. It was from the mid-twelfth century onward the socio-economic forces were strong enough to fight for politico-territorial adjustments. The diverse physiographic base and ethnic character of population facilitated the growth of numerous but independent principalities in the Brahmaputra valley. Rivers and hills acted as boundaries of different kingdoms. Till the advent of the Ahoms in 1228, various kingdoms were there in Assam, of which the Chutiyas, the Barahis, the Morans and the Bhuyans were important. During this period several Muslim invasions were also taking place from the west, which could reach upto central Assam i.e. Nowgong. Most of the time invasions were from the south-western part through Mymenshing-Sylhet or along the north bank of the Brahmaputra via Rangamati above Dhubri (Barua, 1973). The extension of Kamrupa at that time can be ascertained from the composite area of total indigenous principalities which occupied the core of the Brahmaputra valley. It was extending from the east of the territory of Sadiya along the north and south of the river Brahmaputra to the west bordered by the Manah river in the north of Goalpara on the south. The society at that time was in a state of total disintegration which paved the way for the Ahoms entering from the east to establish their kingdom and later become the lord of the land and also became one with the indigenous populations in their fight against invaders from the west.

The Ahoms gradually subjugated all princely rulers of the Brahmaputra valley and became the unchallenged master of the land by 1618. During the Ahom period Aryan culture got special privilege and a privileged class consisting of Brahmins was formed and Hinduism got patronage from the Ahom Court. In the western part of Kamrup was the kingdom of Koch Behar which accepted Mughal vassalege in 1609, which paved the way for the Mughals to occupy Kamrupa Kingdom. The Ahom-Mughal war continued for a long time (1616-1671). The Ahom-Mughal wars had important contribution in unifying different tribal chiefs of the Brahmaputra valley under the overlordship of Ahom to fight against the common enemy, the Mughals. Out of this identity of interests, there sprang-up gradually a civil society, a state, with a common overlord (Baruah, 1985). It is under these circumstances that, a state called"Asam" or"Assam", with a common overlord in the person of the Ahom king and a civil society to be identified as "Assamese" grew up. The growth of the community feeling started with the Bhakti or Neo-vaisnavite movement launched by Sankaradeva and the foundation of a developing feudal society started with the introduction of *Paik* or *khel* system. The vaisnavite institution i.e,the Satras later became very powerful to challenge the Ahom monarchy and consequent Moamariya uprisings invited the circumstances leading to Burmese invasions in Assam. In 1826, the British power was invited by the Ahoms to drive out the Burmese invaders which helped to establish the colonial rule for 120 years. Thus ended the glorious 600 years of Ahom rule in Assam.

Social Formation – First Phase

The adoption of Hinduism by the Ahom kings made the land open for "Brahminic Hindu Civilisation" through the settlement of Brahmins, nobles and other officers of the royal court and making free gifts of land. In other words, the Hindu Brahmin and artisan castes were encouraged by the Ahom royal families to practise settled cultivation which opened a new mode of production. These non-tribal people became the major instrument of change from shifting to settled- plough cultivation. The self-subsistence economy of the tribal peasantry became stratified as it was being transformed to feudalism. This broke the compactness of the tribal territory and a multi-caste and multi-community village organisation and settled communal system of production relation emerged (Census of Assam, 1971). The new state acted as the agent of Brahminic culture and the tribal was undergoing a process of peasantisation. Thus, the blue line of cleavage between tribal and non-tribals, peasants and landlords, underdeveloped and developing emerged in a nascent form at the threshold of the British take over of Assam. In the later period of British rule, this cleavage became more consolidated on the basis of tribes and non-tribes, plains and the hill people as determined by respective geography of the land. The geographical forces can be said to have developed from this dialectic in the early phase of this state making.

Formation of Assam and Adjoining Areas

The period of the British take over in 1826 for about 120 years gave a clear cut geographical identity to its various ethnic elements. During the decade and a half following 1826, the kingdom of Assam and other neighbouring kingdoms along with their dependencies and the petty independent tribal state of the Khasi hills were annexed to the British administration. The North Cachar Hills were organised into a separate administrative unit in 1875. The Garo hills was made a separate district in 1869 and the Lushai hills district was formed in 1898 (Hazarika, 1996). By 1874, the territories of Govind Chandra of Cachar, Ram Singh of Jaintia, Bar Senapati of Muttock, Sadiya Khowa Gohain, the Khamti Chief and the hills inhabited by the Garos, Khasis, Jaintias, Nagas and Lushais were tagged to Assam administration (Bose, 1979). Nevertheless, the British province that came to be known as Assam, took shape more or less by 1873 and the whole area afterwards was officially known as Assam and adjoining areas.

The British province of Assam was divided into fourteen districts viz., the six districts of the Brahmaputra valley, the two districts of Surma valley, the four hill districts and the two frontier tracts (Table 18.1).

Table 18.1
The British territories of Assam

1 Name of District	2 Headquarters	3 Sub - divisions	4 Area in sq. miles	5 Population
Surma Valley Cachar (Plain)	Silchar	Silchar	1,661	359,472
		Hailakandi	287	141,012
Sylhet	Sylhet	North Sylhet	1,035	538,922
		Karimganj	1,069	480,265
		South Sylhet	1,064	418,022
		Habiganj	1,089	592,939
		Sunamganj	1,423	511,193
Total of Surma Valley			7,628	2,541,341
Brahmaputra Valley		-	-	-
Goalpara	Dhubri`	Dhubri	2,865	551,351
		Goalpara	1,089	211,172
Kamrup	Gauhati	Gauhati	2,589	576,156
		Barpeta	1,274	186,515
Darrang	Tezpur	Tezpur	1,828	286,462
		Mangaldai	1,369	191,473
Nowgong	Nowgong	—	3,699	397,921
Sibsagar	Jorhat	Jorhat	1,060	301,752
		Sibsagar	1,028	289,291
		Golaghat	3,114	232,154
Lakhimpur	Dibrugarh	Dibrugarh	2,591	466,205
		North Lakhi-mpur	1,166	122,000
Total of Brah-maputra Valley			23,672	3,812,542
Hill Districts				
Garo Hills	Tura	—	3,140	179,140
Khasi and Jaintia Hills	Shillong	Shillong	3, 936	170,623
		Jowai	2,086	72,640
Naga Hills	Kohima	Kohima	—	90,918

		Mokok Chang	—	70,042
North Cachar Hills	Haflong	—	1,706	26,744
Lushai Hills	Aijal	Aijal	4,701	70,328
		Lungleh	2,526	28,078
Total of Hills Districts			21,210	708,513
Frontier Tracts	–	–	–	–
Sadiya	Sadiya	Sadiya	–	25,949
		Passighat	–	13,582
Balipara	Lokra	—	522	3,819
Total of Frontier Tracts			911	43,350
Grand Total			**53,421**	**7,606,230**

Source : Report on the Administrations of North East India (1921-22), Mittal Publication, Delhi, 1984, pp. 71-72

Policy of Exclusion – Formation of Geopolitical Structures

The clear-cut geopolitical units so formed started expressing their political ideas in purely geographical terms as we shall see later. Moreover, Assam's hill areas were kept by and large outside the scope of provincial autonomy and thereby denied political participation. They were classified under the 1935 Act into two categories: (a) excluded areas and (b) partially excluded areas. They were denied the ordinary competance of provincial legislature. The Balipara and Sadia Frontier Tracts, the Naga Hills district, the Lushai Hills district and North Cachar Hills subdivision were classified as "excluded areas" and went unrepresented in the legislature. On the other hand, the Garo Hills district, the Khasi-Jaintia Hills district (excluding Shillong) and Mikir Hills were classified as "partially excluded areas". These areas enjoyed franchise to elect members to the legislature. These exclusions in political participation which were not there in the Brahmaputra Valley retarted the hill people in political awareness for several decades till 1946. To these exclusions can be attributed the genesis of feelings of separation and aspiration for autonomy in most of the hill areas of Assam as evident from the subsequent developments.

The Garo National Council was formed in March 1946 and in the same year a Khasi-Jaintia political association was set up mainly with the backing of the Siems. In the Mikir hills, the first political organisation of the Mikir (now Karbi Anglong) people, the Karbi Durbar was formed in early 1947 with a modest programme of protection of local customs, consolidation of Mikir area into a single administrative unit and extension of franchise. Thus, all over " the partially excluded areas" the post– war trend was one of a growing desire for local autonomy within Assam and India (Guha, 1977). In the excluded areas of Naga Hills District, Tribal Council was officially sponsored quite ahead of other hill areas in April, 1945. It later became Naga National Councils (NNC), a federation of several tribal councils under the leadership of T.Aliba lmti Ao with its mouthpiece-the Naga Nation. In June 1946, the NNC took a resolution demanding autonomy within Assam and by the turn of the year in February 1947, it formally demanded an interim Government for Nagaland with financial

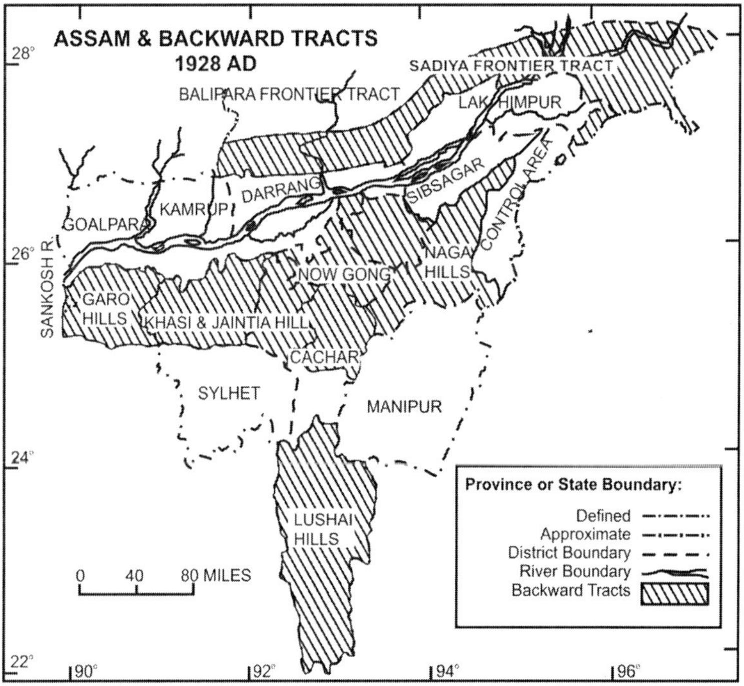

Fig. 18.1

porvisions initially for a period of ten years, at the expiry of which the Nagas would be free to reconsider their political status. Likewise in the Lushai Hills, the Mizo Union was formed in April 1946. It urged for the consolidation of all Mizo areas into a single administrative unit and its right to self-determination within Assam. Moreover, Assam became a Governor's province from 1921, but the hill districts and frontier tracts were declared as" backward tracts" through the modification of Govt. of India Act. (Fig 18.1) The path for political participation was opened with the introduction of legislative council in December 1920. In this process of political participation all the hill areas together got only two member provision against the thirty nine member provision acquired by the plains destrict of Assam. Thus, the divide between the hill and the plains became more distinct and permanent in geopolitical line, as the political aspirations for independence and self - rule were crystalised on the basis of some territorial identity such as Nagaland, Mizoland and in the post-independence period, the Bodoland.

Social Formation--Second Phase

The societal consequences of the colonial rule is also noteworthy. The colonial policies greatly transformed Assam's pre-capitalist economy into a colonial phase. The starting of plantation economy and mineral resource extractions created new relations of production which brought to the surface a conflict of interest between semi-feudal rulers and the mercantile bourgeois. The breakdown of communal system of production and the introduction of private ownership of land ultimately alienated most of the tribal from masses. Moreover, the new technology of wet and dry land farming, commercial jute cultivation etc. to generate surplus agricultural production made the colonialists to bring the land hungry agricultural peasants from the then East Bangal. This "inflow" of peasants paved the way for uncontrolable immigration with far reaching politico-demographic consequences for the next centuries. The political reign of the region went to the hands of Assamese middle class formed by the land owning and/or service holders and "lawyers" families. The bourgeois stratum that was emerging was enlightened as well as enterprising, and it provided leadership, however tenuous it might have been, to the Assamese society in every field (Guha, 1977, p.59). The hill areas and the areas of plains tribal people of Assam lagged behind in this development of a guiding middle class which ultimately affected and alienated them. The formation of the urban elite and the rural poor, which ultimately formed the political super– structure of the region is

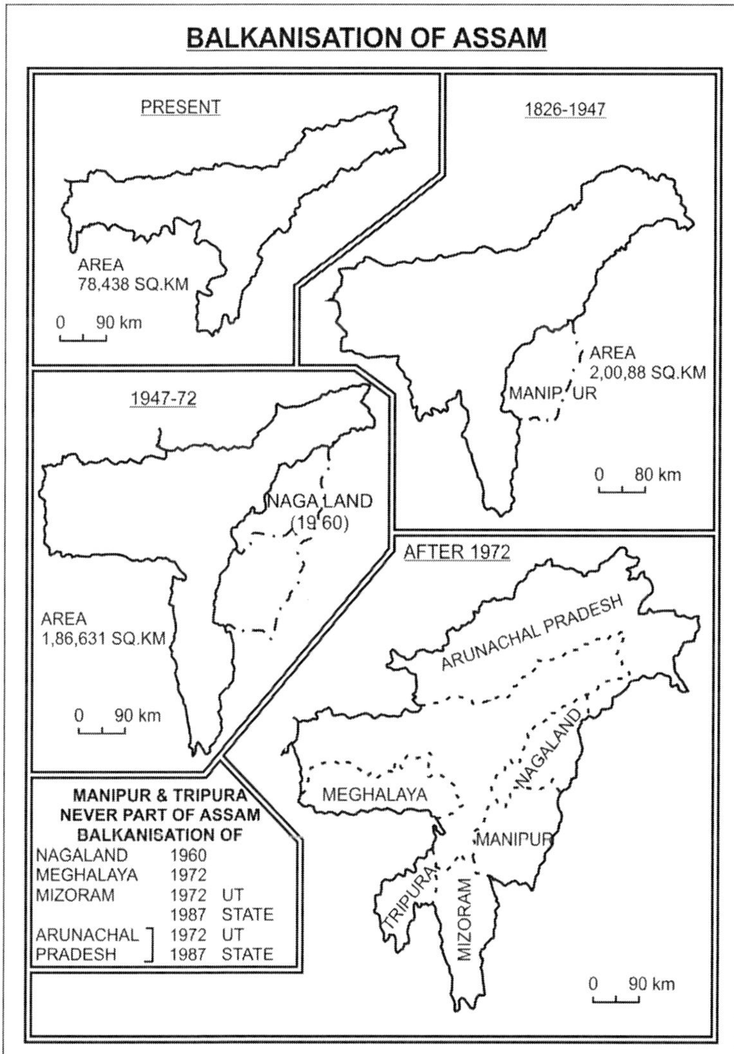

Fig. 18.2

impregnated with this distinctions which proved very hard to erode. Because of these inherent differences in social and economic formations, the hills and valleys could not coordinate in the developmental processes in the post-independence period of India for a unified geographical entity. The root of various geopolitical forces can be said to have emerged from this geographical bearing of political development. (Fig. 18.2).

Epilogue

The Inner Line Regulation which was introduced in 1873 to prevent clashes between colonial rulers and native tribal people of the hills of Assam ultimately alienated the hill people from the mainstream of society, politics and developmental processes. In the post-independence period, when these Inner Lines were transformed to boundaries of newly created states, the problem of inter-state boundary dispute started owing to the under-estimation of economic viability of hill states in defining their own boundaries. In a similar way the undefined Outer Line policy in the northern hill region of Assam (present Arunachal Pradesh), which later in 1914 became the British defined Mc Mahon Line, also drew the two Asian giants- China and India into a boundary war in 1962. This can also be attributed to the following of colonial British boundary policy by the national governments after independence.

The Bodoland movement, the Karbi autonomy and other demands for politico-territorial autonomy status of various insurgent groups are the reflections of relative deprivation, ideology and ethnic class conflict which are crystalised along geographically distinct areas of the lands of Assam. Unless inequality, injustice and exploitation of people are checked the geopolitics of Assam will take its own course, shaping and reshaping history, politics and geography of its people and the land.

References

Barua, K.L., (1973) : *Studies in the Early History of Assam* (ed.), Jorhat-Guwahati, p. 93.

Barua, S.L., (1985) : *A Comprehensive History of Assam*, Munshiram Manohar lal, New Delhi, p. 251.

Bose, M.L. (1979) : *Historical and Constitutional Documents of North-East India,* New Delhi, p. 37.

Guha, Amalendu (1977) : *Planter Raj to Swaraj*, Peoples Publishing House, New Delhi, p. 325.

Hazarika, J. (1996) : *Geopolitics of North East India*, Gyan Publications, New Delhi, p. 70.

Note :

Paik system : During Ahom period (medieval period) vast majority of the population consisted of Paiks, who were the labour class. They formed the peasantry and also the militia of the kingdom. The Paiks were bondsmen by birth, enjoyed limited freedom earned by fulfilling their duties to the state ; a type of feudal system.

19

Place Names

Toponymy is an important branch of **Onomastics.** It is the combination of Greek words '*Topos*' (means 'place') and '*Onoma*' (means 'name)' . So 'Toponymy' is a discipline, which deals with the study of all aspects of place name. It has various branches such as : *Hydronymy* (study of water bodies), *Oronymy* (study of hills and mountains), *Hydonymy* (study of roads, ramparts and embankments), *Chrematonymy* (study of names of things), *Micro-toponymy* (study of smaller places), *Anthroponymy* (study of personal names) and so on. This branch of science flourished in our country as early as 600 B.C. by Sanskrit grammarian Panini in his work 'Astadhyayi'. Panini dealt in length about the nomenclature, classification of place names from Afganistan to Magadh. But the subject 'onoma' or 'namah' in Sanskrit flourished in India since the days of Rig Veda (VIII. 5.7;VIII. 45.1; X.71.1; X.82.2), Aitareya Aranyaka (II.I.6), Satapatha-Brahmana (II.4.21), Ramayana and Mahabharata etc. Later it developed in Greece and people like Aristotle, Plato and others were the pioneers in the subject. In our next door China, toponymy was studied by the Chinese scholars in 223 B C in the regime of Emperor Howang-ti, prior to that also, in Chiao rule this branch of study could draw certain amount of attention in the reign of Emperor Phu-chi. During the modern period, in the western world, this branch was scientifically treated by Fried Wilhelm Lebinitz in A.D. 1768, who is also poularly known as the 'father of modern toponymy'. Leibnitz was follwed by J.J. Eglig, who is known as doyen of English place name studies. In the last two centuries this discipline had expanded to a great extent in the U.K., the U.S.A., Canada, Norway, Sweeden, Germany, Australia, Belgium etc.

In modern India , the work progressed with Ramakamal Sen (1834), B.C. Law, S.K. Chatterjee (1926), H.D. Sankalia, K.P. Goswami etc.

In Assam, the important works are of Banikanta Kakati (Assamese, its formation and development, 1941), R.M. Nath (The background of Assamese culture, 1948), B.K. Barua (A cultural history of Assam, 1951), Benudhar Sarmah (Thai namar atiguri, 1958), Sarbananda Rajkumar (Lik-phan-tai, 1966), Bhuban Chandra Handique (Guide to Rangpur), Hema Buragohain (Lecheri-butala Buranji), Ripunath Buragohain (Nam-talikar Buranji), Maheswar Neog (Pavitra Asom), Lila Gogoi (Buranjie parasa nagar; Buranji, A historical literature of Assam, 1984), Sarat Kumar Phukan (Tai- Ahom element in the place names of Upper Assam, 1975; Asomar Thair Namat Bivinna Janagosthir Prabhav, 1992-93; Place names relating to the battle fields in Assam during Tai-Ahom rule, 1995; An analytical and critical study on the history of place names of Upper Assam as gleaned from old Assamese chronicles, 1996; Bodo-elements in place names of Assam, 1997; Place names of Botanical origin in Assam, 1997; Inscriptional place names of Assam. 1997; Some aspects of cultural elements of Nalbari district through place names study, 1997; Place names as a source of Naga history, 1997; Place names relating to conquest of Assam by the first Tai- Ahom prince Siukapha, 1997; Onomastics, Toponymy and Nomenclature with special reference to Assam, 1998; Toponymy and Cultural Assimilation in the context of North East India, 1998; An etymological and historico-cultural treatment of certain place names of Nocte, 1999 ; A few place names in and around Rangpur, 1999; Mizo element in place names, 2000; A study of biodiversity through toponymy with special reference to Assam, 2000.), P.C. Bhattacharyya (Bodo place names), S. N. Goswami (Place names of Assam: A linguistic analysis, 1988).

According to the United Nations and Geographical Names (UNGN) names are important to each of us as-

Names identify our landscape.
Names express our national identity, our cultural heritage.
Names create our framework of orientation, our keys to the electronic age.
Names promote our awareness of the world around us.

The study of toponymy is worthwhile only when it is studied through inter-disciplinary approach including geography, history, archaeology, linguistics, folk-literature etc. Again any study on history, geography, linguistics, ethnology etc. is incomplete without the study of toponymy as this discipline is important source materials for most of the disciplines. So toponymy is called 'foot-notes of history and fossils of archaeology'.

The **source materials** for place name study are varied in nature. In the case of prehistoric work one has to depend on legendary accounts, folk materials and for the subsequent periods, early literatures, archaeology, coins, temple records, inscriptions, history or chronicles, vamsavalis, foreign travellers' accounts and so on. For the modern period, the maps or atlas, revenue records, gazetteers, district manuals, district census hand books, imperial gazetteers, post office, hospital and police records etc. can be utilised as source material for place name studies.

The **naming process** or nomenclature of place is important in place name study and in fact that is the basic of place name study. According to Mahabharata (I, 2-8), "a place must be named after any of its peculiar features". The naming pattern varies from region to region and from place to place. In studying the naming process one will always confront with the following queries:

a. When an explorer discovers a new geographical feature, or colonist finds a new settlement- what is to be named?
b. When there is already a name, at that juncture, whether a colonist will retain the name or give a new name?
c. If named, whether the letters are bestowed as direct?
d. Whether it is unaltered borrowing?
e. Whether it is transliterated from original one?
f. Whether they were taken over in slightly adopted form?
g. Whether they are prefixed of suffixed to the new name or some other additions?

In this context it will be relevant to mention about the **Canadian Permanent Committee on Geographic Names.** CPCGN and various provincial and territorial names authorities rarely initiate the naming of geographical features. Almost all names are submitted by the public or have been gathered through field survey programmes. In 1898, a primary task for Geographic Board of Canada was to set up standards and guidelines. Originally, the Board approved a set of 13 rules of nomenclature. Changing attitudes and perceptions, in particular, concerning translation, name duplication, language treatment and the handling of native names, have led to periodic revisions and updating of the principles.

The following are the guiding principles and procedures for geographical naming drawn in 1990 by CPCGN which are universially adopted by a number of countries:

1. Names governed by statutory authority-The names of municipalities, territorial divisions, reserves, parks, and other legal entities as created by, or resulting from, legislation by the appropriate government shall be accepted, e.g., the name Newfoundland is statutory, as is the French form Terre-Terre-Neuve. Its capital's name, St. John's is also official, but the French form, Saint-Jean, is not authorized.

2. Names in general public use--First priority shall be given to names with long standing local usage by the general public, unless there are good reasons to the contrary, this principle should prevail.

3. Names given by other agencies-- Names for facilities established by postal authorities, railway companies and major public utilities shall be accepted, if they are in keeping with other principles. Names established for land divisions by federal, provincial, and territorial departments shall also be accepted, if they confrom to the other principles. Active encouragement should be given to such agencies to have open communication with the appropriate name authorities.

4. Naming an entire feature and identifying its extent--A decision on name proposal for a physical or cultural feature should specify the geographical limits of the feature. Future approval of different names with the same generic for a part of what is deemed to be the same feature should be avioded.

5. Use of personal names--A personal name should not be given to a geographical feature unless such application is in the public interest. The person commemorated should have contributed significantly to the area where the feature is located; when such a name is applied, it should normally be given posthumously. The adoption of personal name during the lifetime of the person concerned should be made in exceptional circumstances. Ownership of the land should not in itself be names already in

common local use derived from the names of persons, either living of deceased, here Principle 2 takes precedence.

6. Approving names for unnamed feature-- In approving names for previously unnamed features for which no local names are found to be in use, the following sources are recommended: descriptive names appropriate to the features; names of pioneers; names of persons who died during war service; names associated with historical events connected with the area; and names from native languages formerly identified with the general area.

7. Form and character of names-- Geographical names should be recognizable words, acceptable combination of words, and should be in good taste.

8. Language forms and translation-- A name should be adopted in a single language form, although other form may be accepted where in use and when sanctioned by the appropriate names authority. A name should, where possible, be written in the Roman alphabet. When a name is derived from languages other than English or French, it should be written according to the considered opinion of linguistic specialists acceptable to the appropriate name authorities and to the language communities concerned. Names of selected geographical entities of 'pan Canadian' significance, as established by Treasury Board in 1983, are recognized in both English and French for use on federal maps and in federal texts.

9. Spelling standards in English and French -- The spelling and accenting of names should be according to the rules of the language in which they are written. In English, hyphenation and genetive apostrophe should be approved only when well established and in current use.

10. Uniformity in the spelling of names-- Names of same origin applying to various service facilities in a community should conform in spelling with the official name of the community. Names with the same specific associated features should agree in form and spelling.

11. Duplication-- Where established names are duplicated or are similar in sound or spelling, and tend to cause confusion, local assistance should be obtained to achieve distinction among them.

In giving new names, duplication to the extent that may result confusion in a local community should be avoided.

12. Generic terminology-- A geographical name usually includes both a specific and generic element. The generic term in a newly approved geographical name should be appropriate to the nature of the feature. Its position in the name should be dictated by euphony and usage. The generic term will be recorded in English or in French, by the name authority concerned.

13. Use of qualifying terminology -- Qualifying words may be used to distinguish between two or more similar features with identical specific forms. Such words may be derived from other local names or features or may be terms such as 'upper', 'new', 'west branch', 'nouveau', 'petit' and 'gros'. Whenever possible, however, new names should be distinctive.

14. Names of small features-- Except where local and historic usage dictates, the official approval of a minor feature should be guided by the relative significance of the feature, the familiarity with the name, and the scale of mapping available.

General Procedures (CPCGN, 1990):

I. How to propose a new name or name change? The following information will facilitate prompt decisions: 1. reason for proposal; 2. location by latitude and longitude, either by geographical co-ordinates or universal transverse Mercator grid co-ordinates; 3. identification on a map indicating precise extent; 4. photographs or sketches; 5. origin and meaning of name proposed.

II. Where to inquire about policies and principles on naming, and about information on the origin and use of names and terminology? - Secretariat (at Natural Resources Canada) and CPCGN members.

Every language has its own style of nomenclature or naming process. In western countries one of the renowned toponomist Adrian Room, (Dictionary of world place names derived from British names, 1989, pp. 196-207) put forward seventeen types of naming patterns and classed them as classification of English place names also :

1. Place names - Towns, Counties and Villages etc. 2. Aristocratic titles - Earl, Barons etc. 3. Royal names and titles- Adelaide, Albert, Prince, King, Queen etc. 4. Statesmen and government officers- Addison, Althorpe Islands, Amery Ice Shelf, Walpole Windham etc. 5. Financial baker or sponsor - Barrow, Grenfell, Hutt, Lancaster, Smith etc. 6. Sponsoring farms and other bodies- Banzare Coast, Bridgeport, Society Islands etc. 7. Military leaders - Amherst, Back River, Lawford Islands, Lawrence, Wager Bay etc. 8. Tribute names - Arrowsmith, Bingham, Dover, Dyer, Sidney etc. 9. Battle names- Warterloo, Naseby, Camperdown etc. 10. Explorer and discoverers- Arthur's pass, Hudson Bay, Sturt Desert, Wake Island etc. 11. Ship's captains- Bligh Island, Pennell Coast, etc. 12. Ship names -Alert, Beagle Bay, Griper Bay etc, 13. Founder and settlers - Alden, Harrisburg, Mcallen, Potter County, Wakefield etc. 14. Colonial officials -Baillie Islands, Jesselton, Port Davey etc. 15. Literary names - Avalon, Enid, Selma, Woodstock etc. 16. Name of buildings - Doonside, Melfort, Scarsdale etc. 17. Other names -the places which do not fall in any of the above categories such as in Bairsdale, Invermere, St. George, Union Jack etc.

George R. Stewart in his '*American Place Names*', New York, 1970, pp. xviii-xxviii, made elaborate explanation on naming process of American place names. They are briefly as :

I. Specific-generic-- Generic-river, mountain city, ville, burg etc. Specific -blue, battle, Lincoln etc. II. The article- The Gulf of Maine, The Thousand Islands etc. III. Languages -French, Dutch, Spanish etc. IV Linguistic transfer- Names are transferred from one language to another by four methods, three of them in an oral situation and the fourth in a written situation: A. An informant says the name in his own language, and the recipient renders it by sound (perhaps later in writing) as well as he can according to his own language habits. This process is primitive (existing even before writing and between illiterate) and is especially prevalent in situations where the recipient has little knowledge of the language of the informant. B. The name is translated from the original into the receiving language. Such procedure occurs more commonly when a fair degree of mutual understanding has been established. C. The name is reworked by folk-etymology. D. The transferal is by the written or printed form. The name is then ordinarily pronounced as the spelling suggests, and many of the misprounciations in American names thus arise, e.g., Florida, Berlin. V. Dual and multiple origin -New York, Florida. VI. Abundance vs. rarity - Fish Spring, Fish Creek. VII. Habitations and natural features-New Orleans. VIII. Folk names- Troublesome. IX. Linguistic play, humor- Matrimony, California. X. Evolved names - bestowed names- 'We shall meet at the black rock'. XI. 'Official' name-through official declarations. XII. Naming systems- Classical belts, The numerical system, Alphabetical systems, Grid systems etc. XIII. Literary influence-Shiloh, Bethesda, Mount Hermon etc. XIV. Religious and mythological name- Saint, San, Nephi, Lemphi etc. XV. Railroad stations, mines, brands- Clan Alpine, Tombstone, Ucross, Teedee etc. XVI. Personal names- Wilmington, Albany etc. XVII. Tribal and national names- Shawnee, Chinese, Red Indians. XVIII. Names from nature- Bear, Eagle, Oak, Grass, Coal etc. and XIX. Exotic names-Bogota, Memphis, Hannibal, Cairo etc.

In the Indian cotext, grammarian **Panini (600 B.C.)** in his *Astadhyayi* tells us about the following naming patterns :

I. Place names ending with - *Pura, Nagara, Grama, Kheta, Ghosha, Kula, Suda, Sthala, Karsha, Tira, Rupya, Kachchha, Agni, Vaktra, Gartta, Palada, Hrida, Vaha, Arma, Kanya, Kantha, Kanda etc, (Astadhyayi, IV. 2.84-420).*

II. Place names with addition of *Chaturvarthika* suffixes dealt with in the *Astadhyayi*, IV.2- 67, IV.2-71. The *gana* enumeration in IV. 2-80 has preserved about 228 places names.

III. Place names indicating original home of the people IV.3:90- *Mathura.*

IV. Place names after the finished goods imported in India, e.g., *Kapisa.*

V. Place names known due to control of a particular group of people or *gana* have been enumerated.

VI. Place names under the control of one king have been given by Panini known as *ekarajajanapada (Astadhyayi*, IV.1.168-176).

VII. Place names known after the name of control by a *Samgha* or

Gana existed in the areas of Vahika now-a-days known as Punjab in the territory of India.

In modern times, **H.D.Sankalia** suggested (in *Studies in the Historical and Cultural and Ethnography of Gujarat,* Deccan College, Poona, 1949) :

- Place names after a person–hero, saint, tribal leader,
- Place names after a deity
- Place names after a spirit
- Place names after a tribe
- Place names after an event :
  - Auspicious occasion
  - Bad occasion
- Place names after customs and superstition
- Place names after geographical and physical feature
- Place names after hill, mountain, mound,or elevated places
- Place names after river, stream, lake and ponds
- Place names after forest, steppe, desert
- Place names after flowers, fruit, tree and crop
- Place names after animals
- Place names after bird
- Place names after reptile
- Place names after existing place.

In later date, **Sankalia** modified his work in the line of **S.K. Chatterjee** (in *The Origin and Development of Bengali Language*, Calcutta, 1926, reprint 1971) as :

- Place names from tribes or caste
- Place names of natural feature
- Place names of a religious character
- Place names after person and event
- Place names copied from other place name

Again,

I. Place names, which pertain to geographical and physical feature :

i. Places named after mountains, hills, mounds, rocky surroundings etc.

ii. Places named after all types of water including rivers and rivulets

iii. Places named after different types of lands
iv. Places named after location

II. Place denoting flora :
i. Trees and plants
ii. Part of trees and gardens
iii. Flowers, specified
iv. Flower, general
v. Grains and pulses

III. Place denoting fauna :
i. Animals, wild
ii. Animals, domestic
iii. Birds, specified
iv. Birds, general
v. Reptiles
vi. Fish

IV. Religious place name :
i. Name of the vedic gods
ii. Saivite names
iii. Vaisnavite names
iv. Goddesses both Saivite and Vaisnavite
v. Buddhist
vi. Jain
vii. God general
viii. Goddess general
ix. Philosophical, sacrificial and celestial
x. Sages
xi. Epic heroes

V. Ethnographical name :
i. Castes and professions
ii. Tribes
ii. Races

VI. Place names of historical significance
I. Royal dynasties
ii. Kings and queens
iii. Epithets of kings
iv. Subordinate chiefs and royal officials etc
v. Official designations

VII. Place names of commercial character

VIII. Place names after famous cities

IX. Miscellaneous place names :

i. Names which distinguish between the new and old places of same name
ii. Names indicating two different sizes of places of same name
iii. Places indicating sentiments, aesthetic and superstitions
iv. Places named after metals
v. Places named after colours
vi. Places named after food items

While studying the naming process in **our region**, we have noticed certain peculiarities, which are relatively uncommon in other areas. Prof. G. N. Reddy, in his Survey of Indian Place names- A Case Study, p.30 records that Indian place names are generally analysed into two parts or constituents : specifics and generics. But in our region we have found another prominent class i.e, 'Prefix' like 'Ban', 'Che', 'Di', 'Mung', 'Na', 'Nam', 'Rai' and so on. It is observed that almost all the style of naming process in national level or eastern countries are prevalent in Assam, which is probably due to its location in the greatest migration routes of mankind. Place names with numeral as prefix or suffix is common in our next door West Bengal, which is also a common feature in Assam. The place names containing middle words are also a peculiar feature of this region such as : Pat-kai-cheng-kan, Doi-kao-rong, Ran-che-kham, Nam-dao-phi Nam-ti-lao, Mung-dun-chun-kham and so on. In south India, we have 'in-fix' words, which are not seen here, though some grammarians place the middle word as 'infix'. Thus it can be said, numerous styles of naming pattern of national and international perspective assembled here making the proocess rich, more varied, more beautiful and forming a mini sub-continent. The systematic analysis of the naming process, though interesting, is a difficult task due to conglomeration of different classes or groups of people and their variations in language, culture and thereby creating admixture, corruption in different class of place names. Place names of Assam are the result of a good deal of evolution as result of impact of religious, historical, cultural exposure to continuous migrations and invasions. Nevertheless, here we will try to show some of the important **naming processes in the context of place names in**

Assam :

Place names derived from natural objects-

a. Topography :

* Hills or Mountains
* Mounds or Ting
* Elevated place of Chapori
* Rivers or Streams, Shifting of River courses
* Lake or Ponds or Beels
* Caves or Caverns or Holes
* Gorges
* Waterfalls
* Rapids
* Currents
* Valleys or Plains
* Wetland, Heath and Dry land
* Colour of Soil, Water, Stone, etc.
* Force of Wind

b. Flora :
Surface- Trees, Crops, Pulses, Flowers, Creepers, Grasses, Fruits, etc.

c. Fauna :
Surface Animals- Wild, Domestic, Reptiles etc.
Aquatic Animals- Fish, Reptiles etc.
Ornitho etc.

II. Place names derived from cultural objects-

a. Religious–
Saivic name
Sakta name
Vaisnava name
Islamic name
Saints name
Deity name
Spirit name etc.
Place name after- Dols (Temples)
Place name after-Namghars, Satras
Place name after-Islamic shrines
Place name after-Christian settlements

b. Secular–
Place name after-Persons
Place name after-Existing places
Place name after-Royal titles
Place name after-Important nobles
Place name after-Important offices
Place name after-Tribes or Tribal Groups
Place name after-Khels
Place name after-Mels
Place name after-Clans
Place name after-Community
Place name after-Professional Clans
Place name after-Capitals, Towns, Bahars (Rest camps)
Place name after-Gars (Ramparts)
Place name after-Alis (Roads)
Place name after-Road Crossings or Intersections of Roads
Place name after-Mathauries (Embankments)
Place name after- Pukhuris, Sagars (Tanks)
Place name after-Events
Place name after-War fronts
Place name after-Treaties, Agreements
Place name after-Trades, Commercial characters etc.
Place name after-Markets and Hats
Place name after-Buildings
Place name after-Cultural centres
Place name after-Shifting of professional clans with prefix 'Na' (New).

c. Miscellaneous group :
Place names with- Numerals
Place names after- Food items
Place names with-Metals
Place names with- Colours
Places indicating -- Sentiments, aesthetic and superstitions
Place names indicating- Old and New
Place names showing - Directions
Place names indicating- Two different sizes of the place of the same name etc.

Structurally nomenclatured place names in this region can be seen through - Noun + Noun, Adjective + Noun, Adjective +

Adjective, Noun + Adjective, Onoma-poetic, combinations. In addition-Prefix, Suffix (Generics), Specific and Middle word (Infix) are also common features here. The other linguistic variations are : Corruptions of words, Hybrid formation, Shortening of Vowels and Consonants, Change of Vowels and Consonants, Change of Affixes, Metathesis, Elisions, Sound changes, Consonants Clusters, Assimilation , Droppings, Semantic changes etc.

The common Prefix-formations in the region are (As = Assamese, Aus. = Austric, Bd. = Bodo, Naga = Nagamese, NIA As. = Neo - Indo-Aryan Assamese, T. = Tai, OIA Skt. = Old Indo - Aryan Sanskrit) as follows :

Ban (T. Village)- Ban - chang, Ban-kak, Ban-khen, Ban- para, Ban-phi, Ban-phu, Ban- pur, Ban- rui, Ban- rui-ban, Ban- tung etc. Assamese : Ban-mukh.

Bar (OIA Skt.> NIA As.- Big) - Bardoloni, Barduar, Bargaon, Barghat, Barhat, Barkola, Barnagar, Barpani etc.

Cham (T. -Three)- Chamring > Charing, Chamdang etc.

Che (T.- Town)- Che-hung, Che-mun, Chepon, Che-regua > Chorogua, Cheraidoi > Choraideo etc.

Chong (T. -Two) - Chongtak>Chantak etc.

Dhar (Embankment)- Malidhar, Narayandhar etc.

Di (Bd. River, Water)- Dibong, Dibru, Digaru, Dihing, Dikhow, Dimow etc.

Doi (Bd. Water, T. Mountain, Hill)- Doiyang, Doigrung, Doi-kao-rong, Doi-mung, Doi-kham, Doi-je etc.

Ha (Bd. Land) -Habung, Haflong, Hajo, Hakama etc.

Mung (T. Country)- Mung-dun-chun-kham, Mung-khrang, Mung-klang, Mung-Pakam, Mung- Timicha, Mung- Tiura etc.

Na (T. Field, As. New)- Na-zira, Na-Nagar, Na-pam, Na-gaon, Na-Kachari etc.

Nam (T. River or Water) - Nam-chang, Nam-dang < Deng, Nam-pong, Nam-sai, Nam-ti, Nam-rup<Ruk etc.

Pani (As. Water)- Pani-beel, Pani-Dihing, Pani-Para,

Rai (T. Prominent)- Railung, Raicha etc.

Tai (T. Tai)- Taimung, Tailong, Tai-Khamti etc.

Ti (T. Place) - Timak, Timon, Tiok, Tiphuk, Tiru, etc.

Ting (T. High)- Tingkhang, Tingrai etc.

Tung (T. Field)- Tunkhang.

The common Suffix-words are :
Ali (Road)- Dhaiali, Namtiali, Garali etc.
Ani - Amani, Bakulani, Kathalani etc.
Ati (As. Village)- Auniati, Bengenaati, Haldhiati, Mikirati, Naharati etc.
Bagan - Ambagan, Nilbagan etc.
Baha - Kauribaha, Makhibaha.
Bahar (Rest camp)- Namdangbahar, Numalibahar, Rangalibahar, Tatibahar
Band - Dhuarband, Loharband etc.
Bandha - Jakhalabandha, Komarbandha. Mahebandha etc.
Bar - Dhopahar, Koliabar, Titabar etc.
Bari - (A compound or an enclosure may be of crops, fruits, persons etc)- Adabari, Aghonibari, Bamunbari, Bengenabari, Hemenabari, (old name of Gargaon), Kathalbari, Nalbari, Sarabari, Tamulbari, Tokoubari etc.
Bat (A way, a road)-Panibat
Baushi- Patbaushi
Bazar (A market)-Beltolabazar, Fancybazar, Morabazar, Romanibazar etc.
Beel-Choraibeel, Diparbeel etc.
Bera-Nagarbera
Bhanga-Nangalbhanga, Tekelibhanga, Tuluthabhanga etc.
Bheta-Mikirbheta, Rajabheta etc.
Bhita (A highland)-Japaribhita, Dakurbhita etc.
Bhukti-Pragjyotishbhukti
Bori-Bagharbori, Moinbori
Chala-Barchala
Chali (A shed)-Baruachali
Chapori (A heath)-Baghmorachapori, Jamugurichapori, Rangachapori
Char-Alugutichar, Bhangnamarichar, Tangnamarichar etc.
Chariali-Baihatachariali, Biswanathchariali, Dolomukhchariali etc.
Cherra-Dhalcherra, Durlabhcherra, Katlicherra etc.
Chiga (A breach)-Hiloichiga, Kumaranichiga, Phulpanichiga etc.
Chinga-Alichinga.
Choa-Hahchoa, Kukurachoa, Senchoa etc.
Chora (A rearing place)-Hahchora.
Chuk-(A corner) -Bagharchuk, Jakaichuk, etc.
Dang (T.A store or red colour or check gate)- Chamdang,

Jaikhamdang, Mathadang, Namdang etc.
Danga-Bamundanga, Moudanga etc.
Dar-Bhubandar, Malidar etc.
Desa-Kalitadesa, Mlechadesa
Dhoa-Chauldhoa, Dekdhoa, Nagaldhoa etc.
Dhora (To hold, to catch)-Bangaldhora, Bhatdhora, Dhupdhora etc.
Di-Bardi, Saniadi, Majardi etc.
Dia- Hardia, Khetrihardia, Ramdia, Sundaridia etc.
Doba (To sink)-Dhekeridoba, Bongaldoba etc.
Doi-Bhogdoi, Hamdoi, Khamdoi, Mangaldoi etc.
Dol (A temple)-Bogidol, Khoradol, Namtidol etc.
Dola-Hatdola, Jagdola etc.
Donga-Moudonga, Noudonga, Teliadonga etc.
Dowa or Dewa- Bandardowa, Bordowa etc.
Duar-Chaiduar, Chariduar, Naduar etc.
Dubi - (Depth in water)- Burdubi, Gagaldubi etc.
Ganga-Akasiganga, Burhiganga, Barganga, Basisthaganga etc.
Ganj-(Arabic, Village)-Janiganj, Kabuganj, Karimganj etc.
Gaon (A village)-Athgaon, Baghagaon, Chaygaon, Dergaon, Nagaon etc.
Garh (A rampart)-Hatigarh, Harmatigarh, Jongalbalahugarh, Ladoigarh, Mahgarh, Numaligarh, Vaidargarh etc.
Ghat (A footway to river)-Barghat, Dighelighat, Ghilajharighat, Golaghat, Silghat etc.
Ghopa (An abode)-Hindughopa, Jogighopa, Mushlimghopa etc.
Ghuli (A rearing place)- Hatighuli, Mahghuli etc.
Gola-Bargola, Chorgola etc.
Goma-Dolgoma
Gora (A bank)-Dighalgora, Nagagora, Rangagora etc.
Grama (A village)-Devagrama, Haposhgrama, Panchgrama etc.
Gudam (A store)-Puranigudam.
Guri (A root of a tree, an object)-Ahotguri, Amguri, Chapaguri, Dologuri, Jamirguri, Simaluguri etc.
Hagi-Garehagi, Kauriyehagi etc.
Hat (A market) -Barhat, Bhuyanhat, Dhuyahat, Jorhat, Jamugurihat, Kacharihat, Nagahat, Rupahihat etc.
Hati (A hamlet)-Ambarihati, Guwahati, Koihati etc.
Jan (A rivulet)-Bokajan, Daisajan, Duliajan, Kakojan, Mudoijan, Naojan, Pavajan, Samukjan, etc.
Jhar (A marsey land)-Chipajhar, Kokrajhar etc.

Jonka (Naga. Boundary)-Nagajonka
Juli-Dhekiajuli, Rangjuli, Rupajuli, Sonajuli etc.
Kandi-Ainakandi, Binakandi, Hailakandi etc.
Khal-Dholakhal, Katakhal, Maniyerkhal etc.
Khana-(To dig)-Dhakuakhana, Dhanekhana, Jankhana, Nakhana, Pukhurikhana etc.
Khat (An unit of land or an estate)-Engerakhat, Jahakhat, Jaikhamdangkhat, Pengerakhat etc.
Khati-Sapekhati
Khok-(A shelter or an abode)- Hatikhok.
Khola-(An abode)-Chipahikhola, Choraikhola, Hatkhola, Puranikhola etc.
Khorong (A hollow in a tree)-Choraikhorung, Moinakhorung etc.
Khowa (To enjoy)-Borakhowa, Chapakhowa, Choikhowa, Machkhowa, Morangikhowa, Sadiyakhowa etc.
Khuli-Hatikhuli
Kona -Kharikona, Dargakona, Murikona etc.
Kora-Chaulkora
Kota-(To cut)-Bashkota, Chagalikota, Dangkota, Kukurakota, Lepetkota, Pahukota etc.
Koth-(An encampment)-Baghmorakoth, Narayanpurkoth, Manahkoth etc.
Ksetra(An area)-Hidhuksetra (Sadiya).
Kunchi or Kuchi (A village)-Balikuchi, Banekuchi, Hamchakunchi, Patacharkuchi, Sualkuchi etc.
Makha- Burimakha.
Mandala (Equivalent to modern state)-Kamrupamandala.
Mari (To kill or haunt)-Chitalmari, Ghoramari, Goroimari, Pavamari Rowmari, Sialmari, etc.
Mati (Earth, land)-Kechamati, Namati, Noonmati, Puranimati etc.
Mora (To kill or haunt)- Baghmora, Bengmora, Bhalukmora, Kukurmora, Nagimora, Noramora, Salmora, Sialmora etc.
Mukh (A mouth)-Dhansirimukh, Dibarumukh, Dichangmukh, Goramukh, Jankhanamukh etc.
Mur (An end of a road or rampart) - Alimur, Garamur etc.
Muri (An end)- Sagunmuri.
Nagar (A town) - Barnagar, Champaknagar, Chatainagar, Gotanagar, Patnagar, Rangpurnagar etc.
Nala (Channel) - Bichennala, Chapanala etc.
Nga- Jerenga, Koronga, Mezenga etc.

Noi (River) - Nanoi, Pachnoi etc.
Pahar (A hill)- Bhubanpahar, Burapahar, Hilkundapahar, Mayangpahar, Mikirpahar etc.
Pam (A newly reclaimed field)- Napam, Pathalipam etc.
Panchali (Five road junctions)- Bhadoipanchali.
Pani - Dichangpani, Dighalpani, Lekhapani, Tengapani etc.
Par (A bank)- Bhatiapar, Cherekapar, Dakhinpar, Ghagrapar, Holongpar, Klanchupar, Uttarpar etc.
Para (A hamlet) - Goalpara, Kumarpara, Larupara, Rangapara etc.
Parbat (A mountain)- Bandarparbat, Gabharuparbat, Silaparbat etc.
Pat (A location, a seat)- Dakhinpat.
Pataka (Part of a village)- Arjjapataka, Santipataka, Subhankarapataka
Pathar - (A field) - Ahompathar, Bachapathar, Hamdoipathar, Jerengapathar
Pitha - Kamakhyapitha, Kampitha, Ratnapitha, Swarnapitha, Saumarpitha
Pota (To fix or bury)- Khotipota, Jathipota, Lorapota etc.
Pota- Gaharipota.
Pukhuri (A tank)- Alipukhuri, Gangapukhuri, Jaypalpukhuri, Rajapukhuri, Rajmaopukhuri etc.
Punji (Hamlet) - Barapunji, Khasiapunji, Mikirpunji, Reangpunji etc.
Pur (A town) - Abhoypur, Banpur, Bhabanipur, Jaipur, Lakhimpur, Mathurapur, Narayanpur, Pragjyotishpur, Sonapur, Ratanpur etc.
Puri - Abhoyapuri, Chandrapuri (-visaya) etc.
Ra - Dadara, Dhanora, Korora, Mahora, Nanora etc.
Rastra- Pragjyotishrastra.
Road - Barpetaroad, Dhekiajuliroad, Jagiroad etc.
Sagar (A big tank)- Gaurisagar, Jaysagar, Laksmisagar, Sivasagar etc.
Sal (A sakta shrine)- AmlokhiDeosal, Bakata Deosal, Chatrasal, etc.
Satra (A vaisnavite shrine) - Ahotgurisatra, Auniatisatra, Dakhinpatsatra, Kamalabarisatra etc.
Sindhu - Lauhityasindhu
Ta - (Aus. Land, soil)- Bakata, Chamata etc.
Tari- Bakaitari.

Tek (Aus. Elongated land) - Paglatek, Sialtek etc.
Than - Gopalthan, Norowathan etc.
Ti (Aus., T.) - Badati, Bedeti, Khamti, Namti etc.
Ting (A highland)- Deoriting, Farkating, Haloating, Negheriting etc.
Tiniali - Bogidoltiniali, Tinialibagicha etc.
Tola - Beltola, Kusumtola etc.
Toli - (A base, ground) - Amoratoli, Chinatoli, Eraltoli, Goaltoli, Kathiatoli, Sariahtoli etc.
Tomoni (an abode)- Rotatomoni, Umatomoni
Tul (Aus. A hill) - Dharamtul
Visaya - Chandrapurivisaya, Manjayavisaya etc.

A few endings like-
I - Jerai, Bolai, Romai, Rupai etc.
Ma - Borama, Dolgoma etc.
Nga - Jerenga, Mejenga, Koronga etc.
Ra - Dadara, Jagara, Korara, Nanora etc.

Double suffixes (or primary and secondary suffixes) place names are :
Bada-*ti- ghat,* Bagh- *mora-chapori,* Bah-*bari-pathar,* Bali-*jan-gaon,* Bhangna *-mari-char,* Dhopa- *bar-hat,* Jamu-*guri-chapari,* Jamu-*guri-hat,* Kako *-jan-koth,* Kalia - *bar-hat,* Nagala *-mora-mukh,* Nagini *-mora-hat,* Tangana *-mari-char etc.*

A few Middle- word (in-fix) formation in place names are :
Doi-*kao-* rong, Mung-*dun-chun*-kham, Nam- *dao*-phi, Pat *-kai-cheng-*kan, Ran-*che*-kham etc.

Corruptions of words in certain place names :
Bhokbari > Bhogbari.
Ka-mei-kha> Kamakhya
Maktha > Bakatha > Bakata
Morangkii > Marangi
Namruk or Namhuk >Namrup
Pagar -jo (u) - tic > Pragjyotish

Shortening form of certain place names :
Pat -kai-cheng-kan > Pat-kai
Sung- Saumara> Saumara

Transliterated forms of some of the place names :
Brahmaputra or lohit > or Namtilao or Namdaophi
Dhansiri > Tai. Namtima
Dichang > Tai. Khenamkhun or Lechang
Dihingnagar > Tai-mung
Dikhow> Tai. Khenamchao
Dikhowmukh > Tai. Chupnamchao.
Dilih > Tai. Namkhun
Gajpur> Tai. Che-chang,
Gargaon > Tai, Chehung
Jaji > Tai.Khe- nam - jeng
Rangpur > Tai . Chemun,
Sonai > Tai. Khe-num-kham
Tengapani > Tai. Nam-choom etc.

Acronyms or hybrid formations and language element in place names of the region :
Dharam- tul > As. Dharam + Aus. Tul.
Di -hong > Bd. Di + Aus. Hong.
Di-lao > Bd. Di + Aus. Lao.
Nam- di-ma > T. Nam + Bd. Ti+Ma.
Nam-ti-lao > T. Nam + Bd. Ti (unvoiced form of Di)+ Aus. Lao,
Aus. Svarnasri > Tai. Khe-nam-ti- ma> As. Suvansiri. etc.

Semantic change of place names :
Demow < Dimow, a name of a river to a town.
Kamrupa, once name of a country > contracted to a district name.
Nagaon, anglicised form Nowgong, once a village name of A.D. 17th century > expanded to a district name.
Pragjyotish, once name of a country > contracted to a district name.
Sivasagar, a name of a tank> expanded to the name of the town, sub-division and district and so on.

Structural formations of place names:
Adjective + Noun > Ranga (red) + Jan (rivulet),
Noun + Adjective > Nam, (a classified noun) + Deng > Dang (red).
Noun + Noun > Siva + Sagar,
Onomapoetic name > Bherbheri, Chalchali, Damdama, Domdoma etc.

Isogloss of place names (communitywise) :
Aryan region - Naming of Aryan (Sanskrit) origin,
Austric origin - Naming of Austric origin,
Bodo origin - Naming of Bodo (like Bodo, Rava, Lalung, Dimacha) origin,
Tai origin - Naming after Tai (Tai- Ahom, Tai-Khamti, Tai-Phake. etc.) language etc.

As regards **classifications of place names,** H.D. Sankalia has rightly commented, "Classification of names according to their name contents is indeed difficult. The meaning of which, the dictionaries - Sanskrit and Prakrit -suggest, has also their limitations. Very often a name has more than one meaning. What its exact connotation is, can probably be understood only by a personal visit to the place. For example, Munda 'Sthala', may be after the old non-Aryan tribe of the Mundas of Kols which inhabited this area or it may be after Siva, as in Munda, is one of the names of this God. But this meaning can be varified and proved correct, if there are even ruins of the temple etc. of Siva who was called so. The limitations referred to are that the dictionaries may not be giving really ancient usages, or usages as old as the inscriptional name itself. They at times record the meaning given by later works. So the interpretations suggested here are to be understood as provisional, and not final. They need further collaboration from other sources, but particularly from first hand knowledge of the place itself." (Op.cit. Sankalia, p.80). The most acceptable international classification on place name as referred in Encyclopaedia of Britannica, Vol.24, 1989, pp. 728-729, is :

Place name or Toponymy is basically classified into -

I. Inhabited places- Cities, Towns, Villages etc,
II. Uninhabited places- Hills or Mountains, Forests etc.
III. Hodonymy - Streets, Roads, Ramparts etc.
IV. Hydronymy -Rivers, Lakes, Ponds, Beels etc.
V. Oronymy - Mountains or Hills, Mounds, Elevated places etc.
VI. Chrematonymy - names of things.
VII. Microtoponymy etc.

J.J. Egglig classifies names into two broad categories and then subdivides them as:

A. Natural names and

B. Cultural names. These names are derived from some natural feature or characteristic.

a. Inherent

b. Adherent

c. Relational.

Cultural names - They reflect the cultures:

Material culture : a. Physical, b. Economic, c. Intellectual.

Spiritual culture : a. Moral, b. Religious, and c. Political.

George R. Stewart, in his '*American Place Names*' 1970, pp. xxviii-xxxii put forward a classification of place nemes by Mechanism of Origin :

Place names originate from the need or desire to distinguish one place from others. A place will be named if people frequenting it - I. conceive it as its entity and II. find to be in some way be useful, and therefore worth naming. Primitive people, for instance, rarely had name for a mountain range, since it is too large for them to see all at once and thus to conceive as entity. Even a single mountain may not be named, because there is no advantage in its being so known, though springs and meadows on its slope will have names. To distinguish one place from another of the same kind (generic), the users of a language employ certain devices or mechanisms, ten of which are here discussed. Nearly all (if not, indeed all) of the place-names of USA originate in one of these ways.

I. Descriptive names. II. Association names. III. Possessive names. IV Incident names. V. Commemorative names. VI. Commendatory names. VII. Coined (manufactured) names. VIII. Transfer and shift names, name cluster. IX. Folk - etymology. X. Mistake names.

In this connection, it may be recalled, Adrian Room put forward a naming pattern of English namings which he preferred to call as 'Category of Place names' like —Place names, Aristocratic title, Royal names and titles etc. In the Indian context, S.K. Chatterjee and H.D. Sankalia had drawn a classification of place names on the light of naming pattern which has been already discussed.

In the case of our region, the place names can be categorized under the following heads :

I. Anthropo-Historical

a. Paleolithic origin
b. Megalithic origin
c. Neolithic origin
d. Proto-Historic origin
e. Historical origin

With the existing place names:
I. Historical
II. Geographical
III. Ethnological or Anthropological or Sociological
IV. Linguistic
V. Religio-Cultural
IV. Folk-lore
VII. Miscellaneous.

I. Historical category can be further classifed into (periodwise):
a. Ancient or Pre-historic Era
b. Early Medieval
c. Late Medieval
e. Modern Period

I. Historical category can be divided on the basis of ethnic groups:
a. Austric
b. Negrito
c. Indo-Chinese stock
d. Later Mongoloids
e. Indo- Aryans.

I. Historical classification on the basis of dynastic rules:
a. Danava/Asura: Mairang hill after Mahiranga Danava
b. Naraka-Bhauma (Varmana: 4th century A.D-A.D. 650) : Kaluto (Korotoya), Kamalopo (Kamrupa)
c. Bana, Bhismaka (contemporary to Naraka), Devesvar, Dharmapala, Ramchandra, Sankal or Sankaladvip A.D. 1st century)
d. Salastambha (C.650-990).
e. Palas (A.D. 990-A.D 1138)
f. Devas
g. Kamata
h. Kacharis (Dihing, Dikhow)
i. Bhuyans (Bihpur, Biswanath)

j. Chutiyas (Sonagiri, Chandangiri, Swetagiri)
k. Moran (Moran, Bengmora)
1. Borahi (Bolom, Hallem, Bihali)
m. Mahemmadan
n. Ahom (Tipam, Tirap, Timon, Nampong, Namdang, Namdaophi etc.)
o. Koch (Kocharbhita, Chikangram etc.).

II. Geographical classification : As the theme of geography is a place, so we may have a distinct geographical classification of place names which will cover almost all the place names of the world and remaining names can be placed in miscellaneous category :

1. Physical geography - Hills or Mountains, Plateaus, Valleys, Rivers, Lakes, Beels, Ponds, Caves and Caverns etc.
2. Cultural geography-
   Rural : Names of villages : professional or non-professional, village shrines, village agricultural fields, tanks, temples etc.
   Urban : Capitals, Towns, Growth Centres etc.
3. Economic geography -Place names relating to economic products- Minerals, Agricultural products, Dairy products etc.
4. Political geography - Place names relating to war, treaty, agreements, territorial divisions and the like.
5. Regional geography - Place names relating to regional distributions of human and animal kingdom and the names derived from them.
6. Biogeography - Place names relating to flora and fauna.
7. Historical geography- Place names derived from historical relics, monuments, history connected with the places.
8. Soil geography - Place names connected with characteristics of soil like Rangamati, Engera, Kajalabheti etc.

III. Literary classification : Place names originated in different literary periods can be placed in this categories.

i. Vedas, Upanisadas and Samhitas
ii. Epics
iii. Puranas, Tantras and various other old literary, religious and secular works like Arthasastra, Reghuvamsa, Harshacarita, Yoginitantra, Haragauri Sambada etc.
iv. Epigraphs and Monuments
v. Folk literatures, Folk-lores

vi. Historical literatures -Buranjis, Chronicles, Vamsavalis etc.
vii. Traveller's accounts
viii. Study of coins.

IV. Linguistic classification:

i. Austric
ii. Mongoloid language- Bodo, Rava, Karbi, Lalung, Dimacha, Miching
iii. Aryan language -Sanskrit, Prakrit, Assamese etc.
iv. Later Mongoloids- Tai- Phake, Tai-Khamti, Singpho etc.
v. Others : Muslims, Manipuri, Bengali, Hindi etc.

V. Religio-cultural classification :

i. Hindu shrines -Saiva, Sakta, Vaisnava
ii. Buddha (Vihar) and Jain etc.
iii. Muslim shrines- Dargah, Mosque etc.
iv. Christian- Church.

VI. Miscellaneous : This category includes those place names which are not included in any of the categories.

Comparative Analysis : Though there are certain degree of differences in nomenclature of classification of place names in different parts of the world, yet there is an overall similiarity or uniformity in many respects. As for example, we have number of prefix words like -*Che (Chieng or Xiang,* in Thailand, Vietnam, China, equivalent for city or town), *Khe-nam, Me- nam ,nam* in Thailand, Vietnam, equivalent for river or water), *Mung (Myong* in Yunan, North Myanmar, equivalent for country or land) etc. In western world, some of the nomenclatures of place names are based on : heroic activities of persons, important personalities, war front, tribes name, community name, habitations, religious shrines/leaders, names from nature, food items, minerals, language, building, aristocratic title, royal title etc. which are common to our region also, like- Barbarua, Banamali, Ranchekham, Saraighat, Ghilajharighat, Manah, Bodoland, Michingland, Kacharihat, Ganakkuchi, Kamakhya, Kechaikhati, Bisvanath, Dakhinpat, Naharati, Bhalukmora, Hatichung, Klupet, Mahang, Ranghar chariali, Rajmao, Bogidol, Vaidargarh, Jongalbalahu, Phukanarhat etc. But there are distinct differentiation in namings with our region : colonial officials, ship names, discoverer's names, founders, financial baker's names, christmass names, funny and strange names etc. This is due to basic cultural variations over the centuries.

Now let me throw little light on certain place names of our region:

Athabari : Lat 27⁰-08'-33 N and Long. 94⁰ 1-45'-13 E intersects at Athabari. Athabari, a garden of trees from where adhesive substance or 'gum' had been collected for preparing paste. Siukhampha @ Khora Raja (A.D. 1552-1603) constructed a road from Gargaon to Dihing and while constructing the road one villager brought 8 saplings of plant which would secrete 'gum' and handed over to the king for plantation by the side of the road. The king planted these saplings by the side of the road and it became famous as 'Athabari'.

Brahmaputra : The main river of Assam. A.D. 10th century, this name was conferred by the Indo-Aryans of Assam. Brahmaputra, Brahmar, N. Brahma, the Hindu trinity, + case ending -r, Brahmar + OIA Skt. suffix -putra, son, = Brahmaputra, i.e. son of Brahma (Kalika Purana), here loss of consonant 'r' has taken place. Old name: 1. Lauhitya-sindhu, aryanisation of the Aus. name 'Lao-ti ' or 'Leuti', Lau-hit, (here, change of 'o' to 'u') red + suffix - ya, + suffix - sindhu, big river. sea i.e, a big red river. 2. Lao-ti, Aus. Lao, clear, + suffix -ti, river, i.e.a clear water river. Soften form of Lao-ti to 'Luit' , change of letter, 'a' to 'u', loss of letter 'o' and reverse position of consonant 't' and vowel, 'i' or letters are transposed as 'Laoti' to 'Luit'. 3. Di-lao, Bd. Prefix- Di, river + suffix Aus.- Lao, a case of superimposition. 4. BullungBuththur, Bd. (Rava) - A river flowing with gargling sound (referred by Bishnu Rava). 5. Nam-dao-phi, Tai. Prefix-Nam, a river, water+dao, a star,+phi, god, i.e.a divine water river. 6. Khe-Nam-ti-lao, a hybreed formation of Tai-Bd.- Aus., Tai. Prefix - Khe-Nam, river+ 'ti', unvoiced form of Bd. 'Di' to Tai. 'ti' + suffix Aus. -Lao = Namti-lao, (reference in Tai- Ahom manusript of 13th century A.D, Ahom-Buranji, preserved in the house of Mr. Mukunda Phukan). 7. Greek. Duoinoss, Skt. Hrada, Hradini etc.

Dergaon : In Golaghat district (26⁰-38-08 N.Lat. and 95⁰-54-30 E. Long). A. D. 14th century by the Bodo- Kacharis. N.NIA As. Deor, god+suffix-gaon, a village of god; corruption of word 'Deorgaon' to 'Dergaon', here, mid vowel '0' is lost; known for its famous old Siva temple, later rebulit by the Ahom king Rajeswarsimha and consecrated by Laksmisimha in A.D. 18th century. Old names :

1. Devagram, a name given by the Brahmins of. 'Devagram' of Bengal, a neighbouring state of Assam, a transliterated form of 'Deorgaon'. 2. Sital - Negheri, NIA As. Sital, cold + N, suffix - Negheri, a kind of small birds, i.e.a silent place for 'Negheri' birds; it was royal dockyard of the Ahoms in A.D. 17th century. 3. Kopahtoli, N.NIA As. Kopah, cotton, + suffix -toli, field, a cotton growing area. 4. Phutukatoli (vide Nazira). 5. Tai- Ahom. 'Banphi', a transliterated form of 'Deorgaon'. Semantically, a temple name becomes a name of township of Golaghat.

Gaurisagar : Gaurisagar is about 6 km south-west of Rangpur at the intersection of 26^0-55-07 N. Lat and 94^0-02-50 E Long. Old name, Bezgaon. In the reign of Siutanpha, queen Phuleswari excavated a tank and constructed three temples, Siva, Visnu and Devi (Durga)-dols, at Bezgaon in A.D. 1723-24 and declared the complex as 'Gaurisagar', here 'Gauri' denotes the consort of Siva, 'sagar', a big tank measuring more then 400 bighas of land . Since then this area came to be known as 'Gauri-sagar'. The tank is about 234 Beo in length East-West, 194 Beo in breadth North-South in water area, inclulding the land area : 315 Beo in length 302 Beo in breadth. The Siva temple is about 32 yards in height and 60 yards in peremeter. Visnu-dol is about 45 yards in peremeter and 20 yards in height. Devi-dol is about 40 yards in peremeter and 20 yards in height. The Manipuri princess Kuranganayani (fawn eyed) stationed here prior to her marriage with Rajeswarsimha. It was the fort of the Ahoms against Moamorias from A.D. 1785-1812. This was also encampment of English Captain Mr. Wood along with his wife. In A.D. 1816, Gaurisagar was the station of Col. Richardson against the Burmese in the reign of Chandrakantasimha.

Rangpur : Rangpur rests at the intersection of 26^0-56-10 N Lat and 94^0-37-36 E Long. Rangpur was the fourth permanent capital of the Ahom. The capital of the Ahom was shifted from Gargaon to Rangpur by Siukhrangpha @ Rudrasimaha, in A.D 1702-03 and this capital continued for about a century . The length and breadth of the capital was 12 and 10 miles respectively. The name was kept after the Assamese word, 'Rang-pur', Rang - joy, pur- a city i.e., a city of joy and its Tai-Ahom equivalent is 'Che-mun', Che- a city or town, mun- amusement, joy, happiness etc. The capital had a 'Rangghar', pavillion with highly decorated wood and bamboo works to witness various games and sports of the time, like: elephant-buffalo-crocodile-tiger fights, hawk fight, wrestling etc. Pramattasimha (A.D. 1744-1751), son of Rudrasimha renovated 'Rangghar' complex with the permanent present structure. According to Gunaviram Barua, the capital was about 3 miles in length and one mile in breadth. The capital complex was surrounded by 18 ft. high ramparts all around and covered more than 1000 bighas of land. The actual royal seat was situated towards the eastern side of the town to which there were three approaches. These approaches were known as 'Barduar' in the south, 'Pani-duar' in the east and 'Na-duar' in the west. The road through Barduar leads to Royal palace and the residences of the high officials. A beautiful description of the capital city of Rangpur is found in 'Ananda Lahari' by Ananta Acharya of A.D. 18th century in the reign of Sivasimha. Rudrasimha constructed five temples of 'Ranganatha Siva' in front of the Barduar of the capital Rajeswarsimha (A.D. 1751-1769) founded the royal palace 'Talatalghar'. The descriptions of the 'Talatalghar' and 'Rangghar' can be found in the report of the Archaeological Survey of Bengal Circle, 1902-1903. The capital continued up to the last days of reign of Rajeswarsimha, A.D. 1769, but in the same year Moamoria general Ragha Barbarua forcefully occupied the Ahom capital and Bharathi was made as their king and the displaced Ahom king Laksmisimha took shelter at Sonainagar or Sonarinagar. The Moamorias occupied the cpital intermmitently up to A.D 1791. But the next Ahom king Gaurinathsimha recaptured the capital Rangpur with the help of British Col. Mr. Wood in the same year from the Mataks or Moamorias. Soon after that Gaurinathsimha shifted his capital from Rangpur to Jorhat. In A.D. 1831, the Ahom capital ' Rangpur' was occupied by the Burmese General Mingimaha Tilwa. In A.D. 8131, the Ahom capital 'Rangpur' was occupied by the Burmese General Mingimaha Tilwa. In A.D. 1838 for the last time king Purandarsimha stationed at Sivasagar and Rangpur. The main Royal dockyard of the Ahoms was located near Rangpur at Meteka. The old name for Rangpur was 'Meteka' (residence of Laluksola Barphukan in A.D. 1679, who conspired against the Ahom king in his aspiration to the Ahom throne with the help of a Mughal envoy Govindaram sent by Md. Azam from Bengal) and it was also known as 'Tengabari'.

Suggestion: Regarding place name study and proper documentation of place names the following suggestions have been put forth :

1. As the place names are important tools to reconstruct the history of the past, reasonable care and caution should always be there to accept the correct form of the names (both old and new) and preserve the old and other forms of name.

2. The spelling and application of names of towns, growth centres, villages, land and water features must be clear, accurate, current, and unambiguous. Effective communication depends on appropriate use of these names on maps, in media, and in legal documents, Inaccuracy and misrepresentation may lead to inefficiency and misdirection.

3. Any one and every one gives an etymology of the place names, even a road sider, without proper varification or screening and without involvement of linguists, geographers, historians or folklorists, which will lead to utter confusion. So without varification by the above disciplines, nothing should be accepted as final.

4. To draw any conclusion on place names always an interdisciplinary approach should be adopted.

5. A place name institute or a society is an urgent need of the hour. The components of the institute should consist of a group of Linguists, Geographers, Historians, Archaeologists, Ethnologists and Folk-lorists etc. The institute or society should organise awareness campaigns, trainings, field investigations, collect proper piece of name without any distortion, standardize the name form, draw a classification of place names, screen the naming processes, and incorporate new data, etc. Finally, we should review old records on the face of newly available confirmed data, as we all know that nothing is final about the interpretation of the past.

6. Promote the use of standardized names on maps and in documents.

7. Encourage the dissemination of standardized names through the publication of gazetteers (alphabatical lists of names, with co-cordinates and other data).

8. Promote the development of national automated data processing capabalities and international technology exchange.

9. Encourage collaboration between states on standardization of names for transboundary features. This standardization of geographical names is fundamental to effective communication and essential to our socio-economic development, in such fields as trade and commerce, population census and statistics, property rights and cadastre, urban and regional planning, environmental conservation, natural disaster and emergency preparedness, security strategy, search for rescue operation, maps and atlas production, automatic navigation and tourism etc.

10. To prepare a place name gazetteer or time capsule for the entire state, the following materials will be required-suffix, prefix, specific characteristics, etymology, ethnology, geography, history, archaeology and other cultural data. Similar gazetteer for districts, sub-divisions, circles and mauza can also be thought of. This gazetteer should correspond with the toponymic guidelines and principles of national and international societies or institutes.

For further readings :

J.B. Johnston, *The place names of England and Wales,* 1914; J.E.B. Gover, *The place names of Middlesex,* 1922; J.J. Kneen, *The place names of Isles of Man,* 1925; J.Loth, *Les names des saints Briton,* 1910; Th. Miller, *Place names in the English Bede, 1896;* W.St. Baddeley, *The place names of Gloucestershire, 1913;* H. Abraham Espenshade, *Pennsylvenia place names,* 1925; Henry Gannet, *The Origin of certain place names in the U.S.A.*, 1905; Hemmil Kenny, *West Virginia place names : Their origin and meaning, 1945;* A. Martin, *One thousand and more place names in New South Wales,* 1943; Ernest Scott, *A short history of Australia,* 1920; Alec McEwen, *The English place names of the Galapagos, 1988;* Charles Pettman, *South African Place names : the Past and Present, 1931,* B.F.Holt, *Place names in Transkeian Territory , 1959;* E.S. Dollimore, *The New Zealand Guide 1962;* R.S. Boumphrey and others, *Preliminary Derivations of some Falkland Islands place names, 1950;* B.K. Kakati, 1941, '*Assamese, its formation and development*', S.N. Goswami, *Place names of Assam: A linguistic approach;* Hema Buragohain, *Lecheri Butala*

Buranji; Ripunath Buragohain, *Namtalika Buranji;* Lila Gogoi, *The Buranjis Historical Literature of Assam;* Sarat Kumar Phukan -Place names of Assam, Botanical place names of Assam, Zoological place names of Assam, Anglicised place names of Assam, Place names derived from inscriptions, Place names connected with medieval war fronts, Place names connected with Siukaphas enroute to Assam, Geography of place names, Cultural elements in place names, Tai elements in place names, Bodo elements in place names, Austric elements in place names, Rava elements in place names, Place names connected with Naga history, Karbi elements in place names, A few place names in and around Rangpur, 24 important place names of Assam : A linguistic analysis, Place names as a source of Naga History, Nocte elements in place names, Jinghpho elements in place names of Assam, Place names among the Yubins, Mizo elements in place names, Khasi elements in place names, Anthroponymy of Assam (Personal Names of Assam) and Surnames of Assam.

Index